Generic Drug
Product Development

DRUGS AND THE PHARMACEUTICAL SCIENCES

A Series of Textbooks and Monographs

Executive Editor

James Swarbrick
PharmaceuTech, Inc.
Pinehurst, North Carolina

Advisory Board

Generic Drug Product Development

International Regulatory Requirements for Bioequivalence

Edited by

Isadore Kanfer
Rhodes University
Grahamstown, South Africa

Leon Shargel
Applied Biopharmaceutics, LLC
Raleigh, North Carolina, USA

CRC Press
Taylor & Francis Group
Boca Raton London New York

CRC Press is an imprint of the
Taylor & Francis Group, an **informa** business

CRC Press
Taylor & Francis Group
6000 Broken Sound Parkway NW, Suite 300
Boca Raton, FL 33487-2742
First issued in paperback 2019

© 2010 by Taylor & Francis Group, LLC
CRC Press is an imprint of Taylor & Francis Group, an Informa business

No claim to original U.S. Governme nt works

ISBN-13: 978-0-8493-7785-3 (hbk)
ISBN-13: 978-0-367-38437-1 (pbk)

Library of Congress Cataloging-in-Publication Data

Generic drug product development : international regulatory requirements for bioequivalence / edited by Isadore Kanfer, Leon Shargel.
    p. ; cm. – (Drugs and the pharmaceutical sciences ; 201)
    Includes bibliographical references and index.
    ISBN-13: 978-0-8493-7785-3 (hardcover : alk. paper)
    ISBN-10: 0-8493-7785-4 (hardcover : alk. paper)   1. Generic drugs.
2. Generic drugs–Law and legislation.   I. Kanfer, Isadore.   II. Shargel, Leon, 1941-   III. Series: Drugs and the pharmaceutical sciences ; 201.
    [DNLM: 1. Drugs, Generic–pharmacokinetics.   2. Drugs, Generic–standards. 3. Biological Availability   4. Drug Design.   5. International Coorperation.   6. Legislation, Drug.   7. Therapeutic Equivalency.
W1 DR893B v. 201 2010 / QV 38 G3255 2010]
    RS55.2G454 2010
    615'.19–dc22

                                                                                          2009046567

For Corporate Sales and Reprint Permission call 212-520-2700 or write to: Sales Department, 52 Vanderbilt Avenue, 7th floor, New York, NY 10017.

Visit the Informa Web site at
www.informa.com

and the Informa Healthcare Web site at
www.informahealthcare.com

Preface

HISTORY OF THIS SERIES OF BOOKS ON GENERIC DRUG PRODUCT DEVELOPMENT

The first book in this series, *Generic Drug Product Development: Solid Oral Dosage Forms,* was published in 2005 by Marcel Dekker, Inc., New York, as part of their series in drugs and the pharmaceutical sciences. The objective of the first book was to describe, from concept to market approval, the development of therapeutic equivalent generic drug products, including regulatory and legal challenges. The second volume, *Generic Drug Product Development: Bioequivalence Issues,* was published in 2008 by Informa Healthcare USA, Inc., New York, and focused on problems concerning the scientific demonstration of bioequivalence (BE) of two drug products: a test (T) and a reference (R) product.

ABOUT THIS BOOK

This book, the third in this series on *Generic Drug Product Development*, provides information on the regulatory requirements for bioequivalence in several different countries around the world.

The advent and application of bioequivalence testing has had a significant impact on the process of drug product development both for innovator/brand pharmaceutical manufacturers and also for the generic drug industry. In the former instance, innovator/brand companies' use BE during their formulation development of products prior to embarking on clinical studies on safety and efficacy and subsequently to facilitate postapproval changes in formulation. Generic drug manufacturers, on the other hand, are able to circumvent clinical trials in patients by using BE to obtain market approval for their products.

Although the underlying scientific principles of BE have been well-established and many drug regulatory agencies around the world have largely adopted standard methodology and acceptance criteria, some differences exist in the application of the methodology and interpretation of the outcomes among different countries. Hence the objective of this book is to outline the specific BE requirements of each of the included countries where such information will be useful to prospective generic drug manufacturers intending to obtain market approval in a specific country.

CONTENTS

The bioequivalence regulations and requirements of the following countries and their associated regulatory agencies are included in this book:

- Australasia—Therapeutic Goods Administration of Australia and Medsafe, a business unit of the New Zealand Ministry of Health
- Brazil—The National Agency for Sanitary Surveillance (Agencia Nacional de Vigilancia Sanitaria, ANVISA), General Management of Generic Products (Gerência Geral de Medicamentos Genéricos, GGMED)
- Canada—Health Products and Food Branch (HPFB), Therapeutic Products Directorate (TPD), Health Canada
- European Union—European Medicines Agency (EMEA), Committee for Medicinal Products for Human Use (CHMP)
- India—Central Drugs Standard Control Organization (CDSCO) under the Government of India having certain powers vested in them and each State has its own drug regulatory system having certain powers
- Japan—The Ministry of Health, Labour and Welfare (MHLW) in accordance with the Pharmaceutical Affairs Law
- South America—In accordance with guidelines of PAHO (Pan American Health Organization) through PANDRH (Pan American Network for Drug Regulatory Harmonization)
- South Africa—Medicines Control Council (MCC), Department of Health
- Taiwan—The Bureau of Pharmaceutical Affairs (BPA), Department of Health (DOH)
- Turkey—The General Directorate of Drug and Pharmacy (GDDP, Ministry of Health (MoH)
- United States of America—Food and Drug Administration
- World Health Organization (WHO)

The chapters include background information and history of the development of rules, regulations, and guidelines in a specific country along with patent information and national requirements for interchangeability where such requirements exist. This is followed by bioequivalence requirements including the choice of reference product, bioequivalence metrics used, and acceptance criteria as well as study designs for immediate-release and controlled-/modified-release products for oral administration. Sections also include information on the requirements for the assessment of food effects, metabolites and chiral drugs, use of "add-on" and steady-state studies, conditions for biowaivers, bioanalytical and statistical requirements. Most of the guidelines also include recommendations for the assessment of drug products not intended for absorption into the systemic circulation such as topical products for local/regional action (e.g., dermatological corticosteroid products and other creams, gels, ointments, and lotions as well as nasal sprays/aerosols and oral inhalation products). In addition, the bioequivalence requirements for postapproval formulation changes, biowaivers for additional strengths of the same product, and the use of dissolution data for biowaivers are also included.

Bioequivalence has been acknowledged as a valuable tool and has been used as a surrogate measure to assess the safety and efficacy of generic medicines, also termed multisource products. However, as mentioned previously, in general, the methods used and the bioequivalence requirements of most regulatory agencies have much in common, but in some countries specific issues and approaches differ. This volume provides relatively detailed information required by the regulatory authorities of several different countries and the implications of their different requirements and approaches for the market approval of a generic product, particularly with respect to it's declaration of interchangeability/switchability.

A list of definitions, abbreviations and symbols has been included where many of the terms are commonly used in most countries.

Isadore Kanfer
Leon Shargel

Acknowledgments

The editors wish to express their gratitude to all the authors who have given so generously of their time and expertise to provide these texts and to the organizations, for which they work, which have graciously allowed them to do so.

Disclaimer

The contents of this book represent the views of its authors. They do not necessarily reflect the policies or opinions of any national authorities (regulatory agencies), advisory or regulatory committees, pharmaceutical companies, contract research organizations, consultancy companies, academic institutions, hospitals, etc., with whom the authors may be associated or employed.

Annexure 1
Definitions, Abbreviations and Symbols

AM
    Active moiety (active)
Active moiety (active) is the term used for the therapeutically active entity in the final formulation of a medicine, irrespective of the form of the API. The active is alternative terminology with the same meaning. For example, if the API is propranolol hydrochloride, the active moiety (and the active) is propranolol.
Active pharmaceutical ingredient
A substance or compound that is intended to be used in the manufacture of a pharmaceutical product as a therapeutically active ingredient.

BA
    Bioavailability
Bioavailability refers to the rate and extent to which the API (active pharmaceutical ingredient), or its active moiety (substance), is absorbed from a pharmaceutical product (dosage form) and becomes available at the site of action or biological fluids (usually plasma or plasma) representing the site.

    For drug products that are not intended to be absorbed into the bloodstream, bioavailability may be assessed by measurements intended to reflect the rate and extent to which the active ingredient or active moiety becomes available at the site of action.

    It may be useful to distinguish between the "absolute bioavailability" of a given dosage form as compared with that (100%) following intravenous administration (e.g., oral solution vs. intravenous), and the "relative bioavailability" as compared with another form administered by the same or another nonintravenous route (e.g., tablets vs. oral solution).

BE
    Bioequivalence
Two pharmaceutical (medicinal) products are bioequivalent if they are pharmaceutically equivalent or (pharmaceutical alternatives in some countries) and if their bioavailabilities in terms of peak (C_{max} and T_{max}) and total exposure (AUC) after administration of the same molar dose under the same conditions are similar to such a degree that their effects with respect to both efficacy and safety can be expected to be essentially the same.

    The USA FDA's definition is:

    "the absence of a significant difference in the rate and extent to which the active ingredient or active moiety in pharmaceutical equivalents or pharmaceutical alternatives becomes available at the site of drug action when administered at the same molar dose under similar conditions in an appropriately designed study."

Bioequivalence focuses on the equivalence of release of the active pharmaceutical ingredient from the pharmaceutical product and its subsequent absorption into the systemic circulation.

Comparative studies using clinical or pharmacodynamic end points may also be used to demonstrate bioequivalence.

FDC   Fixed-dose combination
A combination of two or more active pharmaceutical ingredients in a fixed ratio of doses. This term is used generically to mean a particular combination of active pharmaceutical ingredients irrespective of the formulation or brand. It may be administered as single entity products given concurrently or as a finished pharmaceutical product.

FP   Final product
A product that has undergone all stages of production, excluding packaging.

FPP   Finished pharmaceutical product
A product that has undergone all stages of production, including packaging in its final container and labeling.

IPI   Inactive pharmaceutical ingredient
A substance or compound that is used in the manufacture of a pharmaceutical product and does not contribute to the therapeutic effect of the product, but is intended to enhance the consistency, appearance, integrity, stability, release characteristics, or other features of the product.

Usually refers to excipients or other formulation adjuvants/aids.

MSPP   Multisource (generic) pharmaceutical product
Multisource pharmaceutical products are pharmaceutically equivalent products that may or may not be therapeutically equivalent or bioequivalent. Multisource pharmaceutical products that are therapeutically equivalent are interchangeable.

MRDF   Modified-release dosage forms
A modified-release dosage form is one for which the drug-release characteristics of time course and/or location are chosen to accomplish therapeutic or convenience objectives not offered by conventional dosage forms such as solutions, ointments, or promptly dissolving dosage forms.

Delayed-release and extended-release dosage forms are two types of modified-release dosage forms.
- *Delayed-release dosage forms:* A delayed-release dosage form is one that releases a drug(s) at a time other than promptly after administration.
- *Extended-release dosage forms:* An extended-release dosage form is one that allows at least a twofold reduction in dosing frequency or significant increase in patient compliance or therapeutic performance as compared to that presented as a conventional dosage form (e.g., as a solution or a prompt drug-releasing, conventional solid dosage form).

The terms *controlled release, prolonged action,* and *sustained release* are used synonymously with extended release. The term *extended release* is used to describe a formulation that does not release active drug substance immediately after oral dosing and that also allows a reduction in dosage frequency. This nomenclature accords generally with the USP definition of extended release but does not specify an impact on dosing frequency. The terms *controlled release* and *extended release* are sometimes considered interchangeable.

PA   Pharmaceutical alternatives
Medicinal products are pharmaceutical alternatives if they contain the same active moiety but differ either in chemical form (e.g., salt, ester) of that moiety or in the dosage form or strength, administered by the same route of administration but are otherwise not pharmaceutically equivalent.

NB: *Pharmaceutical alternatives do not necessarily imply bioequivalence.*

PDF   Pharmaceutical dosage form (compare with PP)
A pharmaceutical dosage form is the form of the completed pharmaceutical product, e.g., tablet, capsule, injection, elixir, suppository.

PE   Pharmaceutical equivalents
Pharmaceutical products are pharmaceutically equivalent if they contain the same amount of the same API(s) in the same dosage form, if they meet the same or comparable standards, and if they are intended to be administered by the same route.

Pharmaceutical equivalence does not necessarily imply bioequivalence as differences in the excipients and/or the manufacturing process can lead to changes in dissolution and/or absorption.

The USA FDA's definition is:
"drug products that contain identical amounts of the identical active drug ingredient, i.e., the same salt or ester of the same therapeutic moiety, in identical dosage forms, but not necessarily containing the same inactive ingredients, and that meet the identical compendial or other applicable standard of identity, strength, quality, and purity, including potency and, where applicable, content uniformity disintegration times and/or dissolution rates " (21 CFR 320.1(c)).

Products with different mechanisms of release can be considered to be pharmaceutical equivalents or duplicates.

PP   Pharmaceutical product (compare with PDF)
Any preparation for human (or animal) use, containing one or more APIs with or without pharmaceutical excipients or additives, that is intended to modify or explore physiological systems or pathological states for the benefit of the recipient.

Proportionally Similar Dosage Forms/Products (South African Guidelines)

"Pharmaceutical products are considered proportionally similar in the following cases:

- When all APIs and inactive pharmaceutical ingredients (IPIs) are in exactly the same proportion between different strengths (e.g. a 100-mg strength tablet has all API and IPIs exactly half of a 200-mg strength tablet and twice that of a 50-mg strength tablet).
- When the APIs and IPIs are not in exactly the same proportion but the ratios of IPIs to the total mass of the dosage form are within the limits defined by the Post-registration Amendment guideline.
- When the pharmaceutical products contain a low concentration of the APIs (e.g. less than 5%) and these products are of different strengths but are of similar mass."

The difference in API content between strengths may be compensated for by mass changes in one or more of the IPIs provided that the total mass of the pharmaceutical product remains within 10% of the mass of the pharmaceutical product on which the bioequivalence study was performed. In addition, the same IPIs should be used for all strengths, provided that the changes remain within defined limits.

TE   Therapeutic equivalence (equivalents)

Two pharmaceutical products are therapeutically equivalent if they are pharmaceutically equivalent or pharmaceutical alternatives and, after administration in the same molar dose, their effects with respect to both efficacy and safety are essentially the same, as determined from appropriate bioequivalence, pharmacodynamic, clinical or *in vitro* studies.

According to the FDA's Orange Book, the following is stated: "Drug products are considered to be therapeutic equivalents only if they are pharmaceutical equivalents and if they can be expected to have the same clinical effect and safety profile when administered to patients under the conditions specified in the labeling."

FDA classifies as therapeutically equivalent those products that meet the following general criteria: (1) they are approved as safe and effective; (2) they are pharmaceutical equivalents in that they (a) contain identical amounts of the same active drug ingredient in the same dosage form and route of administration and (b) meet compendial or other applicable standards of strength, quality, purity, and identity; (3) they are bioequivalent in that (a) they do not present a known or potential bioequivalence problem, and they meet an acceptable *in vitro* standard, or (b) if they do present such a known or potential problem, they are shown to meet an appropriate bioequivalence standard; (4) they are adequately labeled; (5) they are manufactured in compliance with Current Good Manufacturing Practice regulations. *The concept of therapeutic equivalence, as used to develop the List, applies only to drug products containing the same active ingredient(s) and does not encompass a comparison of different therapeutic agents used for the same condition (e.g., propoxyphene hydrochloride vs. pentazocine hydrochloride for the treatment of pain).* Any drug product in the List repackaged and/or distributed by other than the application holder is considered to be therapeutically equivalent to the application holder's drug product even if the application holder's drug product is single source or coded as nonequivalent (e.g., **BN**). Also, distributors or repackagers of an application holder's drug product are considered to have the same code as the application holder. Therapeutic equivalence determinations are not made for unapproved, off-label indications.

FDA considers drug products to be therapeutically equivalent if they meet the criteria outlined above, even though they may differ in certain other characteristics such as shape, scoring configuration, release mechanisms, packaging, excipients (including colors, flavors, preservatives), expiration date/time and minor aspects of labeling (e.g., the presence of specific pharmacokinetic information), and storage conditions. When such differences are important in the care of a particular patient, it may be appropriate for the prescribing physician to require that a particular brand be dispensed as a medical necessity. With this limitation, however, FDA believes that products classified as therapeutically equivalent can be substituted with the full expectation that the substituted product will produce the same clinical effect and safety profile as the prescribed product.

CV	Coefficient of variation
LOD	Limit of detection
LOQ	Limit of quantification
LLOQ	Lower limit of quantitation
SOP	Standard operating procedure
QC	Quality control (LQC, MQC, HQC: low, medium, high quality control)
GCP	Good clinical practice
GLP	Good laboratory practice
CI	Confidence interval
C_{pd}	Concentration in a predose sample immediately following dosing in a steady state
AUC_t	Area under the plasma/serum/blood concentration–time curve from time zero to time t, where t is the last time point with measurable concentration.
AUC_∞	Area under the plasma/serum/blood concentration–time curve from time zero to time infinity.
AUC_τ	Area under the plasma drug concentration–time curve over one dosing interval at steady state.
	Cumulative urinary excretion from pharmaceutical product administration until time t.
Ae_∞	Amount of unchanged API excreted in the urine at infinite time (7–10 half-lives).
C_{max}	Maximum plasma concentration. It is the maximum drug concentration achieved in systemic circulation following drug administration.
C_{min}	Minimum plasma concentration It is the minimum drug concentration achieved in systemic circulation following multiple dosing at steady state.
$C_{max\,(ss)}$	Maximum plasma concentration at steady state
$C_{min\,(ss)}$	Minimum plasma concentration at steady state

C_{av}	Average plasma concentration
t_{max}	Time to C_{max}
	It is the time required to achieve maximum drug concentration in systemic circulation.
K_{el}	Apparent first-order terminal elimination rate constant calculated from a semi-log plot of the plasma concentration versus time curve
MRT	Mean residence time
$t_{1/2}$	Elimination half-life
	It is the time necessary to reduce the drug concentration in the blood, plasma, or serum to one-half of the initial concentration.
%PTF	$(C_{max\ (ss)}-C_{min\ (ss)})/C_{av}.100$
%Swing	$(C_{max\ (ss)}-C_{min\ (ss)})/C_{min}.100$

Contents

Contributors

Ricardo Bolaños Department of Pharmacology, School of Medicine, University of Buenos Aires; Department of Projects and Plans, Direction of Planification, National Administration of Drugs, Food and Medical Technology (ANMAT), Ministry of Health; and Working Group of Bioequivalence, Pan American Network of Drug Regulatory Harmonization, PAHO, Buenos Aires, Argentina

Márcia Martini Bueno Regulatory Affairs and Pharmacovigilance, Libbs Farmacêutica Ltda, São Paulo, Brazil

Jo-Feng Chi Bureau of Pharmaceutical Affairs, Department of Health, The Executive Yuan, Taipei, Taiwan, Republic of China

Dale P. Conner Office of Generic Drugs, Center for Drug Evaluation and Research, United States Food and Drug Administration, Rockville, Maryland, U.S.A.

Barbara M. Davit Office of Generic Drugs, Center for Drug Evaluation and Research, United States Food and Drug Administration, Rockville, Maryland, U.S.A.

L. A. Folland Zenith Technology Ltd, Dunedin, New Zealand

Silvia Susana Giarcovich Department of Pharmacology, School of Pharmacy and Biochemistry, University of Buenos Aires; and DIFFUCAP-EURAND SACIFI, Buenos Aires, Argentina

John Gordon Division of Biopharmaceutics Evaluation 2, Bureau of Pharmaceutical Sciences, Therapeutic Products Directorate, Health Canada, Ottawa, Ontario, Canada

A. Atilla Hincal IDE Pharmaceutical Consultancy Ltd. Co., Ankara, Turkey

Oliver Yoa-Pu Hu National Defense Medical Center, Taipei, Taiwan, Republic of China

C. T. Hung Zenith Technology Ltd, Dunedin, New Zealand

Noelyn Anne Hung Otago University Dunedin School of Medicine, Dunedin, New Zealand

Isadore Kanfer Faculty of Pharmacy, Rhodes University, Grahamstown, South Africa

Ilker Kanzik   IDE Pharmaceutical Consultancy Ltd. Co., Istanbul, Turkey

F. C. Lam   Zenith Technology Ltd, Dunedin, New Zealand

Subhash C. Mandal   Directorate of Drugs Control, "NALANDA," Fartabad, Amtola, Kolkata, India

Margareth R. C. Marques   Department of Standards Development, U.S. Pharmacopeia, Rockville, Maryland, U.S.A.

Iain J. McGilveray   McGilveray Pharmacon Inc., Ottawa, Ontario, Canada

Li-Heng Pao   School of Pharmacy, National Defense Medical Center, Taipei, Taiwan, Republic of China

Henrike Potthast   Federal Institute for Drugs and Medical Devices, Bonn, Germany

Lembit Rägo   WHO Medicines Prequalification Programme, Quality Assurance and Safety of Medicines, Essential Medicines and Pharmaceutical Policies, World Health Organization, Geneva, Switzerland

S. Ravisankar   GVK Biosciences Pvt. Ltd., Ameerpet, Hyderabad, India

D. Ren   Zenith Technology Ltd, Dunedin, New Zealand

Juichi Riku   Meiji Pharmaceutical University, Tokyo, Japan

Michael F. Skinner   Biopharmaceutics Research Institute, Rhodes University, Grahamstown, South Africa

R. Smart   Douglas Pharmaceuticals, Auckland, New Zealand

Matthias Stahl   WHO Medicines Prequalification Programme, Quality Assurance and Safety of Medicines, Essential Medicines and Pharmaceutical Policies, World Health Organization, Geneva, Switzerland

Sílvia Storpirtis   College of Pharmaceutical Sciences, University of São Paulo, São Paulo, Brazil

Roger K. Verbeeck   School of Pharmacy, Catholic University of Louvain, Brussels, Belgium; and Faculty of Pharmacy, Rhodes University, Grahamstown, South Africa

Roderick B. Walker   Faculty of Pharmacy, Rhodes University, Grahamstown, South Africa

Joelle Warlin   Federal Agency for Medicines and Health Products, Brussels, Belgium

1 Introduction

Isadore Kanfer

Faculty of Pharmacy, Rhodes University, Grahamstown, South Africa

INTRODUCTION

The introduction and application of bioequivalence testing as a surrogate indicator of safety and efficacy has had an enormous impact on the development and formulation of solid oral dosage forms in particular, not only for generic product manufacturers but also for the innovator or "brand" companies. Using bioequivalence as a tool to establish therapeutic equivalence between a test (generic) and a reference product certainly forms the basis for market approval of multisource drug products, the latter being products marketed by more than one manufacturer and containing the same active pharmaceutical ingredient (API) in the same dosage form intended to be administered by the same route of administration. The reference product is usually the comparator product developed and marketed by a so-called "innovator" or brand company, where their product had been approved on the basis of clinical trials to support claims for safety and efficacy. It is however a little-known fact that many innovator products that eventually become available on a particular market were in fact approved on the basis of a bioequivalence study (1) between the formulation used in the clinical trials (as reference) and an amended formulation of the same active(s), that is, the test product. Hence, it is apparent that using bioequivalence methodology and sound scientifically based acceptance criteria to show therapeutic equivalence overcomes the need to redo clinical trials that are expensive and time consuming, often taking several years to produce outcomes. As mentioned previously, it is in the area of solid oral dosage forms where the major benefits of bioequivalence testing occur since plasma drug concentrations can be measured following oral administration of a test and reference product in a cross-over fashion in healthy human subjects. This procedure is feasible and scientifically sound for drugs, which are intended to be absorbed into the systemic circulation since the underlying principle is that a link exists between drug concentrations in the blood and therapeutic effect. However, for other dosage forms, particularly those intended for other routes of administration and more specifically those which are not intended to be absorbed into the systemic circulation, generally no such link exists between drug concentrations in blood and effect. Although some specific methodologies have been developed and validated, such as the human skin blanching assay (HSBA) for topical corticosteroid products, to date relatively few surrogate measures to assess bioequivalence of such dosage forms have been accepted by regulatory agencies. It is thus apparent that the main objective of bioequivalence testing is to compare the formulation performance between a test and reference product with the premise that if bioequivalence is proven, there is unlikely to be any significant differences in clinical outcomes between such products.

Each country has specific regulatory requirements to approve these products as therapeutic equivalents and therefore considered interchangeable with the innovator/brand drug product. Regulatory approval of therapeutic equivalents generally requires that the generic drug product meets certain requirements including: (*i*) they are approved as safe and effective; (*ii*) they are pharmaceutical equivalents in that they (a) contain identical amounts of the same active drug ingredient in the same dosage form and route of administration and (b) meet compendial or other applicable standards of strength, quality, purity, and identity; (*iii*) they are bioequivalent in that (a) they do not present a known or potential bioequivalence problem, and they meet an acceptable in vitro standard, or (b) if they do present such a known or potential problem, they are shown to meet an appropriate bioequivalence standard; (*iv*) they are adequately labeled; (*v*) they are manufactured in compliance with Current Good Manufacturing Practice regulations. Therapeutically equivalent products can be substituted with the full expectation that the substituted product will produce the same clinical effect and safety profile as the prescribed product[a]. As previously mentioned, globally, regulatory approval for interchangeable multisource generic drug products is not identical in all countries and the regulatory agency for each country may differ in its regulatory requirements for the demonstration of bioequivalence. Moreover, there is no universal reference listed drug to be used as the comparator for the proposed generic drug product. Each country establishes its own reference listed drug product that is commonly available in its own domestic marketplace. In addition, the statistical criteria for bioequivalence may vary in each country where the brand drug product is marketed. Consequently, a generic drug product that has established bioequivalence to the reference listed drug product in the United States may not be bioequivalent to the reference listed drug product in another country. The following is a summary of regulatory requirements by the relevant countries included in this book.

AUSTRALASIA (AUSTRALIA AND NEW ZEALAND)

The chapter on Australasia includes the Australian requirements according to the Therapeutic Goods Administration (TGA) of Australia controlled by the Therapeutic Goods Act 1989 (2) and the New Zealand requirements as set out by Medsafe, a business unit of the New Zealand Ministry of Health, in accordance with the Medicines Act 1981, Section 20 (3). It was initially intended that from July 1, 2006, all medicines were to receive regulatory approval from a proposed Trans-Tasman Joint Regulatory Authority. This Joint Agency was to have responsibility for the control of medicines in both Australia and New Zealand. However establishment of the Joint Agency has been postponed and there has been no indication if and when such an agency will come into being. Currently, the Australian and New Zealand registration requirements are largely aligned with the Committee of Proprietary Medicinal Products (CPMP) guidelines of the

---

[a] Approved Drug Products with Therapeutic Equivalence Evaluations," Orange Book, (www.fda.gov/downloads/Drugs/DevelopmentApprovalProcess/UCM071436.pdf).

European Medicines Agency (EMEA) (4). Prescription medicines for Australia are evaluated by the Drug Safety and Evaluation Branch (DSEB) of the TGA and generic prescription medicines in both Australia and New Zealand are required to conform to the (European Union) EU's Common Technical Document (CTD) format although, unlike the EU, an abbreviated format is acceptable.

BRAZIL

In Brazil, the National Policy for Drug Products published in 1998 created the National Agency for Sanitary Surveillance (Agencia Nacional de Vigilancia Sanitaria, ANVISA), which is responsible for the approval of the law and the publication of the technical guidances for the registration of generic products (5–8). The law and guidances for the registration of generic products in Brazil are based on the regulations of countries such as Canada, United States of America, and the EU. Brazil was the first country in South America to implement the evaluation of pharmaceutical equivalency and bioequivalence studies for the registration of generic products.

CANADA

Canadian regulators, under the Health Products and Food Branch (HPFB), Therapeutic Products Directorate (TPD), Health Canada, were the first to apply pharmacokinetics (PK) to safety and efficacy risk assessment of generic drug products as a consequence of 1969 amendments to the Patent Act (9) (compulsory licensing). Guidelines [Reports A, B, and C (10–12)] were only published in the 1990s by an Expert Advisory Committee (EAC), currently referred to as the Scientific Advisory Committee (SAC). However, Canada is governed as a Confederation of Provinces and the regulations and guidelines for bioequivalence are federal, leading to a Notice of Compliance (NOC) to sponsors for marketing in Canada. Although an application may lead to a Declaration of Bioequivalence for specific products, the various Canadian provinces and associated Provincial formulary committees may not accept the Federal decision of bioequivalence (which also evaluates quality) to be sufficient to list a particular product as interchangeable.

Revision of the Canadian guidelines was mentioned during presentations at the 2008 and 2009 annual meetings of the Canadian Society of Pharmaceutical Sciences (CSPS) held in Banff during May 2008 and in Toronto during June 2009 (13).

EUROPEAN UNION

The EU has established requirements which must be met by a generic drug product to receive marketing authorization (14). The EU offers four routes for the registration of generic drug products (15):

1. National Procedure
2. Mutual Recognition Procedure
3. Decentralized Procedure
4. Centralized Procedure.

The *national procedure* (NP) is strictly limited to medicinal products that are not authorized in more than one member state and may lead to marketing authorization of the generic drug product in the concerned member state. The *mutual recognition procedure* (MRP) makes provision for the extension of marketing authorizations granted to one member state, the so-called reference member state (RMS), to one or more member states identified by the applicant. The *decentralized procedure* (DCP) was implemented from November 2005 where a submission can be made to each of the member states, where it is intended to obtain a marketing authorization with a choice of one of them as the RMS. The *centralized procedure* (CP) has been in use since 2004 for marketing authorization of medicinal products in the EU. Here a single application is introduced for evaluation that is carried out within the Committee for Medicinal Products for Human Use (CHMP) of the EMEA and is valid throughout the EU with the same rights and obligations in each of the member states.

The first European bioequivalence guidelines were published in 1991 (16) and assessment was based on the principles published in the scientific literature, FDA guidelines, and the first European guidelines on pharmacokinetic studies in man (17). In 2001, the EMEA's CPMP published the current version of the Note for Guidance on the Investigation of Bioavailability and Bioequivalence (18). A draft version of revised bioequivalence guidelines, entitled Guideline on the Investigation of Bioequivalence, was made publicly available in August 2008 on the EMEA Web site and a modified version is due to come into effect in 2010 (19).

INDIA

Bioequivalence assessment for generic medicines in India was instituted by the incorporation of Schedule Y of the Drugs and Cosmetics Act in 1988, followed by subsequent amendments of Schedule Y in 1989 and 2005 (20).

In India, generic medicines are those medicines, which are labeled with their generic names. There is no separate law for registering generic medicines. Drugs and drug products are classified as either "new drugs" or drugs other than new drugs. However, in general "generic drugs" refer to those drugs than are no longer subject to patent protection and are being marketed by their generic name. In fact, it's interesting to note that India only started observing product patents from January 1, 2005 (21,22).

India has a two tier regulatory system, the Central Drugs Standard Control Organization (CDSCO) under the Government of India and each State has its own drug regulatory system having certain powers.

JAPAN

In Japan, data from 2006 indicated that generic drug products accounted for as little as 16.9% of the market by volume 5.7% by value. In the following year, the Japanese government announced a specific program to increase the use of generic products to more than 30% by the year 2012 and a system for generic substitution was targeted for implementation in April 2008.

Approval to manufacture and market generic drug products in accordance with the Pharmaceutical Affairs Law in Japan is granted by the Minister of Health, Labour and Welfare (MHLW). An independent administrative organization, the Pharmaceutical and Medical Devices Agency (PMDA), under the auspices of the MHLW, undertakes the review and assessment of bioequivalence data as part of the "Equivalence and Compliance Review."

The first guideline for bioequivalence studies for generic dugs was released in 1971, in which large animals, such as dogs and rabbits, could be used in bioequivalence studies, but humans were not required. Subsequently in 1997, the MHLW published a revised Guideline "The Guideline for Bioequivalence Studies of Generic Products 1997 (23) was introduced. Several newer guidelines and revisions have been published such as the Guideline for Bioequivalence Studies for Different Strengths of Oral Solid Dosage Forms 2000 (24), Guideline for Bioequivalence Studies for Additional Dosage Forms of Oral Solid Dosage Forms 2001 (25), Guideline for Bioequivalence Studies of Generic Products for Topical Dermal Application 2003 (26), Guideline for Bioequivalence Studies for Formulation Changes of Oral Solid Dosage Forms 2000 (27), Draft Guideline for Bioequivalence Studies for Changes in Manufacturing of Oral Solid Dosage Forms: Conventional, Enteric Coated and Prolonged Release Products, 2003 (28), and Revision of Guideline for Bioequivalence Studies of Generic Products, 2006 (29).

For oral drug products, in Japan, dissolution testing of these dosage forms is required together with bioequivalence studies and plays an important role in selecting appropriate subjects for the in vivo study. In fact, it is noteworthy that the use of dissolution testing for BA/BE in Japan clearly marks a distinct difference between Japanese BA/BE guidelines and those of most other countries.

SOUTH AFRICA

The Medicines Control Council (MCC) in South Africa is a statutory body that was established in terms of the Medicines and Related Substances Control Act (MRSCA), 101 of 1965, to oversee the regulation of medicines in South Africa. To facilitate the registration process for generic medicines, guidelines have been prepared to serve as a recommendation to applicants wishing to submit data in support of the registration of such medicines (30–33).

In the early 2000s, the requirements for the registration and market approval of generic medicines were published as official guidelines. Prior to that time, proof of safety and efficacy of generic medicines were based on requirements described in "official" notices or circulars issued by the MCC. In many instances, only in vitro dissolution testing was required based upon Circular 14/95 that was first issued in the early 1990s and subsequently updated in 1995 (34).

During 2002 (35,36), legislation was introduced to make provision for generic substitution whereby dispensers of medicines were mandated to inform patients of the benefits of the substitution for a branded medicine by an interchangeable multisource medicine and to recommend accordingly. The final decision, however, has been left in the hands of the patient.

SOUTH AMERICA (EXCLUDING BRAZIL)—PAN AMERICAN HEALTH ORGANIZATION

In South American countries (Brazil has been dealt with previously) harmonization efforts have been carried out by different economic integration groups, such as TLCN (*Tratado de Libre Comercio de Norteamérica*) in Canada, Estados Unidos, and México; MERCOSUR in Argentina, Brazil, Paraguay, and Uruguay (Chile and Bolivia participate without being members); SICA (*Sistema de la Integración Centroamericana*) in Guatemala, El Salvador, Honduras, Nicaragua, and Costa Rica; CAN (*Comunidad Andina de Naciones*) in Bolivia, Colombia, Ecuador, Perú, and Venezuela; CARICOM (Caribbean community) in Caribbean Islands; and by PAHO (Pan American Health Organization) through PANDRH (Pan American Network for Drug Regulatory Harmonization). Generally, most South American countries, apart from Brazil, do not have regulations for the registration of generic products as such. However, proof of bioequivalence and an inference of therapeutic equivalence and/or declaration of interchangeability (through either in vitro or in vivo methodology) are required as a condition either for registration or commercialization of generic (i.e., noninnovator) products.

TAIWAN
The Bureau of Pharmaceutical Affairs (BPA) within the Department of Health (DOH) is responsible for the regulation of medicinal products in Taiwan. In 1984, the BPA outsourced the drafting of BA/BE guidelines and Taiwan guidelines on BA/BE for generic medicines were issued in 1987. Generic products are reviewed and approved within the BPA at the DOH. The applicant needs to certify to the DOH that the patent in question is not infringed by the generic product. Once the BPA staff have completed the filing review of the submission and have verified that the application has met all the necessary regulatory requirements, the application is then assigned to technical review which focuses on BE data and chemistry and manufacture quality.

TURKEY
Marketing authorization for medicinal products for humans are issued by the Ministry of Health (MoH) in Turkey, but the entire procedure is managed by the General Directorate of Drug and Pharmacy (GDDP). Turkish licensing regulations for all pharmaceutical products, innovator/brand and generics, came into force on June 30, 2005 (37). This legislation brought Turkish law in line with that of the EU and covered all aspects of the drug registration procedure and all new Turkish regulations are intended to be, as far as possible, compatible with those of the EU. However, bioequivalence became compulsory for a generic drug product to receive a marketing license after the publication of a regulation on May 27, 1994 (38). In general, the design and conduct of a BE study should follow Turkish and/or EU regulations on Good Clinical Practice (GCP).

UNITED STATES OF AMERICA
The Food and Drug Administration (FDA) in the United States of America introduced the Abbreviated New Drug Application (ANDA) procedure in 1984[b], which is used for the marketing approval of generic drug products. The ANDA pathway for generic drug product approval was enacted due to earlier concerns in the 1960s that there may be differences in bioavailability between two chemically equivalent products (39). In 1974, a report entitled "Drug Bioequivalence" (40) was published and in 1977, the FDA published its Bioavailability and Bioequivalence regulations that facilitated the rational development of generic drug

---

[b] The Drug Price Competition and Patent Term Restoration Act of 1984, informally known as the "Hatch-Waxman Act" is an amendment to the U.S. Food and Drug Law that established the Abbreviated New Drug Application (ANDA) as a pathway for approval of generic drug products.

products (41), which was revised in 1992 (42). The current FDA guidance for industry was posted in 2003 and updates recommendations for documentation of bioavailability and bioequivalence required for regulatory submissions (43). Although this guidance indicates that it provides recommendations for orally administered drug products, the guidance also applies to nonorally administered drug products where reliance on systemic exposure measures is suitable to document bioavailability/bioequivalence such as for transdermal and some rectal and nasal products. Several separate guidances have also been published and which relate to topical dermatologic corticosteroids (44), food effects (45), and guidance on waivers for in vivo studies for immediate-release solid oral dosage forms on the basis of the Biopharmaceutics Classification System (BCS) (46). On Friday, January 16, 2009, "Final rule" on the requirements for submission of Bioequivalence Data was published in the Federal Register (47). Several other guidances that have been published are given here.

CDER Guidance for Industry: BE Recommendations for Specific Products (48) and CDER Guidance for Industry: Individual Product Bioequivalence Recommendations (49). These latter two guidances provide information on specific drug products relating to the design of BE studies to support ANDAs and information on individual products. Another draft, Guidance on the Submission of Summary Bioequivalence Data for ANDAs (50), relates to the FDA's requirement that ANDA applicants submit data from *all* BE studies that the applicant conducts on a drug product formulation including studies that do not demonstrate bioequivalence of their generic product submitted for approval. This guidance also provides information on the format for summary reports of BE studies and types of formulations that the FDA considers to be the same drug product formulation for different dosage forms, based on differences in composition.

The ANDA is submitted to the Office of Generic Drugs (OGD)[c] and upon filing acceptance the application is assigned for bioequivalence review, where the review process assesses the bioequivalence data comparing the generic product and the reference listed drug (RLD) indicated in the Orange Book (51).

WORLD HEALTH ORGANIZATION

World Health Organization (WHO) has developed a document entitled "937, WHO Expert Committee on Specifications for Pharmaceutical Preparations," 40th Report (52) in which annex 7 relates to bioequivalence. WHO, which is the directing and coordinating authority for health within the United Nations system, is responsible for providing leadership on global health matters, shaping the health research agenda, setting norms and standards, articulating evidence-based policy options, providing technical support to countries, and monitoring and assessing health trends. In particular, the intention is to help national and regional authorities (in particular drug regulatory authorities), procurement agencies, as well as major international bodies and institutions, to combat problems of substandard medicines and underpin important initiatives. Importantly, since the overall tendency is that resource-constrained or resource-poor countries are less likely to control the quality of products on the market, due to the absence

---

[c] New BE Review Organization: Divisions 1 and 2; http://www.fda.gov/AboutFDA/CentersOffices/CDER/ucm119462.htm. Accessed June 8, 2009.

of properly resourced and functioning regulatory authorities, WHO publications and guidelines are intended to fill an important void by providing appropriate information in the quest to ensure medicinal product quality.

ISSUES OF SEMANTICS AND DIFFERENCES IN INTERPRETATION

Several semantic issues prevail in the official documents and guidelines published by various regulatory authorities, which appear to have resulted in confusion and misinterpretation of criteria and conditions for the declaration of bioequivalence and consequently establishing the interchangeability/switchability of a generic product for and innovator/brand product.

The definition of the term "bioequivalence" itself gives rise to a degree of misunderstanding when related to generic substitution. For example, the Orange Book (51) provides the following definition for bioequivalence, viz:

> the absence of a significant difference in the rate and extent to which the active ingredient or active moiety in pharmaceutical equivalents or pharmaceutical alternatives becomes available at the site of drug action when administered at the same molar dose under similar conditions in an appropriately designed study.

Importantly, in this definition, provision is made for "pharmaceutical alternatives" to be declared bioequivalent to the listed reference product in the Orange Book. However, pharmaceutical alternatives are not considered as therapeutic equivalents and therefore are neither interchangeable nor switchable.[d]

According to the Orange Book (51), the *following is stated:*

> Drug products are considered to be therapeutic equivalents only if they are pharmaceutical equivalents and if they can be expected to have the same clinical effect and safety profile when administered to patients under the conditions specified in the labeling.

Consequently, in terms of the FDA's requirements, only products classified as therapeutically equivalent can be substituted with the full expectation that the substituted product will produce the same clinical effect and safety profile as the prescribed product. FDA specifically excludes a "pharmaceutical alternative" dosage form from attaining therapeutic equivalent status. In contrast, this specific requirement for therapeutic equivalence is either overlooked or deemed unnecessary in several countries where pharmaceutical alternatives, once shown to be bioequivalent are considered to be interchangeable. Reference to the WHO document (52) includes the following definitions/descriptions where the term, pharmaceutical alternatives is mentioned. For example, a *multisource (generic) pharmaceutical product* is described as pharmaceutically equivalent or pharmaceutically alternative products that may or may not be therapeutically equivalent. This intention is unclear since what would be the basis of market approval for a pharmaceutically equivalent or pharmaceutically alternative generic product that has not been shown to be therapeutically equivalent?

---

[d] Pharmaceutical alternatives may be a capsule and a tablet containing the same active drug substance in the same dose strength. However, the U.S. FDA does not allow capsule formulations to be interchanged for tablet formulation of the same drug even if both products are bioequivalent.

COMPARATOR (REFERENCE) PRODUCTS

A further confusing issue also arises from some of the requirements stated in the bioequivalence WHO document (52). Specifically, mention is made in that document of a "Comparator Product" under section 3 entitled *Choice of Comparator Products*. This section describes how to identify the correct Comparator Product against which the proposed multisource pharmaceutical product (MPP or generic) must be compared. Generally, the comparator or reference product for use in a bioequivalence study needs to be the "nationally authorized innovator," usually the innovator/brand product that has been approved and available on that domestic market. However, the WHO document makes provision for some options of choice regarding the comparator product by including the permitted use of an innovator product currently available on the market in a well-regulated country. This brings into consideration the use of a foreign or nondomestic reference product. Although such a choice is permitted only when a nationally authorized innovator may not be available (i.e., countries where no innovator/brand product is available on their domestic market), this constraint appears to have been dismissed by some countries where in spite of the availability of a nationally authorized innovator, use of a foreign reference product is permitted. The WHO comparator product was instituted specifically as an effort to assist national drug regulatory authorities and pharmaceutical companies in selecting appropriate comparator products from a list of comparator products derived from information collected from drug regulatory authorities and the pharmaceutical industry. The WHO document includes the following:

> Note: a product that has been approved based on comparison with a non-domestic comparator product may or may not be interchangeable with currently marketed domestic products.

Hence, it is important to be aware that should a nationally authorized innovator product not be available as the comparator product, products approved on the basis of comparison to the chosen comparator product may or may not be interchangeable with other products currently available within that particular market. In other words, "generic substitution" per se cannot be recommended unconditionally.

GENERIC SUBSTITUTION (INTERCHANGEABILITY)

In some countries, where the market share of a new generic product may be relatively small, sponsors tend to submit bioequivalence studies where the reference product from another country (foreign/nondomestic) was used. Therefore, bioequivalence studies that were performed in compliance with the requirements of one country are submitted in support of approval of that generic in another country. The intent for such practice is to reduce costs and avoid the necessity of performing an additional bioequivalence study. However concerns have been raised about such practice in view of the fact that different formulations of the same product are often used in different countries and may have significant consequences with respect to bioavailability.

A unique situation currently exists in South Africa. The government, a few years ago, decided that when a generic medicine has been approved for marketing by the national medicines regulatory authority, the MCC, prescribers and dispensers of medicines must inform the patient that such prescribed medicines may be available at a lower price than the innovator/brand product and recommend accordingly. In other words, where an approved generic medicine exists, the generic medicine will be substituted for the innovator/brand product unless the dispenser is prohibited to do so by the patient. The mandatory instruction is based on the premise that approved generic medicines have been assessed by the MCC and deemed to be interchangeable. However, in terms of the national guidelines, most generic medicines approved and marketed in South Africa do not comply with the usual internationally accepted requirements for interchangeability since most have not been assessed by comparison with the innovator/brand product available on the South African market. The reason for the "noninterchangeability" notion is that provision has been made in the local guidelines to permit bioequivalence assessment to be conducted using a "foreign" reference product. This means that the "foreign" reference product, although being supplied by the same innovator/brand company, may not be the same (identical formulation, manufacture, etc.) as the innovator/brand product being sold on the South African market.

There are many instances where the innovator/brand products are formulated differently for different markets. For example, Tegretol XR® tablets, a prolonged action carbamazepine product is marketed in the United States as a nondisintegrating dosage form using the OROS® mechanism. The same innovator is listed as the manufacturer of prolonged action carbamazepine dosage forms in various other countries where those dosage forms are also tablets but which disintegrate in aqueous fluid and are marketed as Tegretol CR® in South Africa. The release mechanisms and formulation of the United States reference listed product and the product marketed by the same innovator in South Africa are clearly different. In some instances this is done intentionally due to patents. In other cases, there may be unintended differences in release of the active ingredient(s), due to various factors such as the manufacturing process. Bioavailability differences between products can be due to factors such as sources of raw material and synthesis (nature) of the API including particle size and crystal forms (polymorphs, crystal shapes and degree of hydration or salvation, etc.), use of different methods of manufacture and manufacturing equipment, amongst others. All of these factors can have significant effects on bioavailability with consequent implications for their bioequivalence. Hence, in the absence of specific confirmatory data, a nondomestic innovator/brand product used as the reference product in a bioequivalence study involving a generic medicine intended for a particular domestic market cannot be assumed to be bioequivalent to the domestic innovator/brand product. In spite of the foregoing, the only data required by the MCC to show that the "foreign" reference product is the "same" as the reference product marketed in South Africa are dissolution profiles comparing f_2 values between the foreign and domestic reference product conducted in three different dissolution media at pH 1.5, 4.5, and 6.8. These comparisons are not constrained to any particular class of medicine or properties such as the BCS (53) or drug use, potency, therapeutic index (e.g., narrow) amongst others. Risk assessment has apparently neither been done nor is required?

Whereas it should be noted that a generic product which has been shown to be bioequivalent to a reference product purchased in a nondomestic market will likely provide a spectrum of safety/efficacy associated with the included active ingredient, it is important to stress that such a generic product cannot be deemed to be equivalent to the innovator/rand product available on another (domestic) market and vice versa, unless appropriate data have been obtained to show bioequivalence. In other words, a generic product that has been shown to be bioequivalent to a nondomestic reference product may well be usable or "prescribable" but in the absence of the necessary comparative bioavailability data showing bioequivalence between that same generic product and the domestic reference product, that generic product cannot be considered to be interchangeable.

From the foregoing discussion it appears that the MCC, by making provision for a foreign reference product to be used in a bioequivalence study (i.e., using a nondomestic innovator/brand product as the reference) has inadvertently and naively created a two-tiered system for the approval of generic medicines in South Africa. The top tier can therefore be considered to consist of generic products approved on the basis of comparison with the domestic innovator/brand product as the reference, whereas another (second or lower) tier includes those generic products approved on the basis of a comparison of the generic with a nondomestic innovator/brand as the reference in a bioequivalence study. In addition, the latter tier probably also includes all other generic medicines which have been approved on the basis of in vitro testing only, apart from those products which incorporate an API classified as Class 1 according to the BCS (53,54).

It also appears that a similar situation of ill-conceived interchangeability for approved generic controlled/modified release dosage forms and also nonoral dosage forms such as topical products for local use, inhalation products and various other such generic products which are not intended to be absorbed into the systemic circulation.

In summary, it should be emphasized that only when a generic medicine has been shown to be bioequivalent with the domestic innovator/brand product used as reference, will substitution be acceptable and appropriate. On the basis of "similar" bioavailability, as would the case be when the generic has been shown to be bioequivalent to a nondomestic reference product, a clinical decision may be justified to declare a "second tier" generic product "prescribable" for a patient who is naïve to that particular medicine, but certainly interchangeability of such a product is highly questionable.

BIOWAIVERS

The main intentions of a biowaiver is to provide a cost-effective approach to improve the efficiency of drug development and the review process by recommending a strategy for identifying expendable clinical bioequivalence tests.[e] The BCS published by Amidon et al. in 1995 (53) finally lead to consideration of the possibility of waiving in vivo bioequivalence studies for certain immediate-release drug products in favor of specific comparative in vitro testing to conclude bioequivalence without requiring an in vivo bioequivalence study. This approach is meant to reduce unnecessary in vivo bioequivalence studies and the use of human subjects but is, however, restricted to noncritical drug substances in terms of solubility, permeability, and therapeutic range, and to noncritical

---

[e] http://www.fda.gov/AboutFDA/CentersOffices/CDER/ucm128219.htm.

pharmaceutical forms. Several countries, however, such as Canada, Japan, Taiwan, and Brazil, amongst others, do not make provision for BCS biowaivers, even for Class 1 drugs whereas the WHO have proposed an extension of the BCS for biowaivers for some Classes 2 and 3 drugs (52). In general, BCS-based biowaivers appear to be relatively under-used due to uncertainties by both pharmaceutical companies and regulatory authorities (55) although some countries have permitted the indiscriminate use of BCS dissolution conditions for certain biowaivers.

SUMMARY

Most countries appear to have adopted the basic tenets and approaches used by those countries that pioneered (Canada and United States) the introduction and application of this particular tool for the assessment of bioequivalence of different formulations of the same drug product. Apart from its general use, both by innovator and generic drug companies, there still appears to be a certain degree of naivety and ignorance associated with BE testing where the notion persist that it is specifically a tool for use only in the assessment of generic drug products.

However, some causes for concern still remain, and those relate to issues of semantics and interpretation of the main objectives such as, for example, the association between bioequivalence, therapeutic equivalence, and generic substitution. In particular, statements such as,*"may or may not be interchangeable,"* or *"may or may not be bioequivalent or therapeutically equivalent,"* which appear in some formal definitions such as the following are disconcerting.

- Note: a product that has been approved based on comparison with a nondomestic comparator product *may or may not* be interchangeable with currently marketed domestic products (Ref. 52, 6.5.2, p. 364).
- Multisource pharmaceutical products (52).
  Pharmaceutically equivalent or pharmaceutically alternative products that *may or may not* be therapeutically equivalent. MPP that are therapeutically equivalent are interchangeable.
- Pharmaceutical alternatives (52).
  Pharmaceutical alternatives deliver the same active moiety by the same route of administration but are otherwise not pharmaceutically equivalent. They *may or may not* be bioequivalent or therapeutically equivalent to the comparator product.

In addition, the term *"essentially similar"* (56) as used in EU guidances and documents creates confusion with respect to considerations of bioequivalence, therapeutic equivalence, and interchangeability. The issue of *pharmaceutical alternatives* remains inconsistent between countries with respect to permission for market approval with interchangeability status.

Finally, the issue and questionable use of an "acceptable reference product" or comparator product for generic substitution (AB rating) differs between countries. In the United States it is quite clear that the reference product for use in a bioequivalence study must be a reference-listed drug (RLD) as published in the Orange Book (51) whereas in many other countries, including the EU, a nondomestic reference product may be used to establish bioequivalence, therapeutic equivalence, and interchangeability of an MPP.

In spite of the foregoing, the applications of the principles of average bioequivalence and its related methodologies have clearly served the international pharmaceutical industry very well over the years since its introduction. BE testing has proved to be an extremely valuable tool, expedient, efficient, and relatively inexpensive in comparison with the costs to conduct clinical trials. It serves both innovator/brand companies during their formulation development and post-approval changes of products initially approved on the basis of clinical safety and efficacy studies and generic manufacturers to obtain market approval for their MPP.

REFERENCES

1. Benet LZ. Understanding bioequivalence testing. Transplant Proc 1999; 31(suppl 3A):7S–9S.
2. Therapeutic Goods Act 1989—Australian Government Publishing Service, Canberra.
3. Medicines Act 1981 (Reprint 2006) (published under the Authority of the New Zealand Government).
4. Note for Guidance on the investigation of bioavailability and bioequivalence (CPMP/EWP/QWP/1401/98). CPMP Guidance—As adopted by the TGA—with amendment. April 10, 2002. http://www.tga.gov. Accessed May 30, 2009.
5. Health Minister. National Health Surveillance Agency. MS/GM Ordinance no. 3916 of October 30, 1998. Approval of the National Drug Policy. Brazil. Brasília, DF; 1998. http://e-legis.bvs/leisref/public/showAct.php?id=751. Accessed May 30, 2009.
6. Brazil. Law No. 9782 of January 26, 1999. Definition of the National Health Surveillance System, establishment of the National Health Surveillance Agency, among other provisions. Brasília, DF; 1999. http://e-legis.bvs/leisref/public/showAct.php?id=182. Accessed May 30, 2009.
7. Brazil. Law N. 9787 of February 10, 1999. Amendment of Act No 6360 deals with health surveillance and provides for the use of generic names in pharmaceutical and other provisions. Brasília, DF; 1999. http://e-legis.bvs/leisref/public/showAct.php?id=245. Accessed May 30, 2009.
8. Brazil. Health Minister. National Health Surveillance Agency. Resolution N. 391 of August 9, 1999. Technical regulation for generic drugs. Brasília, DF; 1999. http://e-legis.bvs/leisref/public/showAct.php?id=251. Accessed May 30, 2009.
9. Bill C-102 An Act to Amend the Patent Act, the Trademarks Act and the Food and Drugs Act. See www. curc.clc-ctc.ca/lang. Accessed May 30, 2009.
10. Guidance for Industry. Conduct and Analysis of Bioavailability and Bioequivalence Studies – Part A: Oral Dosage Formulations Used for Systemic Effects. Health Canada 1992. http://www.hc-sc.gc.ca/dhp-mps/alt_formats/hpfb-dgpsa/pdf/prodpharma/bio-a_e.pdf. Accessed May 30, 2009.
11. Guidance for Industry. Conduct and Analysis of Bioavailability and Bioequivalence Studies – Part B: Oral Modified Release Formulations, Health Canada 1996 http://www.hc-sc.gc.ca/dhp-mps/prodpharma/applic-demande/guide-ld/bio/bio-b-eng.php. Accessed May 30, 2009.
12. Expert Advisory Committee on Bioavailability, Health Protection Branch, December 1992. Report C: Report on bioavailability of oral dosage formulations, not in modified-release form, of drugs used for systemic effects, having complicated or variable pharmacokinetics. Health Canada, 1992. http://www.hc-sc.gc.ca/dhp-mps/prodpharma/applic-demande/guide-ld/bio/biorepc_biorapc-eng.php. Accessed May 30, 2009.
13. Presentations by Eric Ormsby (Health Canada) to the Canadian Society of Pharmaceutical Scientists Annual Meetings, (i)Banff, Alberta, May 24, 2008 (ii)Toronto, Ontario, June 6, 2009.
14. Chen M-L, Shah V, Patnaik R, et al. Bioavailability and bioequivalence: An FDA regulatory overview. Pharm Res 2001; 18:1645–1650.

15. EC Notice to Applicants, volume 2A: Procedures for marketing authorisation. The Rules Governing Medicinal Products in the European Union, Chapter 1: Marketing Authorisation, November 2005. http://ec.europa.eu/enterprise/pharmaceuticals/eudralex/vol-2/a/vol2a_chap1_2005–11.pdf. Accessed May 30, 2009.
16. CPMP Note for Guidance on the Investigation of Bioavailability and Bioequivalence, Brussels, December 1991.
17. Pharmacokinetic Studies in Man, 1987. http://www.emea.europa.eu/pdfs/human/ewp/3cc3aen.pdf. Accessed May 30, 2009.
18. EMEA Note for Guidance on the Investigation of Bioavailability and Bioequivalence, London, July 2001, CPMP/EWP/QWP/EMEA. http://www.emea.europa.eu/pdfs/human/qwp/140198enfin.pdf. Accessed May 30, 2009.
19. EMEA Guideline on the Investigation of Bioequivalence, EMEA, London, July 24, 2008, CPMP/EWP/QWP/1401/98 Rev. 1. http://www.emea.europa.eu/pdfs/human/qwp/140198enrev1.pdf. Accessed May 30, 2009.
20. The Drugs and Cosmetics Act 1940 and The Drugs and Cosmetics Rules 1945, Ministry of Health and Family Welfare, Government of India, 2005:425–459.
21. The Patents (Amendment) Act 2005, The Gazette of India: Extraordinary, Part II-Sec. I, 1–18, 2005.
22. The Patents (Amendment) Rules, 2005, The Gazette of India: Extraordinary, Part II-Sec. 3 (ii), 60–108, 2004.
23. Ministry of Health, Labour and Welfare. Guideline for Bioequivalence Studies of Generic Products, 1997. http://www.nihs.go.jp/drug/be-guide(e)/Generic/be97E.html. Accessed May 30, 2009.
24. Ministry of Health, Labour and Welfare. Guideline for Bioequivalence Studies for Different Strengths of Oral Solid Dosage Forms, 2000. http://www.nihs.go.jp/drug/be-guide(e)/strength/strength.html. Accessed May 30, 2009.
25. Ministry of Health, Labor and Welfare. Guideline for Bioequivalence Studies for Additional Dosage Forms of Oral Solid Dosage Forms, 2001.
26. Ministry of Health, Labor and Welfare. Guideline for Bioequivalence Studies of Generic Products for Topical Dermal Application, 2003. http://www.nihs.go.jp/drug/DrugDiv-E.html. Accessed May 31, 2009.
27. Ministry of Health, Labor and Welfare. Guideline for Formulation Changes of Oral Solid Dosage Forms, 2000. http://www.nihs.go.jp/drug/DrugDiv-E.html. Accessed May 31, 2009.
28. Ministry of Health, Labor and Welfare. Draft Guideline for Bioequivalence Studies for Changes in Manufacturing of Oral Solid Dosage Forms: Conventional, Enteric Coated and Prolonged Release Products.
29. Ministry of Health, Labour and Welfare. Revision of Guideline for Bioequivalence Studies of Generic Products, 2006.
30. Pharmaceutical and Analytical: Medicines Control Council, Department of Health, 2.02 P&A, Jun07, v2, 2007. http://www.mccza.com/showdocument.asp?Cat=17&Desc=Guidelines%20-%20Human%20Medicines. Accessed April 29, 2009.
31. BIOSTUDIES: Medicines Control Council, Department of Health, 2.06 Biostudies, Jun07, v2, 2007. http://www.mccza.com/showdocument.asp?Cat=17&Desc=Guidelines%20-%20Human%20Medicines. Accessed April 29, 2009.
32. DISSOLUTION: Medicines Control Council, Department of Health, 2.07 Dissolution, Jun07, v2, 2007. http://www.mccza.com/showdocument.asp?Cat=17&Desc=Guidelines%20-%20Human%20Medicines – Accessed 29 April 2009
33. Generic Substitution: Medicines Control Council, Department of Health, December 2003. http://www.mccza.com/showdocument.asp?Cat=17&Desc=Guidelines%20-%20Human%20Medicines – Accessed 29 April 2009
34. Circular 14/95, Data Required as Evidence of Efficacy, Annexure 13, Medicines Control Council, Department of Health, October, 2, 1995
35. Medicines and Related Substances Control Act, 1965 (Act No. 101 of 1965) as amended by Act No. 90 of 1997 and Act No. 59 of 2002. http://www.mccza.com. Accessed April 29, 2009.

36. Medicines and Related Substances Control Act, 1965 (Act No. 101 of 1965) as amended by Act No. 59 of 2002. http://www.mccza.com. Accessed April 29, 2009.
37. Regulation on Licensing for Medicinal Products for Human Use, Official Gazetta No: 25725/19 January 2005
38. Regulation on the Evaluation of Bioavailability and Bioequivalence of Medicinal Products Official Gazetta of Turkey No: 21942. http://rega.basbakanlik.gov.tr/. Accessed May 31, 2009.
39. Conner DP, Davit BM. Bioequivalence and drug product assessment, in vivo. In: Shargel L, Kanfer I, eds. Generic Drug Product Development: Solid Oral Dosage Forms. New York: Marcel Dekker, 2005:229.
40. Office of Technology Assessment. Drug Bioequivalence. A Report of the Office of Technology Assessment Drug Bioequivalence Study Panel. Washington, DC: US Government Printing Office, 1974.
41. 42 Fed Regist 1648, Jan. 7, 1977.
42. 57 Fed Regist 17950, Apr. 28, 1992.
43. US Dept of Health and Human Services, Food and drug Administration, Center for Drug Evaluation and Research. Guidance for Industry. Bioavailability and Bioequivalence Studies for Orally Administered Drug Products—General Considerations. March 19, 2003.
44. US Dept of Health and Human Services, Food and Drug Administration, Center for Drug Evaluation and Research. Guidance for Industry: Topical Dermatological Corticosteroids: In Vivo Bioequivalence. March 6, 1998.
45. US Dept of Health and Human Services, Food and Drug Administration, Center for Drug Evaluation and Research. Draft Guidance for Industry: Food—Effect Bioavailability and Bioequivalence Studies. December 2002.
46. US Dept of Health and Human Services, Food and Drug Administration, Center for Drug Evaluation and Research. Guidance for Industry: Waiver of In Vivo Bioavailability and Bioequivalence Studies for Immediate-Release Solid Oral Dosage Forms Based on a Biopharmaceutics Classification System. August 31, 2000.
47. US Dept of Health and Human Services, Food and Drug Administration, 21 CFR Parts 314 and 320, /Vol. 74, No. 11 //Rules and Regulations, [Docket No. FDA–2003–N–0209] (Formerly Docket No. 2003N–0341) RIN 0910–AC23, Requirements for Submission of Bioequivalence Data; Final Rule
48. US Dept of Health and Human Services, Food and Drug Administration, Center for Drug Evaluation and Research Guidance for Industry: BE Recommendations for Specific Products. http://www.fda.gov/downloads/Drugs/GuidanceComplianceRegulatoryInformation/Guidances/ucm072872.pdf, Accessed June 6, 2009
49. US Dept of Health and Human Services, Food and Drug Administration, Center for Drug Evaluation and Research Guidance for Industry: Individual Product Bioequivalence Recommendations. http://www.fda.gov/cder/guidance/bioequivalence/default.htm. Accessed May 31, 2009.
50. US Dept of Health and Human Services, Food and Drug Administration, Center for Drug Evaluation and Research Draft Guidance for Industry: Submission of Summary Bioequivalence Data for ANDAs. http://www.fda.gov/downloads/Drugs/GuidanceComplianceRegulatoryInformation/Guidances/UCM134846.pdf. Accessed June 10, 2009.
51. Approved Drug Products with Therapeutics Equivalence Evaluations, Electronic Orange Book. Approved Drug Products with Therapeutic Equivalence Evaluations, Current through April 2009, Department of Health and Human Services, Public Health Service, Food and drug Administration, Center for Drug Evaluation and Research, Office of Information Technology, Division of Data Management and Services. http://www.accessdata.fda.gov/scripts/cder/ob/default.cfm. Accessed June 10, 2009.
52. Multisource (Generic) Pharmaceutical Products: Guidelines on registration requirements to establish interchangeability. In: WHO Expert Committee on Specifications

for Pharmaceutical Preparations, Fortieth Report. Geneva: World Health Organization, 2006 (WHO Technical Report Series, No. 937, Annex 7): 347–390.

53. Amidon GL, Lennernas H, Shah V, Crison JR. A theoretical basis for a biopharmaceutics classification: the correlation of in vitro drug product dissolution and in vivo bioavailability. Pharm Res 1995; 12:413–420.

54. Guidance for Industry, Waiver of the in vivo Bioavailability and Bioequivalence Studies for Immediate-release Solid Oral Dosage Forms Based on a Biopharmaceutics Classification System. Rockville, MD: Food and Drug Administration, 2000. http://www.fda.gov/downloads/Drugs/GuidanceComplianceRegulatoryInformation/Guidances/UCM070246.pdf. Accessed June 10, 2009.

55. European Medicines Agency. London, 24 May 2007, Doc. Ref. EMEA/CHMP/EWP/213035/2007.

56. EC Notice to Applicants, volume 2A: Procedures for marketing authorisation. The Rules Governing Medicinal Products in the European Union, Chapter 1: Marketing Authorisation, November 2005. http://ec.europa.eu/enterprise/pharmaceuticals/eudralex/eudralex_en.htm. Accessed June 10, 2009.

2   Australasia

C. T. Hung, D. Ren, L. A. Folland, and F. C. Lam
Zenith Technology Ltd, Dunedin, New Zealand

Noelyn Anne Hung
Otago University Dunedin School of Medicine, Dunedin, New Zealand

R. Smart
Douglas Pharmaceuticals, Auckland, New Zealand

INTRODUCTION

History

In Australia, as in many other countries, registration of medicines prior to their sale became a legal requirement soon after the problems with thalidomide (1) were identified. At present, the registration of medicines is controlled by the Therapeutic Goods Act 1989 (2). Under this Act, all medicines sold across state borders, prescription medicines and any nonprescription medicines intended for inclusion in the Pharmaceutical Benefits Schedule (PBS), cannot be sold until they are either registered or listed on the Australian Register of Therapeutic Goods (ARTG) (www.tga.govt.au). Responsibility for inclusion or otherwise of medicines on the ARTG rests with the Therapeutic Goods Administration (TGA) based in Canberra. Prior to the Therapeutic Goods Act 1989, only prescription medicines and nonprescription medicines intended for inclusion in the PBS were evaluated. Nonprescription medicines not intended for inclusion in the PBS and medicines manufactured and sold within a single state or territory did not need to receive formal regulatory approval before being sold except in some states such as the state of Victoria, where registration dubbed "Reg. Vic" by the Victorian Ministry of Health was required. Hence, in most states, no controls on the quality, safety, or efficacy of such medicines were enforced. "Reg. Vic" requirements related to good manufacturing practices (GMP) compliance of the finished product manufacturer along with evaluation of the quality and labeling of the finished product. With the coming into effect of the Therapeutic Goods Act 1989 on the January 15, 1991, these "Reg.Vic" requirements developed some time later into the OTC compliance branch of the TGA. From July 1, 2006, it was intended that all medicines, irrespective of scheduling, site of manufacture or distribution receive regulatory approval from a proposed Trans-Tasman Joint Regulatory Authority before being permitted for sale. The Joint Agency was to have responsibility for control of medicines in both Australia and New Zealand such that both countries were to be viewed as a single regulatory market, without being a single sales market. However in July 2007, the Labor party in the New Zealand Government at that time, failed to achieve sufficient votes to ratify the legislation designed to give authority to the agency. Establishment of the

Joint Agency has, therefore been postponed and no indication is presently available as to when such an agency will come into being. Therefore until such time as negotiations recommence, the status quo applies with the Therapeutic Goods Act 1989 being the applicable legislative authority. Continuation of registration or listing on the ARTG requires payment of an annual fee and no evaluation occurs with such payment. In relation to the sale of generic medicines, substitution, at the discretion of the dispensing pharmacist or at the request of the end user, on prescription is legal in Australia.

In New Zealand, as in Australia, government approval prior to the sale of a medicine also became a legal requirement soon after the problems with thalidomide (1) were identified. New Zealand, however, does not have either a registration or licensing system. Instead, companies who wish to market medicines in New Zealand must apply for "Consent to Distribute" (Medicines Act 1981, Section 20) (3) and cannot begin to market the medicine until such consent is notified in the New Zealand Gazette. Responsibility for recommending "Consent to Distribute" rests with Medsafe, a business unit of the New Zealand Ministry of Health based in Wellington (www.medsafe.govt.nz). Prior to the Medicines Act 1981 coming into force in 1984, Government approval of medicines was controlled by the Food and Drug Act 1969. Unlike Australia, as New Zealand is a single country rather than a federation, all medicines irrespective of scheduling, site of manufacture, or distribution must have "Consent to Distribute" before sales can commence. "Consent to Distribute," once notified, remains in force until 5 years after either the date of the last sale of the product by the company responsible for placing the product on the market or the date of the last regulatory activity such as submission and approval of a Changed Medicine Notification, the equivalent of a variation in other regulatory systems. No annual fees apply. In relation to the sale of generic medicines, substitution on prescription may only occur where the doctor either writes on the prescription that substitution of another brand of the same active ingredient is allowed or the prescription is written using the generic name only of the intended medicine.

In both Australia and New Zealand, medicines may be scheduled as either prescription medicines, pharmacist medicines, pharmacy only medicines, or general sale medicines. In Australia, the scheduling is performed by the Subcommittee on the Uniform Scheduling of Drugs and Poisons (SUSDP), while in New Zealand, scheduling is the responsibility of an advisory committee to the minister of health known as the Medicines Classification Committee (MCC). The SUSDP and MCC have worked together closely in the past few years to promote harmonization of the schedules in each country.

Regulatory Format for Marketing Applications

In both Australia and New Zealand, the preferred format for submission of an application to market a new generic medicine is the common technical document (CTD). In line with the use of the CTD in other countries, the information in Modules 2 to 5 is product related and identical to the information intended for submission in other countries. Module 1 is administrative data and at present and until such time as the Joint Agency is established, the information required to be included in Module 1 is unique to each country.

An application for a generic prescription medicine in Australia must be accompanied by 75% of the evaluation fee described in the fees schedule published on the TGA Web site (www.tga.gov.au). However, in New Zealand the application must be accompanied by 100% of the fee described in the fees schedule published on the Medsafe Web site (www.medsafe.govt.nz).

In Australia, the information that is required to be included in an application for registration of a generic medicine on the ARTG is detailed in the Australian Regulatory Guidelines for Prescription Medicines (ARGPM) June 2004 (4) and Notice to Applicants CTD-Module 1 TGA Edition September 2007 (5) (www.tga.gov.au). As stated earlier, Modules 2 to 5 are identical to Modules 2 to 5 prepared for submission elsewhere in the world. The following information is required for Module 1.

1. Letter of application
2. Comprehensive table of contents
3. Application form: The template for this form can be downloaded from the TGA Web site.
4. Australian labeling and packaging
5. Information about the experts
6. Specific requirements for different types of application
7. Drug and Plasma Master Files and Certificates of Suitability of Monographs of the European Pharmacopoeia: Where products involve manipulation of blood at some stage, applications must be accompanied by a file, the "Plasma Master File," describing that manipulation and the controls applied. For medicines, the Master file in Module 3.2.S of the application should include either a full description of the synthesis and controls applied in manufacturing the active pharmaceutical ingredient (API) as is usual for applications for new chemical entities (NCEs) or for generics, either a current Certificate of Suitability issued by the European Directorate for the Quality of Medicines (www.edqm.eu) or a Drug Master File describing the synthesis and controls applied in the manufacture of the API. The information must be accompanied by a specific authorization to allow the TGA to access the information provided by the API manufacturer on behalf of the applicant.
8. Good Manufacturing Practice: The TGA requires both the API and finished product manufacturers located outside Australia to comply with the same GMP standards as manufacturers resident in Australia. The evidence of such compliance that is accepted by the TGA is detailed in the 16th edition of the Guidance on the GMP Clearance of Overseas Manufacturers (6) available on the TGA Web site.
9. Meetings: A meeting can be arranged between the TGA and an applicant to discuss and resolve possible issues that are important for a marketing application or that may arise during evaluation of an application. The intention is to reduce barriers to approval of an application
10. Individual patient data: In Australia, individual patient data are not required to be included as part of a product application except for the plasma/blood/serum or urine concentrations and derived data from bioavailability studies. Such data should however, be held by the applicant in a form suitable for submission in the European Union (EU) or the United States and be able to be submitted to the TGA on request.

11. Overseas regulatory status: The TGA requires applicants to provide a list of overseas countries where the product has been submitted, the product information for those countries, any differences between the applications submitted in those countries and a statement about the regulatory status in Canada or the United States.
12. Summary of biopharmaceutical studies (defined as bioequivalence and/or bioavailability studies): Where an application contains biopharmaceutical studies, the application should include a summary of the study. The appropriate template for the summary can be downloaded from the TGA Web site and covers details such as study design, analytical validation, and pharmacokinetic results obtained. If no biopharmaceutical study is included in the application, justification for not providing such data is required. The information to be included can be found in Appendix 15 of the ARGPM.
13. Pediatric development program

Besides the above information there are three additional annexures to be completed:

1. Environmental risk for non-GMOs (genetically modified organisms) containing medicines.
2. Antibiotic Resistance Data.
3. Overseas Evaluation Reports.

An application in Australia for a generic product may be classified as either a Category 1 or a Category 2 application. A Category 1 application is an application for a generic product that is not supported by copies of evaluation reports from two or more overseas regulatory authorities. The Therapeutic Goods Act 1989 requires that the TGA complete the evaluation of a Category 1 application within 255 working days otherwise the remaining 25% of the evaluation fee is not paid to the TGA. Where an application for a generic product is supported by two or more independent evaluation reports of Module 3 data from overseas regulatory authorities, the application is a Category 2 application, and must be evaluated within 120 working days. The TGA currently accepts evaluation reports from Canada, Sweden, the Netherlands, the United Kingdom, and the United States. The evaluation reports should be included in Annex 3 to Module 1.

In New Zealand, there is no formal listing of the requirements for inclusion in Module 1. In general however, the following information should be included.

1. Application form: The same form is used irrespective of whether or not the medicine is a generic or NCE. Appendices to this form include evidence of compliance with GMP for the active ingredient manufacturer, finished product manufacturer and finished product packer.
2. Generic medicine check list: This is a unique form that must be completed by an applicant requesting approval of a generic medicine, but not by an applicant requesting approval of a NCE.
3. Comprehensive table of contents for the whole application.
4. Data sheet: New Zealand has a specific format for presentation of quality, safety, and efficacy information to health professionals; however, New Zealand will also accept such information in the form of an SPC (summary of product characteristics) or Australian Product Information. If a New Zealand specific data sheet is prepared, it must be accompanied by a completed data

sheet checklist and declaration. The format of the data sheet, checklist, and declaration may be found in the New Zealand Regulatory Guidelines for Medicines (NZRGM) (7) (www.medsafe.govt.nz).

5. Proposed labeling: The requirement is that it should be in the form of either finished label(s) or finished artwork ready for printing the label(s) or as draft artwork showing design, color, and wording proposed to be used on the label for each dose form, strength and pack size. It is emphasized that in New Zealand, the term "label" refers to the words and designs that are attached to or part of the container in which the medicine is packed. It does not include the data sheet or consumer medicine information (CMI).

6. Consumer Medicine Information: The provision of CMI is not a mandatory requirement; however, it is recommended and if prepared it must be accompanied by a declaration of content. The proposed format for CMI and the declaration to be completed may be found in the New Zealand Guide to the Registration of Medicines.

In Australia, generic medicines that are scheduled as prescription medicines are evaluated by the Drug Safety and Evaluation Branch (DSEB) of the TGA. The information required to be included in an application is detailed in the Australian Guide to the Registration of Drugs (AGRD) Volume 1 (www.tga.gov.au) Generic medicines that are scheduled as pharmacist medicines, pharmacy only medicines, or general sale medicines are evaluated by the over-the-counter (OTC) branch of the TGA. The information required to be included in an application for a pharmacist medicine, pharmacy only medicine, or general sale medicine is detailed in the Australian Regulatory Guidelines for OTC medicines (8) (www.tga.gov.au).

In New Zealand, there is no distinction made between the four medicine schedules as to which regulatory agency will evaluate an application.

As already indicated, applications for Generic Prescription Medicines in both Australia and New Zealand should be in CTD format as is required for submission of an application in the EU. However, unlike the EU, an abbreviated format is acceptable. The abbreviated format consists of Module 1, Module 2.3, Module 3 (3.2.S and 3.2.P), and Module 5.3.2 (bioequivalence). In Australia, applications for pharmacist medicines, pharmacy only medicines, or general sale medicines may also be in an abbreviated CTD format that comprises: Module 1, Module 3.2.P, and Module 5.3.2 for specified medicines. Applications for pharmacist medicines, pharmacy only medicines, or general sale medicines in New Zealand require Module 1, Module 3 (3.2.S and 3.2.P), and Module 5.3.2 for specified medicines.

An application for a generic prescription medicine submitted to Medsafe in New Zealand or to the TGA in Australia, should preferably contain a bioequivalence study that has been performed using either the New Zealand or Australian innovator product as the reference product, respectively. However, where justified by appropriate additional in vitro comparative studies, both agencies will accept bioequivalence studies where the innovator product used as reference was sourced from outside their respective countries. The in vitro studies required are detailed in either Section 14.2 of the NZRGM or Appendix 15 (Section 7) of the ARGPM.

BIOEQUIVALENCE REQUIREMENTS

Ethical Review and Approval Processes

Ethical review of research involving humans has been a requirement of the Australian and New Zealand health research system since the 1950s. The ethical principals outlined in the *Nuremberg Code* (9) and *Declaration of Helsinki* (10) formed the basis of the development of the ethical review process in both countries. These historic documents (along with their subsequent amendments) were fundamental in guiding the ethics for human research practice.

New Zealand

In New Zealand the current system was developed largely in response to the Cartwright Inquiry, which was concluded in 1988 (11). This inquiry formed the basis of an independent ethical review process. Further refinements occurred as a result of the Gisborne Cervical Screening Inquiry in 2001 (12), which recommended that the current process for ethical review be re-evaluated.

The primary role of all ethics committees is the protection of the rights, health, and well-being of consumers and research participants. Ethics committee approval is required for all research involving human participants (whether health or disability support services consumers, healthy volunteers, or members of the community at large) and where the research falls under one of the specified categories for matters requiring ethical review. Ethics committees consider whether the research meets ethical guidelines and considers such matters as to how participants are selected for the trial and whether they are provided with enough information to enable them to make an informed decision about participating. An ethics review is required for each study.

For human ethics, the following committees have been established in statute, under the New Zealand Public Health and Disability Act 2000:

- The Health Research Council Ethics Committee
- The National Advisory Committee on Health and Disability Support Services Ethics
- The Ethics Committee on Assisted Reproductive Technology
- Six Regional Health and Disability Ethics Committees and the Multiregion Ethics Committee

The regional ethics committees consider applications for research that is to be carried out entirely within just one of New Zealand's four ethics committee regions. The multiregion ethics committee considers applications for research that is carried out in more than one of the ethics committee regions.

Ethics committees are also formed by other organizations and gain accreditation through the Health Research Council Ethics Committee (HRCEC). These include institutional ethics committees (e.g., Universities) and private sector ethics committees (e.g., Zenith Biomedical Ethics Committee). These committees may have restrictions upon them as to what type of research they can review (e.g., healthy volunteer studies only). Any research, however, that involves the use of patients in the health sector has to be reviewed and approved by a health and disability ethics committee (or the Multiregion ethics committee if the proposal covers more than one region).

The accreditation of ethics committees by the HRCEC is a formal process. Every accredited ethics committee is required to provide an annual report plus other relevant information as required in the *HRC Guidelines for Ethics Committee Accreditation* (13). All accredited ethics committees are required to provide independent, competent and timely review of research studies. All members of the ethics committee are required to be independent, having no connection with any research proposals that are reviewed, which may cause a conflict of interest.

Ethics committees generally operate in accordance with the ministry of health's *Operational Standard for Ethics Committees* but can also have their own operational procedures. There is generally a turnaround time of 1 to 2 months from the ethics committee submission date to obtain full approval.

All applications for research proposals to accredited ethics committees are completed by way of a formal application process although the exact requirements can vary between committees. A requirement of all applications however is the submission of the study using the *National Ethics Application Form*.

It is the ethics committee's responsibility to ensure that adequate compensation provisions are in place for injuries suffered by participants in a clinical trial. Trials that are sponsored by a pharmaceutical company principally for the benefit of the manufacturer or distributor are not covered by the Accident Compensation Corporation (ACC). Compensation is provided by the pharmaceutical company to the extent described in the *Researched Medicines Industry Guideline* (RMI) (14). The "no-fault" principle forms the basis of these guidelines and the sponsor company should pay compensation to participants suffering bodily injury in accordance with these guidelines.

A separate approval process is required for clinical trials involving new and unregistered medicine formulations. The Ministry of Health's Standing Committee on Therapeutic Trials (SCOTT) reviews this research in parallel with an ethics committee application. SCOTT will initially issue a "recommended for approval" letter, which means it will give final approval once the ethics committee approval has been obtained. The SCOTT approval process takes a maximum of 45 days but is generally less. Final approval only takes a few days once ethics committee approval has been given.

The cost for each SCOTT application is approximately NZ $10,000.

Australia

In Australia, clinical trials of medicines and devices are subject to government regulations that are administered by the (TGA). Approval for a clinical trial is gained through a Human Research Ethics Committee (HREC). Only ethics committees that are constituted and operate in accordance with the National Health and Medical Research Council's (NHMRC) *National Statement on Ethical Conduct in Research Involving Humans* can approve a clinical trial. HRECs must follow the guidelines outlined in the *National Statement* regarding their composition, appointment of members and various aspects of their operational procedures. The terms of reference and working procedures of each ethics committee must be documented. All HRECs must also report annually to the NHMRC through its principal committee, the Australian Health Ethics Committee. Clinical trials must be approved by an institutional HREC with jurisdiction at the site where the study is to be conducted (e.g., hospital, medical center, etc.). In addition, there is a private ethics committee registered with the NHMRC that can be used for some sites.

There are two schemes under which clinical trials involving therapeutic goods may be conducted, the Clinical Trial Notification Scheme (CTN) and the

Clinical Trial Exemption (CTX) Scheme. The CTN scheme is a notification scheme whilst the CTX scheme is an approval process. The vast majority of clinical trials that are conducted in Australia are approved through the CTN process.

The CTN scheme involves the protocol being submitted directly to the HREC who is then responsible for assessing the scientific validity and safety of the project. The TGA does not undertake any review of the study but the sponsor company is required to submit the CTN form to the TGA along with the appropriate fee (presently AU $240) once ethical approval has been granted.

The CTX scheme is intended to assist HRECs when technical, scientific, or medical data are lacking or where further advice is required. Under this scheme the sponsor company submits data to the TGA for evaluation and comment. The CTX application can be made under a 50- or 30-day review and has associated costs that are substantially higher than a CTN review. Under the CTX scheme, the TGA will focus primarily on the study in relation to safety issues. An HREC must also approve the study and the sponsor must wait for written approval from the TGA before they are permitted to commence their research. All CTN and CTX trials must have an Australian sponsor.

Compensation for participants for person injury must be provided by the sponsor company in a form no less favorable than the current version titled Medicines Australia Form of Indemnity for Clinical Trials (15).

Although clinical trials conducted in New Zealand and Australia are not required to be registered, registration is recommended. In May 2006, the World Health Organization (WHO) recommended that all international clinical trials should be registered. Registration can be completed through the Australian New Zealand Clinical Trials Registry (www.anzctr.org.au).

There is an intention to establish a Joint Agency that will replace the TGA and the New Zealand Medicines and Medical Devices Safety Authority (Medsafe) and be accountable to the Australian and New Zealand Governments.

Participant Selection

Studies should normally be performed using healthy volunteers with the inclusion/exclusion criteria clearly outlined in the study protocol. Participants may belong to either sex, with any risk to woman of childbearing potential considered on an individual basis.

Generally participants should be aged 18 to 55 years and have a weight within the normal range according to body mass index (BMI) or other appropriate accepted weight range indicators (e.g., Metropolitan Life Tables). In New Zealand and Australia, where the population is multicultural, no restrictions are placed on the selection of race when conducting studies.

Genetic Phenotyping

Although to date, phenotyping of participants has, to the best of our knowledge, not been requested by either Medsafe or the TGA, it is worth considering specifically for parallel design studies where the medicine is known to be subject to major genetic polymorphism. Phenotyping would allow fast and slow metabolizers to be evenly distributed in the two groups of participants. Genetic phenotyping may also be considered for multiple dosing to steady state, where slow metabolizers are known to present a significantly longer $t_{1/2}$ of the active compound. Under these circumstances slow metabolizers should not be included in the study since the dosing period would have to be prolonged to accommodate their inclusion.

Informed Consent

In obtaining and documenting informed consent, investigators should adhere to GCP (good clinical practice) and the Declaration of Helsinki. Prior to commencement of any study, written approval of (i) information for participants and (ii) consent to participate documentation should be obtained from an independent ethics committee (IEC). Any amendments to this documentation, which are relevant to participant consent, should be approved by the IEC and revised copies provided to the participants. The language used in this documentation should be in lay terms, both practical and understandable to participants.

Participants must give their consent willingly and without coercion or undue influence from the investigator or study staff. Prior to obtaining consent from any participant the investigator must ensure that each participant is made aware of the procedures to be carried out during the study. Participants are required to attend a meeting at which the information for participants form is read aloud. Participants are encouraged to ask questions relating to the study medicine, possible side effects, and study procedures.

Prior to informed consent being obtained, subjects are given time to inquire about details of the study and to decide whether or not to participate. A trial physician, listed in the study protocol, must be present at each meeting to answer any medical questions that participants may have and to witness that informed consent was given freely by each participant. Written informed consent should be signed and personally dated by the participant and trial physician as well as an independent witness.

Informed consent should be filed as part of each individual participant study record.

Participant Screening

The screening and consent of volunteers involves many challenges. In the case of healthy volunteers, it involves both medical and psychological aspects. Most participants with a health science background will request or appreciate knowing their laboratory results. The most frequently requested result in our experience is blood type, but obviously this is not a routinely performed test for studies.

Social History and Exclusion Criteria

A questionnaire detailing social history will usually identify issues regarding substance abuse, high-risk travel, or high-risk behaviors for infectious disease such as HIV infection.

Tobacco smoking is not permitted currently or within the previous 6 months for most studies. Beverages containing high levels of caffeine are also frequently used especially in young adults, and may require a withdrawal period for some participants who use these daily.

Blood donations should not have been made within the previous 60 days of participating in the study, or within a certain period after the study depending on test drug half-life, test drug metabolites, or possibly lowered hemoglobin levels.

Recognition of cultural values is also essential in some instances. In New Zealand, for example, Maori have certain requirements concerning the body and examination thereof, which must be respected. Food restrictions are required for participants from Hindu, Jewish, or Islamic faith, for example.

Medical History

A full medical history is essential and a list of medical conditions forms the basis of screening. It acts as an aide-memoir for the healthy volunteer who may ignore or minimize previous illnesses, and can be completed in a relaxed environment before meeting the examining physician.

Asthma is a commonly experienced condition in childhood, which may extend into young adulthood, and which a participant will often deny or regard as a "nonillness," especially if it is controlled by intermittent inhaler use and avoidance of precipitating factors. At least a 5-year period free of inhaler use or symptoms to confirm the participant is suitable for inclusion in a study is required for study participation.

The medical history should include previous hospital admissions, allergies, adverse reactions, drug reactions, and complications with anesthesia, in addition to family medical history. The examining physician also has the opportunity to assess the participants understanding of the Protocol and procedures, understanding of English, and answer any further questions about the study. A multilingual staff and access to independent interpreters are essential in some cases to ensure informed consent, understanding of the procedures, and cooperation with the study requirements.

Provocation Screening

In some instances, especially antibiotics, where the risk of anaphylaxis is higher than with other drugs, skin testing is performed. The test drug is placed on broken skin and the response observed and recorded immediately and over the ensuing days. A full medical response must be in place to treat anaphylaxis. Usually, any rash is mild and resolves without need for treatment.

Physical Examination

Blood pressure (supine and sitting), height, and weight are recorded. A rise in systolic blood pressure to about 140 mm Hg is typical for some participants, who promptly return to a lower pressure when relaxed. Occasionally, a high blood pressure is found, particularly in older participants (>50 years), females on the oral contraceptive, and overweight male participants who are then referred to their family practitioner. Frequently, in the young adult, blood pressure is less than 100 mm Hg systolic, and this becomes a concern for studies involving blood pressure lowering drugs. A systolic pressure around 90 mm Hg should exclude participants from blood pressure lowering studies.

The routine physical examination involves the visual assessment of skin features (scars, wounds, body fat, peripheral edema) respiratory and cardiac auscultation, abdominal examination, and lymph node examination.

The exclusion of cardiac valve abnormalities is particularly important as the skin barrier is breached multiple times for venipuncture and intravenous (IV) line access. An enlarged spleen may be asymptomatic, as may be a cirrhotic liver or enlarged lymph node.

The forearm skin should be checked for scarring, especially from burns, which might preclude suitable IV line access. Long forgotten abdominal scars may also indicate previous cholecystectomy (perhaps high lipid levels) or the recent onset of numerous seborrheic keratoses (Leser–Trélat sign of internal malignancy).

Electrocardiograph

A 12-lead electrocardiograph (ECG) is useful for determining the risk of sudden cardiac death from prolonged QT interval, short PR interval, or cardiac arrhythmias, for which a participant will be excluded from a study. A copy of the ECG will be given to the participant along with a referral letter to the family medical practitioner for further counseling.

In some instances the "stress" of the screening process will be associated with supraventricular ectopic events, for which a repeat ECG the following day is recommended.

Laboratory Tests

Medical screening that includes basic hematology, biochemistry, pregnancy tests, and, drugs of abuse, is designed not only to identify potential medical problems but also to provide a baseline for possible adverse events.

A hematology screen including full blood count, red cell indices, and ESR is extremely useful. For example, the young iron deficient vegetarian female participant, with a mild anemia may not demonstrate any of the more obvious clinical features of anemia such as pallor, glossitis, stomatitis, or heart failure, but on prompting will often admit to nonspecific lethargy and weakness. In the process of venipuncture during intensive sampling over a short study period, the hemoglobin may decrease by 10 to 15 g/dL. This can be rapid enough to produce nonspecific symptoms such as lethargy. Iron replacement therapy may be prescribed or dietary advice given. Also, participants should be within the recommended hemoglobin range to avoid anemia from borderline iron stores (assessed with a ferritin measurement).

A platelet count is useful to identify idiopathic thrombocytopenic purpura in the undiagnosed participant. As this is a relatively common disorder in young or middle-aged adults, medical screening is likely to identify such problems.

Although the ESR is a nonspecific test it has some value if a particularly high value is obtained (e.g., >100 mm/hr). In the early nonspecific phases of collagen vascular disorders and malignancy (especially mediastinal Hodgkin's lymphoma and multiple myeloma), it may provide an early indicator. While not diagnostic, it invites referral to the family medical practitioner and further testing, and exclusion from the study.

Biochemistry including plasma sodium and potassium levels provides a nonspecific overview of homeostatic mechanisms that include renal and nonrenal causes of potential abnormalities. Most commonly a raised potassium result reflects delayed processing or refrigeration of the specimen. Many drugs likely to be studied (e.g., angiotensin-converting enzyme inhibitors) also run the risk of hyperkalemia, and normal baseline levels are useful in those instances. In rare instances the bulimic anorexic participant will present with hypokalemia, in addition to other factors such as BMI should be helpful.

A screen of liver enzymes can produce small elevations in aspartate aminotransferase (AST), alanine aminotransferase (ALT), or gamma-glutamyltransferase (gGTP), which can rapidly disappear on repeat testing within days of the initial test. If ALT and/or gGTP are raised, in our experience a history of alcohol intake 1 or 2 days previously is commonly elicited. Very occasionally the mean red cell hemoglobin is also raised implicating more chronic use (or abuse) of alcohol. A raised AST level may reflect skeletal muscle derived enzyme following exercise and this is readily resolved by creatine kinase measurements and history of exercise. The AST can also be falsely elevated by delay in plasma separation, and release of AST from red cells.

In some instances the residual effects of glandular fever remain, or, the participant has a high BMI and fatty liver. In either case the participant would not be entered into the study.

Alkaline phosphatase is occasionally mildly elevated in the late, male teenagers, reflecting possible bone injury in extreme sports or a late growth spurt.

An unconjugated hyperbilirubinemia between 20 and 40 μmol/L, and occasionally up to 60 or 80 μmol/L is a relatively common finding (up to 10%) in the general population, and, in the absence of hemolysis and other hepatic disorders, reflects Gilbert's syndrome. Gilbert's syndrome may become evident at any age but is often seen in young adulthood associated with menstruation, fasting, intercurrent illness, or dehydration. Fluctuating plasma bilirubin levels are typical, and the syndrome should not be interpreted as "disease." Dubin–Johnson and Rotor syndromes are likewise benign conditions that produce a conjugated hyperbilirubinemia due to the failure of excretion of conjugated bilirubin.

An important adjunct to hepatic laboratory testing is, of course, the clinical examination and history, as early cirrhosis may produce near normal or even normal results for the above-mentioned tests.

Serological testing for hepatitis A, B, and C exclude participants with liver damage from hepatitis B and C. Hepatitis A is diagnosed by the detection of IgM anti-HAV (hepatitis A virus) during the acute illness, while IgG anti-HAV may reflect a previous resolved episode of hepatitis A or vaccination.

A panel of antibody and antigen tests for hepatitis B and C will determine the hepatitis status adequately in most instances, even in the early stages of disease. A note of caution is added, however, in participants who may have been recently vaccinated, as we have found that vaccine antigens can be detected by sensitive newer tests.

HIV testing must be performed with the knowledge that professional HIV counseling is available should an unexpected positive test result occur, even though HIV positive cases do not participate in a study. Counseling may also be required for inconclusive first-line testing for other tests such as hepatitis B and C, before confirmatory testing is available.

Renal function testing is an important part of screening to rule out renal impairment that may affect drug excretion or drug handling by the body. Athletic or body-building participants are particularly prone to the use of dietary supplements such as proteins and creatinine, with consequent increases in plasma urea and creatinine values. Conversely, significant renal impairment is possible while plasma creatinine values remain within the reference range.

Adhering to conditions of testing is important. Glucose and cholesterol levels need to be performed in the fasting state for reliable results. Furthermore, a regular blood taking time, such as 8 AM, helps to negate circadian rhythm and diurnal variation for some tests.

Interpretation of Laboratory Results

Minor deviations from the reference ranges can cause considerable consternation, especially for participants who may be health science students early in their education, or the non–health science educated participant. Frequent use of the internet by participants for information concerning the results, and to check and countercheck protocol information, is a common occurrence, and minor variations outside reference ranges need to be put into perspective by the trial physician. These need to be explained clearly and precisely to alleviate concerns of ill health. Screening results of any type should not, unnecessarily, create a "patient" out of a healthy volunteer.

By convention, reference ranges include 95% of the population or about 2 standard deviations from the mean, and are usually determined by individual laboratories having investigated their own local healthy population with their own analytical instruments. About 2.5% of the results at either extreme of a normally distributed Gaussian distribution will be outside the reference range. Reference ranges will vary between laboratories even within the same locality, as different laboratories use different assay temperatures, substrates, pH conditions, assays, and instruments. Hence a result outside of the reference range is not necessarily abnormal, and reference ranges are not always the same at the same laboratory or between laboratories. Local laboratories can also influence reference ranges by using a preponderance of hospitalized patients compared to ambulant community-based patients. Protein fractions in particular are influenced by whether the blood was drawn from a recumbent hospital patient with a minor illness in the early hours of the day or an ambulant participant. The ambulant participant, or participant who is somewhat dehydrated by the end of a busy workday, is likely to have a higher protein fraction, that may even exceed the stated reference ranges.

Further variation of laboratory results can also occur between participants on the basis of age, sex, and ethnicity. High-density lipoprotein cholesterol, for example, is higher in premenopausal women than in men. Plasma creatine kinase may also be higher in those of African descent compared with those of Caucasian descent.

Repeat Examinations

While it is incumbent upon the clinical investigator to minimize participant discomfort and time use, some tests and procedures need to be repeated to fully determine their significance. These tests must be carefully followed and reported to the participants (and/or family medical practitioner), especially if they have been excluded from the study because of the initial abnormality.

Poststudy Testing

Laboratory tests are repeated at the end of the study. If abnormalities are detected the test is usually repeated until it returns to normal (or nonsignificant) levels, or, referred to the family medical practitioner for follow-up.

A Team Approach

Staff cognizant of the need for confidentiality initially greet and enroll participants and perform the social histories, ECG, blood pressure, height, and weight recordings.

The participant will meet several physicians during the reading, consent, examination stages, and study periods. This provides the participant with a variety of opportunities for discussion, in both group and individual situations, and with both physicians and nonphysician staff. In cases where doubt is cast on a participant's ability to understand the protocol or be willing to adhere to the conditions of the study, a consensus of opinion is useful.

Study Design

Generally, if only two products are to be compared, a two-period, two-sequence cross-over study design should be used. However, alternate, well-established designs could also be considered such as a parallel design for substances with a very long half-life (>100 hours).

For highly variable medicines (refer to Highly Variable Drugs and Drug Product section), a replicate cross-over design should be considered. For example, the following four-period, four-sequence cross-over design has been successfully employed for TGA registration of alendronate. In this study, the urine concentration of alendronate was measured and the bioavailability of alendronate, as summarized by the accumulated excreted amount, was compared between the two formulations.

Four-Period, Four-Sequence Cross-Over Design

Sequence	Period 1	Period 2	Period 3	Period 4
1	A	B	B	A
2	B	A	A	B
3	A	A	B	B
4	B	B	A	A

Treatment A represents the test formulation while treatment B is the reference formulation. Many different designs can be constructed for two formulations with four periods and four sequences; see Jones and Kenward (16) for some examples. However, the above replicated cross-over design has the smallest error variance among all four sequences, four periods, and two formulations cross-over design (17). The intrasubject variation for testing the formulation effect is only a quarter of that of a standard two-formulation, two-period, two-sequence cross-over design. Recently a three-period, reference-replicated cross-over design, with sequences of TRR, RTR and RRT has been proposed by the US FDA. A simulation study on this design has been performed but the results are yet to be published.

Steady-State Studies

For most formulations, single-dose studies should generally suffice. However there are instances where a steady-state design will be (*i*) required by the TGA and Medsafe, for example, dose- or time-dependent pharmacokinetics and modified release products or (*ii*) considered, for example, where single dose administration does not allow adequate sensitivity for precise plasma concentration determination or where it is intended to reduce intraindividual variability by using a steady-state design. However, in some countries, steady-state studies are no longer "considered" for the very reason that it dampens "true" variability of a product as well as the ethical implications of multiple dosing to healthy participants. Consideration of a steady-state design should also be given when the study population is patients. For example, with some pharmaceutical ingredients, the effects of a normal dose in healthy volunteers could be intolerable or toxic, hence in such a case either a lower dose should be used or if the active compound is not dose proportional, a steady-state design in patients should be considered. An example of this would be the antipsychotic medicine clozapine, which can be given to healthy volunteers in a low dose (12.5 mg or less). However, such compounds can also be studied in the appropriate patient population by using a steady-state design with no washout period, if recommended for submission by the regulatory body, for example, clozapine for US registration, cytotoxic compounds for TGA, and Medsafe or any compound that can potentially cause severe untoward side effects to healthy volunteers. In addition, for the measurement of compounds in healthy volunteers where there is an inherent plasma concentration, such as calcitriol, the use of a steady-state design for the determination of bioequivalence, in our experience, may be acceptable for TGA and Medsafe registration.

Fasted and Fed Studies

The use of fasted and fed study designs in bioequivalence determinations is dependent on two factors: (*i*) whether the formulation is immediate release or modified release and (*ii*) formulations that are indicated for administration with food.

Studies on immediate release formulations are generally conducted using a fasted state design, unless the product is known to cause severe gastrointestinal disturbance or is indicated for administration with food. Under these circumstances a fed state design would be used without a fasting study. A high-fat content, standardized meal consisting of, for example, two pieces bacon, one hash brown, one english muffin, 20 g cheese, 20 g butter, 240 mL milk, 240 mL water should be consumed over a 30-minute period immediately prior to dosing. The dose should be administered no more than 5 minutes after completing the meal.

Studies on modified release formulations include fasted, fed, and steady-state studies typically at the highest marketed strength. However, in formulations where the highest strength is known to cause intolerable adverse events and relevant proof of dose proportion between strengths can be provided, that is, the active compound and metabolite exhibit linear pharmacokinetics, a lower dose could be administered.

As an alternative to conducting two single-dose studies, a four-way, four-sequence, cross-over design could also be employed. The design allows for the conduct of both the fed and fasted studies by using a single group of subjects, typically with a 32 to 48 subject sample size. Each subject is randomly assigned a sequence as follows:

Four-Way, Four-Sequence, Cross-Over Design

Sequence	Period 1	Period 2	Period 3	Period 4
1	A	D	B	C
2	C	B	D	A
3	B	A	C	D
4	D	C	A	B

Treatments A, B, C, and D can be either test (modified release) fed, test (modified release) fasting reference modified release fed or reference modified release fasting.

The advantage of adopting a four-way design is the ability to compare, for example, any food effect on the pharmacokinetic parameters of the test formulation with that of the results from the reference formulation

Sample Size

The number of subjects should be determined statistically, however, generally the minimum number of subjects starting the study should not be less than 12. The minimum number of completing subjects should be determined and outlined clearly in the study protocol. In a cross-over design, the total variation consists of sequence variation, subject (sequence) variation, period variation, formulation variation, and error variation. For testing the formulation effect, the error variation or intrasubject variation (18) is used. Therefore the sample size required for a bioequivalence study should be estimated using the intrasubject variation. In particular, since the logarithmic transformed AUC and C_{max} values are used in constructing the 90% confidence intervals (CIs), the intrasubject variation based on the lognormal distribution (19) must be used. However, this intrasubject variation (or coefficient of variation, CV) can only be estimated if the data of a previous bioequivalence study on the same formulation is available. In most cases sample size is estimated from literature data. This typically consists of only summary statistics such as average and standard deviation for the untransformed AUC and C_{max} values. Sample size estimation is highly dependent on the standard deviation of the above pharmacokinetic parameters and the standard deviation data provided in the literature do not usually provide information on the intrasubject variation. In general, the CV calculated from published data on standard deviation would be much larger than the intrasubject CV based on the logarithmic transformed values. Therefore the sample size estimated from these summary statistics is generally extremely conservative.

Data obtained from one of our studies on tramadol is used here to illustrate this discrepancy. In this study, the mean and standard deviation for $AUC_{0-\infty}$ (ng hr/mL) of the reference formulation were 2997.8 and 1259.2 respectively. The CV estimated using only this information is 42%. With a CV of 42%, at a 5% level of significance and 80% power, a minimum sample of 76 subjects would be required even if the reference and the test formulation means were the same, and a sample size of 94 would be required when the two means differed by 5% (18). However, the mean squares error, obtained from the ANOVA table for the logarithmic transformed $AUC_{0-\infty}$ values, was 0.03997637. By using the lognormal distribution, the intrasubject CV was estimated to be 20% and a sample size of 20 was large enough for a 5% difference in the two means (19). Twenty-four subjects were included in this study and the resulting 90% CI for logarithmic transformed $AUC_{0-\infty}$ was (0.975, 1.074). It is therefore obvious that a sample size of 94 is excessive.

Washout Periods

Treatment periods should be separated by an adequate washout period. The interval between study days should be long enough to ensure elimination of the previous dose. The interval should be no less than five terminal elimination half-lives of the active compound or metabolite. The interval between treatments

should not exceed 3 to 4 weeks. If a longer washout period is indicated a parallel design should be considered.

Sampling Schedules

The sampling schedule should provide an adequate estimation of C_{max} and cover the plasma concentration versus time curve long enough to provide a reliable estimate of the extent of absorption. The Medsafe and TGA adopted guidance indicates this is achieved if AUC_{0-t} is at least 80% of $AUC_{0-\infty}$. However, typically, AUC_{0-t} is at least 90% of $AUC_{0-\infty}$.

Long Half-Life Compounds

For medicines with a long half-life, relative bioavailability can be adequately estimated using truncated AUC. In this instance the sample collection period should be adequate to ensure comparison of the absorption process. Although this is specified by MedSafe and the TGA (20), such an approach, to our knowledge, has never been requested by either a sponsor company or the regulatory authorities, for compounds with long half-lives (>100 hours). Furthermore, with the continuing advances in LC-MS/MS (liquid chromatography mass spectrometry) technology providing greater sensitivity and lower detection limits for most compounds, the necessity for truncated AUC to overcome assay sensitivity issue is questionable. It is our opinion that the either $AUC_{0-\infty}$ or AUC_{0-t} should be used as the primary parameter for the determination of bioequivalence. The conditions under which a study is conducted should be standardized as far as possible to reduce the variability of factors external to the formulations/products being tested.

Test and Reference Products and Foreign Data

Test products are generally compared with the corresponding dosage form of an innovator product (reference product). Both Medsafe and TGA prefer an application for registration of a generic product to include a bioequivalence study versus a leading brand purchased from within their own country. However, in certain circumstances where extensive in vitro testing of innovator products has been performed and comparative excipient analysis, the similarity should be justified by dissolution profiles using buffers at three different pHs. The similarity factor, f_2, is evaluated and an application containing foreign bioequivalence data may be accepted by both Australian and New Zealand authorities provided that the following conditions have been met:

1. A minimum of three time points excluding zero.
2. The time points should be the same for the two formulations.
3. Twelve individual values for every time point for each formulation.
4. Not more than one mean value of >85% dissolved for any of the formulations.
5. The relative standard deviation of coefficient of variation of any product should be less than 20% for the first point and less than 10% from second to last time point. Test products must be prepared according to the GMP-regulations. Oral solid forms for systemic action should usually originate from a batch of at least 10% of an intended full production batch or 100,000 units, whichever is greater. Other dosage forms of test products, such as suspensions, should originate from a batch of at least 10% of a production batch.

Highly Variable Drugs and Drug Product

Highly variable drugs (HVDs) have been defined as active pharmaceutical ingredients associated with a within-subject variability of $\geq 30\%$ in terms of the ANOVA-CV (21). Therefore, proving bioequivalence of products containing HVDs is problematic because the higher the ANOVA-CV the wider the 90% CI. Consequently the sample size required to achieve adequate statistical power in these compounds is typically large. There are several approaches currently being applied to this problem (22). Medsafe and the TGA indicate that in rare cases a wider 90% CI may be acceptable for AUC if it is based on sound clinical justification and that the C_{max} interval may be widened if it is predefined in the protocol, for example, from 0.75 to 1.33 and justified with respect to safety and efficacy concerns of patients switching between formulations. In our experience, however, it is unlikely that an AUC falling outside the general acceptance criteria of 0.8 to 1.25 would be accepted by either authority. It is also our experience that by using a replicated cross-over design, as indicated in the previous section (Study Design), bioequivalence of HVD products can be demonstrated without widening the generally acceptable 90% CI.

Combination Products

Combination generic products should be compared with an equivalent innovator combination product. However, in cases where there is no marketed innovator product, separate products administered alone can be used for comparison. The sample schedule should adequately provide for the determination of the pharmacokinetic parameters of all active analytes and statistical analyses should be conducted for all active analytes using the 90% CIs for all the active components to fall within the usual acceptance ranges. However, in such cases, the combination product will be considered under a new medicine application (i.e., a new formulation).

Over-the-Counter Products

The bioequivalence requirements for oral OTC products are the same as for prescription only products However, both the TGA and Medsafe will consider a bioequivalence study where the reference product is not sourced from Australia or New Zealand provided that comparative dissolution testing and comparative excipient analysis demonstrates that the reference products are essentially similar (refer to Test and Reference Products and Foreign Data section).

Add-On Studies, Outliers, and Dropouts

The method of statistical analysis should be clearly defined in the protocol. The protocol should include provisions for subject withdrawal (dropouts) and for biologically spurious data (single-point outliers), for example, the sample shows no peak where a peak is expected; the sample shows a peak where not expected; unexpected low concentration around the C_{max} region; unexpected high concentration in the elimination phase or data that produces a peculiar concentration/time curve. The most effective method of dealing with subject dropouts is to include an additional 5 to 10% of subjects above the number required for completion (e.g., 26 subjects with 24 completing). A definitive statement regarding replacement subjects should also be included in the protocol. As a general guideline, subjects should only be replaced when the number of completing subjects falls below the minimum. All subject samples should be analyzed, and only those completing all phases of the study should be included in the final pharmacokinetic and statistical analyses. Exclusion of entire subject data (outliers) is not generally accepted, except when a biological justification can be proved, such as vomiting within twice the T_{max} after dosing (23). Add-on studies as a rule are not permitted unless included in the original protocol. However, revised analysis is allowed if the original statistical plan (e.g., $AUC_{0-\infty}$) proves to be invalid. For example, if the results obtained in the study are markedly different from the results referenced in published data or from previous studies, that is, the sampling profile used in the study is not long enough to allow estimation of $AUC_{0-\infty}$ or the accurate estimation of $AUC_{0-\infty}$ is difficult due to an erratic elimination phase and consequent difficulty in determining $t_{1/2}$, most likely the result of endohepatic recycling. The original and revised results should be presented in the final report (20).

Sample Analysis
All analytical methods used to determine the concentrations of the analyte(s) and/or metabolites in the biological matrix must be fully validated and the results documented (validation report). Validation procedures, methods, and acceptance criteria must be specified in relevant standard operating procedures (SOPs). A calibration curve should be constructed for each analyte in each analytical run and should be used to calculate the concentration of the analyte(s) in the unknown (subject) samples. The calibration curve should consist of a blank sample (drug-free plasma sample processed without internal standard), a zero sample (drug-free plasma sample processed with internal standard), and an appropriate number of nonzero plasma samples, including LLOQ (lower limit of quantification) to cover the expected range of concentrations from the subject samples. Quality control samples [at three concentrations over the range of the calibration curve, that is, one within $3 \times$ LLOQ (low QC sample), one at the geometric or arithmetic mean of the low and high QC sample (QC sample), and one close to the upper boundary of the standard curve, that is, 80% of ULOQ, (high QC sample)] should be included in the run in duplicate (six samples per run). All the samples for each individual subject (all periods) should preferably be medium analyzed in the same analytical run.

Analytes to be Measured

Parent Compound Versus Metabolite(s)
The analyte to be measured (23) in biological fluids collected in bioequivalence studies is either the parent compound or when appropriate, its active metabolite(s) (24). Measurement of only the parent compound rather than the metabolite is generally recommended. The rationale for this recommendation is that the concentration-time profile of the parent is more sensitive to changes in formulation performance than a metabolite, which is more reflective of metabolite formation, distribution, and elimination. However, the following are exceptions to this general approach. Measurement of a metabolite may be preferred when the concentrations of the parent compound are too low to allow reliable analytical measurement in blood, plasma, or serum

If there is a clinical concern related to efficacy or safety of the parent compound, it is also recommended that sponsors and/or applicants contact the appropriate regulatory review division to determine whether the parent compound should be measured and analyzed statistically. For example, following administration of simvastatin (an inactive lactone), the drug is hydrolyzed to simvastatin acid. Simvastatin acid is the principal metabolite, which acts to lower cholesterol levels (25). If the metabolite contributes meaningfully to safety and/or efficacy, it is recommended that the metabolite as well as the parent compound be measured. Accordingly, the metabolite data should therefore also be subjected to the usual statistical analysis and meet the usual acceptance criteria for bioequivalence.

Chiral Compounds
For bioequivalence studies, the guidance recommends measurement of the racemate using an achiral assay. Measurement of individual enantiomers in bioequivalence studies is recommended only when all of the following conditions are met: (i) the enantiomers exhibit different pharmacodynamic characteristics, (ii) the enantiomers exhibit different pharmacokinetic characteristics, (iii) primary efficacy and safety activity resides with the minor enantiomer, and (iv) nonlinear absorption is present (as expressed by a change in the enantiomer concentration ratio with change in the input rate of the medicine) for at least one of the enantiomers. In such cases, it is recommended that bioequivalence factors be applied to the enantiomers separately. For example, bicalutamide, a racemate with activity being almost exclusively in the R-enantiomer. Bicalutamide undergoes stereospecific metabolism, with the S-enantiomer cleared rapidly relative to the R-enantiomer (26). In this case, the enantiomers should be analyzed separately and determination for bioequivalence based primarily on the R-bicalutamide with S-bicalutamide treated as a secondary parameter since R-bicalutamide is active while S-bicalutamide is inactive.

Prodrugs
There is no recommendation from the regulatory authorities about measurement of prodrugs and their metabolites in bioequivalence studies. On the basis of our experience, we recommend measuring the metabolite(s) rather than the prodrug if the following conditions are met: the prodrug is well absorbed and rapidly and almost completely converted to its active metabolite(s), peak plasma concentrations of prodrug are less than 10% and occur in a very short time (e.g., less than 30 minutes), and are at or below the limit of quantification within a short time (e.g., 2 hours) after dosing. In the absence of specific guidance information pertaining to the study design for the determination of bioequivalence, the sponsor should consult the applicable regulatory body.

Urine Data
Assessment of bioequivalence in urine can be used in compounds where there is very little systemic absorption (<5%) and the dose absorbed is exclusively eliminated in urine, for example, alendronate. When urine samples are utilized to determine bioequivalence, cumulative urinary recovery (Ae), and maximum urinary excretion rate (R_{max}) are employed in the statistical analysis instead of AUC and C_{max}. A replicate cross-over design, as indicated in section (Study Design), is recommended for low systemic absorption medicines based on urinary data.

Data Analysis

All pharmacokinetic analyses should be performed using validated computer programs. Individual concentrations, AUC, C_{max}, T_{max}, and $t_{1/2}$ calculations should be presented along with individual and mean plasma drug concentration–time curves on both linear/linear and log/linear scales for each completing subject. In addition, calculations for mean, geometric mean, standard deviation, coefficient of variation, and ranges must also be presented. Spreadsheets for each individual subjects' raw data should be submitted. These should include the following information: chromatogram identification, method file, date and time of collection, retention times for the active compound(s) and internal standard(s), peak areas (heights) for the active(s) and internal standard(s); calibration data, regression data and actual calculated drug concentrations. The analytical report should also present a summary of data for each individual subject run, that is, mean and standard deviation of peak retention times, regression data, and "spiked" sample concentrations. Samples that require re-analysis, according to a prespecified protocol, should be clearly identified and reasons for re-analysis must be provided. The analytical report should also contain the validation data. The method used to determine the active compound and/or metabolites in a suitable matrix should be characterized, fully validated and documented (20). The main objective of the validation is to demonstrate the acceptability and reliability of the analytical results from the study and should include the following information: (*i*) stability of stock solutions and of the analyte(s) in the biological matrix under processing conditions and during the entire storage period, (*ii*) specificity, (*iii*) accuracy, (*iv*) precision, (*v*) limit of quantification, (*vi*) matrix effect when relevant, and (*vii*) dilution effect. The validation report should include certificates of analysis for the analyte(s) and internal standard(s) and representative chromatograms.

Calibration Curves

Calibration curves must be generated for each analyte in each analytical run and used to calculate the concentration of the analyte in all subject samples. The standard curve fitting is determined by applying the simplest model that adequately describes the concentration-response relationship by using appropriate weighting and statistical tests for goodness of fit. Both $\frac{1}{x^2}$ weighting and $\frac{1}{y^2}$ weighting are commonly used in regression calculations. Pateman (27) recommends a weighting of $\frac{1}{x^2}$ by assuming that "the SD of response tends to be proportional to the concentration (*x*)." However, if no assumption is made on the SD except that the CV is constant over the majority of the calibration range then the appropriate weight will be $\frac{1}{(\alpha+\beta x)^2}$ and the regression model is $y = \alpha + \beta x + \varepsilon$. The random error, ε, is unknown but the mean of the random error, $E(\varepsilon) = 0$ so it is reasonable to replace ε by 0. This results in a weight of $\frac{1}{y^2}$, which is also called the "empirical weight" (28). To compare the difference between the unweighted, $\frac{1}{x^2}$ weighting and $\frac{1}{y^2}$ weighting a simple simulation can be performed by adding 10% to a single point in an ideal linear standard curve. This process is repeated at each concentration and the absolute percentage deviation from the ideal concentration of the entire calibration curve standards is calculated. The results of this simulation are presented in the table on page 38.

Standard[a] concentration (µg/mL) 10% added	Unweighted[b] sum of abs.% deviation of all standards	Abs.% deviation of the lowest and the top standards[c]	$1/y^2$ sum of abs.% deviation of all standards	Abs.% deviation of the lowest and the top standards[c]	$1/x^2$ sum of abs.% deviation of all standards	Abs.% deviation of the lowest and the top standards[c]
1	9.97	(8.596,0.001)	11.04	(2.545,0.586)	11.08	(2.201,0.612)
2	12.75	(−2.802,0.002)	15.17	(−−3.066, −0.272)	16.04	(−3.582, −0.318)
4	18.27	(−5.575,0.004)	16.69	(−1.239, −0.653)	17.95	(−1.473, −0.777)
8	29.13	(−11.033,0.006)	16.68	(−0.358, −0.845)	17.93	(−0.425, −1.004)
16	50.16	(−21.598,0.010)	16.65	(0.082, −0.943)	17.88	(0.097, −1.118)
32	89.45	(−41.316,0.010)	16.94	(0.302, −0.992)	18.22	(0.357, −1.174)
64	156.94	(−75.070, −0.021)	17.08	(0.412, −1.017)	18.37	(0.488, −1.203)
128	247.30	(−119.658, −0.209)	17.15	(0.467, −1.029)	18.45	(0.552, −1.217)
256	248.90	(−116.960, −1.077)	17.18	(0.495, −1.036)	18.48	(0.585, −1.224)
460.8	260.77	(131.533, −3.722)	17.19	(0.507, −1.039)	18.50	(0.560, −1.227)
512	463.45	(236.620,4.898)	17.19	(0.509,8.857)	18.50	(0.601,8.649)
Mean	144.28		16.27		17.40	
S.D	146.12		1.83		2.21	
CV%	101.27		11.24		12.71	

[a]Ten percent increase in the ideal peak area (height) ratios (PAR, range between 0.01 and 5.12 in this simulation) at one concentration each time while the ideal PARs of the remaining concentrations are unchanged.

[b]Sum of the absolute% deviation from the estimated concentrations to the stated concentrations for all standards.

[c]Absolute% deviation of the lowest and the highest standards.

By examining the percentage deviation it is obvious that the unweighted regression should not be used. A 10% variation at the lowest standard will cause only 0.001% variation of the highest standard, whilst an increase of 10% in the highest standard will cause over 200% variation of the LLOQ. Although the sum of the absolute% deviation for the $\frac{1}{x^2}$ and $\frac{1}{y^2}$ weighting are very similar, $\frac{1}{y^2}$ weighting fit is always smaller than that obtained from the $\frac{1}{x^2}$ weighting fit, indicating that, the $\frac{1}{y^2}$ weighting provides a better fit to the data. This is not surprising as using the $\frac{1}{y^2}$ weighting, the weighted least squares method minimizes the quantity,

$$\sum \left(\frac{y_i - \hat{y}_i}{y_i} \right)^2,$$

that is, it minimizes the relative errors.

Statistical Analysis

Pharmacokinetic parameters derived from the measures of concentration, for example, AUC and C_{max} should be analyzed using ANOVA. The data should be transformed prior to the analysis by using a logarithmic transformation. The statistical model includes the factors, sequence, subject (sequence) (i.e., subjects nested within sequences), period, and formulation. The sequence effect should be tested using the subject (sequence) mean square as the error term while the other main effects should be compared to the mean square error obtained from the ANOVA. Ninety percent CIs should be constructed for the ratio between the test and reference formulation averages based on the logarithmic transformed AUC and C_{max} values using the Schuirmann's two one-sided tests procedure (29).

The pharmacokinetic parameters to be tested, the procedure for testing and the acceptance criteria for each parameter should be outlined in the protocol. The acceptance intervals for the primary pharmacokinetic parameters are as follows:

AUC (AUC_{0-t}, $AUC_{0-\infty}$, or AUC_T for steady-state designs)—the 90% CI should lie within an acceptance interval of 0.80 to 1.25.

C_{max}—the 90% CI should lie within an acceptance interval of 0.80 to 1.25. As previously discussed, in certain cases a wider interval may be acceptable, but must be justified and predefined in the protocol.

T_{max}, $t_{1/2}$, and DF ($C_{max}-C_{min}$)/($AUC_{0-t/t}$) for a steady-state design] are considered to be secondary parameters and hence would not be used for evaluating bioequivalence. However, 90% CI, based on nonparametric technique, for untransformed data of T_{max} and 90% CIs for the untransformed data of the other secondary parameters are also presented for completeness.

Biowaivers

As previously indicated, an application for registration of a generic product in either New Zealand or Australia should generally include a bioequivalence study versus a leading brand (market leader) obtained from within the relevant country. However, in certain well-defined circumstances, both authorities may accept submission of foreign bioequivalence data, provided it can be shown that the local product and foreign product are identical, that is, supported by extensive comparative dissolution testing and comparative excipient testing.

If an application pertains to several strengths of an active ingredient, a bioequivalence study using only the highest strength may be acceptable. In this case Medsafe and the TGA may grant a biowaiver for the other strengths provided the strength chosen for investigation has been justified and the following conditions have been met:

1. The products are manufactured by the same manufacturer and using the same process.
2. The active ingredient has demonstrated linear kinetics over the therapeutic dose range.
3. The qualitative compositions of the different strengths are the same.
4. The ratio of the amounts of active ingredient and excipients is the same, or where the concentration of the active ingredient is low (less than 5%), the ratio of the amounts of excipients is similar.
5. The similarity should be justified by dissolution profiles covering at least three time points, in three different buffers covering the physiological range, that is, normally pH 1 to 6.8 (e.g., 1.0, 4.5, and 6.8), where the f_2 similarity factor is >50 for the additional strengths and the strength of the batch used in the bioequivalence study.

Lifespan of Bioequivalence Data

In general, bioequivalence data and study reports do not have a "use by date." In our experience, bioequivalence reports produced and submitted for local registration in a particular country can be re-submitted to other countries, for example, a study undertaken for Australian registration can subsequently be submitted, for example, in South America or Asia. As the submission process varies for each country, this may occur anywhere from 1 to 10 years after the study report has been finalized and released. The local regulatory body will assess the bioequivalence data according to the relevant guidance available at the time of submission. In this case, given the ever improving and evolving nature of tests and testing methods, the sponsor/CRO may be asked to clarify results or present additional information depending on the requirements of the reviewing regulatory body.

Topical Dosage Forms

Topical Corticosteroid products

Apart from topical corticosteroid products, currently Medsafe and the TGA prefer that topical generic dosage forms be compared using a clinical end point study design. We understand that sponsor companies have submitted bioequivalence applications by using the human skin blanching assay (HSBA) to both authorities on the basis of chromometer studies for topical corticosteroids according to the FDA guidance (30). Use of the HSBA requires a pilot study which determines the dose duration–response relationship of a marketed reference product followed by a pivotal equivalence study. In the FDA guidance only the marketed reference product is employed in the pilot study. In our experience, the pilot study should also include the test formulation as it can provide the sponsor an estimate of the likelihood that the test product will be bioequivalent in the pivotal equivalence study.

Safety Studies
When comparing topical corticosteroid formulations using a clinical end point study, the safety of the product following application must be considered. For example, one of the most obvious side effects of prolonged use of steroids is hypothalamic–pituitary–adrenal (HPA) axis suppression. Typically this can be addressed by using a parallel study design that monitors the plasma cortisol levels before and after prolonged application (e.g., 4 weeks) of the test and reference topical formulations. The area under the plasma cortisol concentration versus time curve (AUC_{0-24}) on Day 1 should be compared with the cortisol AUC_{0-24} on Day 29 to determine any significant difference between each treatment. The test formulation will be deemed safe if the 90% CI is below the limit of 125%, when compared to the reference formulation. If a toxicity study is required a randomized, single topical application, irritancy study using synchronized application and synchronized removal may be adopted. This design can be adopted to show that there is no irritancy effect following the administration of a topical formulation.

Sample Size Determination for Human Skin Blanching Studies
An analysis of variance for parallel group design is performed with only the data on the "detectors" that was obtained in the pilot study. The mean square error in the ANOVA and the mean AUEC for the reference formulation should be used to calculate the CV. The sample size required for a parallel group study, n_p, should be computed using the method specified by Hauschke et al. (31).

However, a randomized block design should be used for the pivotal in vivo bioequivalence study. It can be shown that the relative efficiency between the randomized block design and the parallel group design is $100/(1 - \rho)$, where ρ is the correlation coefficient between the AUEC (reference) and AUEC (test) from the same subject (32). Therefore the sample size for the randomized block design, n_B, is equal to the sample size for the parallel group design $\times (1 - \rho)$, that is, $n_B = n_p \times (1 - \rho)$. Since ρ is expected to be positive, n_B is typically smaller than n_p.

Finally, the sample size obtained for the randomized block design will be adjusted upward by the percentage of "detectors," p_d, observed in the pilot study. Therefore the required sample size for the pivotal in vivo bioequivalence study will be $n = n_B / p_d$.

Nasal Sprays and Inhalers
At present Medsafe and the TGA require data to confirm bioequivalence of nasal sprays and inhalers using clinical studies.

Nasal Sprays
At present, there is no guidance from the Authorities in either New Zealand or Australia for demonstrating bioequivalence between nasal sprays for locally acting products. However, protocols have been proposed by Sponsors to address the efficacy and safety issues of nasal sprays for allergic rhinitis. The efficacy study design is based on the draft guidance proposed by the FDA (33), which

is a sequential, randomized, double-blind, double-dummy (one active test, one inactive reference cross-over to one inactive test, one active reference), placebo-controlled, parallel group study of 14 days duration (each for test and reference) preceded by a 7-day placebo run-in period. Volunteers with seasonal allergic rhinitis are required to be administered doses from the nasal spray product, either the test or reference formulation, according to the dosing schedule. The lowest labeled adult recommended dose should be employed in the study to increase the sensitivity of detecting potential differences between products. A scoring system based on the common nasal symptoms of allergic rhinitis is used to assess and compare the study results. A total nasal symptom score (TNSS) is used, that includes a 4-point scale with signs and symptoms ordered in severity from 0 (no symptoms) to 3 (severe symptoms). About 20% of the volunteers will be identified as placebo responders or who will not meet the minimum qualifying criteria during the 7-day placebo period. For the equivalence and efficacy analyses, the *primary* endpoint involves reflective scores for the 12-hour pooled TNSS over the 2-week randomized portion of the study. The suppression of eosinophil count (34) in the nasal lavage has also been used as a more objective endpoint in addition to the TNSS in the final analyses.

The safety issue, for example, HPA axis suppression resulting from nasal spray administration can be determined by measuring serum cortisol levels in healthy volunteers as indicated for the topical corticosteroid studies (35).

Inhalers

Both Medsafe and the TGA have developed guidance documents to demonstrate bioequivalence of inhaled medications, Medsafe in Section 16 and TGA in Appendix 19 of their respective guidelines. In general, the following protocols address the requirements for exhibiting the bronchodilation and bronchoprotection effects of inhalers.

Bronchodilator Inhalers

Typically, therapeutic equivalence of bronchodilators, such as short-acting and long-acting β_2-agonists, should be compared by their bronchodilatation potency and efficacy to protect against bronchoconstriction caused by stimuli such as methacholine (36), histamine (37), hypertonic saline (38), or exercise (39). A double-blind, double-dummy, placebo-controlled, six-period, four-sequence cross-over study, similar to that employed by Lavorini et al. (40) has been proposed for the assessment of bioequivalence of salbutamol inhalers. Bioequivalence is assessed by calculating the relative potency of two salbutamol formulations by using either Finney's 2×2 parallel regression analysis or the E_{max} model (AUC$_{0-60}$ of FEV1 vs. time). The relative potency is calculated and expressed as the ratio of the estimated doses of salbutamol delivered by the two formulations to achieve similar AUC$_{0-60}$ values with Finney's regression. If the E_{max} model is used then the relative potency is estimated by the ratio of the ED$_{50}$ values obtained from the models. Two formulations are considered bioequivalent if the 90% CI of the relative potency is between 0.67 and 1.5. Whether this protocol can be applied to long-acting β_2-agonists, for example, salmeterol is not known. It is our experience that the FEV$_1$ is relatively insensitive to dose differences and there is evidence that the commonly used doses of salmeterol may be higher than required to produce a maximal effect. Other parameters such as PEFR (peak expiratory flow rate) heart rate, QTc interval, plasma potassium concentration and tremor in addition to bronchial reactivity can be measured to provide unequivocal clinical effectiveness of the medicinal product (41).

Steroid Aerosols

The following protocols have been proposed for the evaluation of the bioequivalence of inhaled steroids.

Cross-Over Designs

A randomized, double-blind, double-dummy, cross-over design with a 2- to 3-week run-in period followed by a 4-week treatment period has been used. Each treatment consists of either the test product and placebo as reference, or the reference product and the placebo as test. PD20 (36), after methacholine challenge testing, is measured at the start and at the end of each treatment period.

1. Start of treatment period 1 (PD20_S1)
2. End of treatment period 1 (PD20_E1)
3. Start of treatment period 2 (PD20_S2)
4. End of treatment period 2 (PD20_E2)

For each subject, the difference of PD20 within each treatment is calculated:

$$D1 = PD20\_E1 - PD20\_S1 \text{ and } D2 = PD20\_E2 - PD20\_S2$$

An ANOVA is performed on the logarithmic (base 2) transformed differences $\log_2$ (D) and a 95% CI is calculated on the $\log_2$-scale. The formulations are considered to be bioequivalent if the 95% CI is within the range $(-1, 1)$.

Parallel Design

A randomized, double-blind, double-dummy, parallel group design with a 2-week run-in period followed by a 6-week treatment period can be performed. PD20 after methacholine challenge testing is measured at the start and at:

1. Start of run-in period (PD20_R)
2. End of run-in period (i.e., Start of treatment) (PD20_S)
3. After 3 weeks of treatment (PD20_M)
4. End of treatment period (PD20_E)

For each subject, the following difference will be calculated:

$$D = PD20\_M - PD20\_S \text{ or } D = PD20\_E - PD20\_S$$

Analysis of covariance (with PD20_S as the covariate) is performed on the untransformed differences D and a 90% CI is calculated on the untransformed scale. The formulations are considered bioequivalent if both values of the 90% CI are 0.8 or above (i.e., noninferior).

Side effects such as HPA axis suppression by inhalers can be determined by measuring serum cortisol levels in healthy volunteers as described for topical studies (35).

Combined (Steroid Plus β2_Agonists) Aerosols

A combination of the above-mentioned study protocols may be useful for bioequivalence evaluation of these products.

Antiallergic Aerosols

Antiallergic aerosols such as sodium cromoglycate and nedocromil sodium are generally less potent bronchodilators than β_2-agonists and less potent bronchoprotectors than steroids. Common side effects are limited to throat irritation on inhalation and unpleasant taste. A combination of the above-mentioned study protocols may be useful for bioequivalence evaluation of the antiallergic aerosols. However, these medicines are not effective in every subject and so the demonstration of their effectiveness, on the provoked response, is a prerequisite for entry into such studies (42).

REFERENCES

1. McBride WG. Thalidomide and congenital abnormalities. Lancet 1961; 2:1358.
2. Therapeutic Goods Act 1989—Australian Government Publishing Service, Canberra.
3. Medicines Act 1981 (Reprint 2006) (published under the Authority of the New Zealand Government).
4. Australian Regulatory Guidelines for Prescription Medicines, June 2004. http://www.tga.gov.au.
5. Module 1—Administrative Information and Prescribing Information for Australia. Notice to Applicants CTD-Module, 1 September 2007.
6. Guidance on the GMP Clearance of overseas medicines manufacturers, 16th ed. March 2008. http://www.tga.gov.au/manuf/gmpsom.htm.
7. New Zealand Regulatory Guidelines for Medicines, Vol 1, 5th ed. 2001. http://www.medsafe.govt.nz.
8. Australian Regulatory Guidelines for OTC Medicines, July 2003. http://www.tga.gov.au.
9. The Nuremberg Code from Trials of War Criminals before the Nuremburg Military Tribunals under Control Council Law No. 10. Nuremberg, October 1946-April 1949. Washington, D.C.: U.S. G.P.O, 1949–1953.
10. World Medical Association Declaration of Helsinki-Ethical Principals for Medical Research involving Human Subjects 1964 (current version 2000).
11. Cartwright Inquiry. 2008. http://www.womens-health.org.nz/cartwright/cartwright.htm.
12. Gisborne Cervical Screening Inquiry. 2008. http://www.csi.org.nz.
13. The Health Research Council of New Zealand, HRC Guidelines for Ethics Committee Accreditation, June 2008.
14. New Zealand Researched Medicines Industry Guidelines on Clinical Trials Compensation for Injury Resulting From Participation in an Industry Sponsored Clinical Trial. 2008. http://www.rmianz.co.nz.
15. Medicines Australia Form of Indemnity for Clinical Trials. 2008. http://www.medicinesaustralia.com.au.
16. Jones B, Kenward MG. Design and Analysis of Cross-Over Trials. London: Chapman and Hall, 1989.
17. Haider SH, Davit B, Chen ML, et al. Bioequivalence approaches for highly variable drugs and drug products. Pharm Res 2008; 25:237–241.
18. Chow SC, Liu JP. Design and Analysis of Bioavailability and Bioequivalence studies, 2nd ed. New York: Marcel Dekker, 2000.
19. Hauschke D, Steinijans VW, Diletti E, et al. Sample size determination for bioequivalence assessment using a multiplicative model. J Pharmacokin Biopharm 1992; 20: 557–561.
20. Note for Guidance on the investigation of bioavailability and bioequivalence (CPMP/EWP/QWP/1401/98). CPMP Guidance-As adopted by the TGA-with amendment. 10 April 2002. http://www.tga.gov.
21. Blume HH, Midha KK. Bio-International 92, Conference on bioavailability, bioequivalence and pharmacokinetic studies. J Pharm Sci 1993; 2:1186–1189.

22. Guidelines on registration requirements to establish interchangeability, Multisource (Generic) Pharmaceutical Products: WHO, Draft Revision, QAS/04.093/Rev.4.
23. Guidance for Industry, bioavailability and bioequivalence studies for orally administered drug products — general considerations, CDER, U.S Dept of Health and Human Services, FDA, March 2003 (Revision 1).
24. Title 21—Food and Drugs. Code of Federal Regulations. Part—Section 320.24 (b)(1)(i).
25. Simvastatin Prescribing Information. eMIMS MIMS, April 2008.
26. Bicalutamide Prescribing Information. eMIMs MIMS, April 2008.
27. Pateman, J. Bio-International 92. In: Blume HH, Midha KK, eds. Bioavailability, Bioequivalence and Pharmacokinetic Studies, Stuttgart: Medpharm Scientific, 1995:399–403.
28. Miller RG. Beyond ANOVA: Basics of Applied Statistics. New York: Wiley, 1986.
29. Schuirmann DJ. A comparison of the two one-sided tests procedure and the power approach for assessing the equivalence of average bioavailability. J Pharmacokin Biopharm 1987; 15:657–680.
30. Guidance for Industry, Topical Dermatologic Corticosteroids: In Vivo Bioequivalence, CDER, U.S Dept of Health and Human Services. FDA, 2 June 1995.
31. Hauschke D, Kieser M, Diletti E, et al. Sample size determination for proving equivalence based on the ratio of two means for normally distributed data. Stat Med 1999; 18:93–105.
32. Fleiss JL. The Design and Analysis of Clinical Experiments. New York: Wiley, 1986.
33. Guidance for Industry, Bioavailability and Bioequivalence Studies for Nasal Aerosols and Nasal Sprays for Local Action. FDA, 2003.
34. Beclomethasone 50 µg/dose nasal spray—Clinical Equivalence Study. C93-586-LBB, April 1994, Zenith Technology Corporation Ltd. Study Archive.
35. A Parallel, 4 arm, Safety Study to evaluate cortisol levels following continuous application of topical corticosteroids in healthy subjects. Zenith Technology Corporation Ltd. Study Archive.
36. Creticos PS, Adams WP, Petty BG, et al. A methacholine challenge dose-response study for development of a pharmacodynamic bioequivalence methodology for albuterol metered-dose inhalers. J Allergy Clin Immunol 2002; 110:713–720.
37. Ahrens RC, Harris JB, Milavetz G, et al. Use of bronchial provocation with histamine to compare the pharmacodynamics of inhaled albuterol and metaproterenol in patients with asthma. J Allergy Clin Immunol 1987; 79:876–882.
38. Delvaux M, Henket M, Lau L, et al. Nebulised salbutamol administered during sputum induction improves bronchoprotection in patients with asthma. Thorax 2004; 59:111–115.
39. Anderson SD, Lambert S, Brannan JD, et al. Laboratory protocol for exercise asthma to evaluate salbutamol given by two devices. Med Sci Sports Exerc 2001; 33:893–900.
40. Lavorini F, Geri P, Camiciottoli G, et al. Agreement between two methods for accessing bioequivalence of inhaled salbutamol. Pulm Pharmacol Ther 2008; 21:380–384.
41. Wong CS, Williams J, Britton JR, et al. Bronchodilator, cardiovascular, and hypokalaemic effects of fenoterol, salbutamol, and terbutaline in asthma. Lancet 1990; 336:1396–1399.
42. Wong BJO, Hargreave FE. Bioequivalence of metered-dose inhaled medications. J Allergy Clin Immunol 1993; 92:373–379.

3 Brazil

Margareth R. C. Marques
Department of Standards Development, U.S. Pharmacopeia, Rockville, Maryland, U.S.A.

Sílvia Storpirtis
College of Pharmaceutical Sciences, University of São Paulo, São Paulo, Brazil

Márcia Martini Bueno
Regulatory Affairs and Pharmacovigilance, Libbs Farmacêutica Ltda, São Paulo, Brazil

INTRODUCTION

The publication of the National Policy for Drug Products in 1998, the creation of the National Agency for Sanitary Surveillance or Agencia Nacional de Vigilancia Sanitaria (ANVISA), the approval of the law, and the publication of the technical guidances for the registration of generic products dramatically changed the pharmaceutical market in Brazil. New concepts such as pharmaceutical equivalency, therapeutic equivalency, bioavailability, and bioequivalence were introduced (1–4).

The law and guidances for the registration of generic products in Brazil were developed based on the regulations from countries or regions, such as Canada, United States of America and the European Union, with a great deal of experience with these type of products.

Brazil was the first country in South America to implement the evaluation of pharmaceutical equivalency and bioequivalence studies for the registration of generic products and nowadays is considered a model for the other countries in the region.

A "similar" drug product is a product that contains the same drug(s) in the same concentration, dosage form, route of administration, strength, and therapeutic indication as the reference drug product registered at the federal agency in charge of sanitary surveillance, being allowed to differ only in characteristics related to size and shape of the dosage form, expiry date, packaging, labeling, excipients, and vehicles. It is always identified by its trade (branded) name and not by the generic name.

Using the regulations for generic products (3,4) as a model, in 2003 new regulations and guidances were published for the registration of new similar products and for similar products already in the market. These regulations included the evaluation of pharmaceutical equivalency, relative bioavailability, and good manufacturing practices (GMP).

Similar products registered from May 2003 up to date must be in accordance with the same regulations as generic products with the following exceptions:

Characteristics	Generic	Similar
Interchangeability with the reference product[a]	Yes	No
Use of a trade name	No, only generic designation	Brand name is mandatory together with the generic designation
Number of approved suppliers for active pharmaceutical ingredient (API)[b]	Not more than three suppliers per API	No limits

[a]Interchangeability between similar products and the reference product is not allowed by the Brazilian regulations because of the similar products registered before 2003. The deadline for similar products to meet the registration criteria according to the current regulations is 2014. If similar products registered before 2003 comply with current requirements for the registration of generic products, ANVISA may approve the interchangeability between similar and reference products. In this case, the product is designated as a branded generic.
[b]All API suppliers must be GMP certified by ANVISA. The company must submit stability studies for each API from different suppliers. In the case of dosage forms, the company must demonstrate that the dissolution profiles of products manufactured with API from different suppliers are equivalent. The dissolution profile comparison must include the reference product approved by ANVISA.

GENERIC PRODUCTS IN BRAZIL

Definitions and History

According to the Brazilian regulations, a generic drug product contains the same active substance(s), it has the same strength(s), same pharmaceutical dosage form, same route of administration, dosing, and therapeutic indications as the reference product. The differences could be size and shape of the product, its expiry date, packaging material, and excipients. It is intended to be interchangeable with a reference product and can be manufactured after the patent expiration or after transfer of the patent rights to the appropriate organizations. In other words, a generic product as opposed to a similar product is intended to be interchangeable—it cannot use a branded name but must state the drug substance name. It must be efficacious, safe, and manufactured under current GMP. Its name should be in accordance with the Brazilian Common Nomenclature (Denominação Comum Brasileira, DCB) or with the International Drug Name (3).

The reference product must be registered at ANVISA, supported by documentation related to its efficacy, safety, and quality, and it must be sold in the Brazilian market (3).

Brazilian Regulations

The use of the generic name of the drug substance(s), according to the Brazilian Common Nomenclature, together with the brand name on the packaging material of products marketed in Brazil has been mandatory since 1980 (5,6).

The Law 793/93, published on April 5, 1993, defines the information on the packaging materials of generic products. The generic name should be written using a font with a size three times bigger than the brand name. The main objective of this law was to stimulate competition on the drug product market with subsequent reduction of product price. Unfortunately, the text of the law was not very clear, with the main problem being the lack of demonstration of therapeutic equivalency between generic and reference products (7).

On May 14, 1996, Law 9279 establishing the regulations regarding intellectual property was approved. This law was a significant step for the introduction of new pharmaceutical dosage forms in the Brazilian market because there were no regulations in Brazil for intellectual property rights for medicines and drug products, allowing the registration of drug products, based only on its similarity. This law was the first step toward regulations for the registration of generic products similar to other countries with established and more modern regulatory systems (8).

The regulations for generic drug products are part of the National Policy of Drug Products, with the registration of this type of product being made according to the following:

- Definition of criteria to demonstrate therapeutic equivalency, mainly regarding bioavailability.
- Training and infrastructure of local laboratories to perform bioequivalence studies.
- Incentive to produce generic drug products, laws and guidances for the marketing, prescription and dispensing of generic products in the Brazilian market (1).

On February 10, 1999, a new approved Law, 9787 (3), defined the concept of a generic drug product and established the conditions for the use of the generic drug name for all products. According to this law, the responsibilities of ANVISA are

- to establish criteria and conditions for the registration and quality assurance of generic products,
- to establish criteria for bioavailability studies of any drug product,
- to establish criteria for the verification of therapeutic equivalency through bioequivalence studies to allow their interchangeability,
- to establish criteria for the dispensing of generic products. The decision to replace the reference or branded product with a generic version is at the discretion of the practitioner (9). Items 3 and 5 of this law promote the prescription and acquisition of generic drug products by all clinics and dispensaries in the network maintained by the federal government, government actions to facilitate the registration, sales and marketing of these products and facilitation of the information and education of the public and funding of special programs to improve the quality of drug products. As a consequence of Law 9787/99 (3), a group of Brazilian experts in the areas of quality control, pharmacology, and pharmaceutics was created to write the technical guidances for the registration of generic products in Brazil. All the texts were revised by a consultant with a great deal of experience on bioequivalence studies from the University of Texas. As a result of the work of this team, on August 9, 1999, Resolution RDC 391 (4) was approved. This document defined the technical parameters for the registration of generic drug products and had the following six annexures:
  1) Guidance for stability studies.
  2) Guidance for protocol and technical reports, bioequivalence, or bioavailability studies.
  3) Guidance for the validation of analytical methods.

guidance for templates for pharmaceutical equivalency studies.
ﾉ) Guidance for biowaivers.
6) First list of reference products.

To implement all these new guidances and regulations and to review the new dossiers submitted for the registration of this new category of products, a new department within ANVISA, the General Management of Generic Products (Gerência Geral de Medicamentos Genéricos, GGMED) was created in September 2000. The major actions undertaken by this new department were the following:

1. Elimination of similar products without brand name or similar products with generic names to avoid the confusion created by some pharmaceutical companies that were promoting similar products as if they were generic products;
2. Inclusion of a yellow strip containing the letter G in blue and the term "Medicamento Genérico" (generic product) on the packaging materials of generic products;
3. Use of billboards in public areas and broadcasting of special advertisement on TV to explain and promote the use of generic products together with special leaflets distributed to clinics and dispensaries;
4. Training of practitioners on the prescription of generic drug products;
5. Introduction of a nation-wide program to monitor the quality of generic products on the market. This activity is under the coordination of the National Institute of Quality Control in Health Products and Services (Instituto Nacional de Controle de Qualidade em Saúde, INCQS).
6. Special funding for manufacturers of generic products by the Banco Nacional de Desenvolvimento Econômico e Social (BNDES).
7. Publication of a new law, Decreto 3675, on November 28, 2000 (10) [reprinted with modifications as Decreto 3841 (11) on May 11, 2001]. This law established the process of registration of generic products already approved as generics in the United States, Canada, and some European countries with similar regulations for generic products. This registration would be valid for one year and, during this period of time, the company should present a bioequivalence study done in accordance to the Brazilian regulations. After 8 months following registration approval, the pharmaceutical company should provide their indication of intent to undertake manufacture of the product in Brazil.

Besides all these activities, ANVISA promoted and funded the installation of Pharmaceutical Equivalency Centers and Bioequivalence Centers in the country. The former were accredited by ANVISA
In 2000, ANVISA introduced the Chamber of Regulations for Medicines Market (Câmara de Regulação do Mercado de Medicamentos, CMED) to be responsible for the management of the prices of generic drug products, maintaining them at least 35% below the price of the reference product (12).
Since the introduction of generic products in Brazil, their sales have been increasing at a very fast pace, and currently they represent about 16% of medicines sold in Brazil (13).

Bioavailability

ANVISA defines bioavailability as the rate and the extension of absorption of a drug product in a dosage form, based on its concentration/time curve in the systemic circulation or its excretion in urine (14).

Relative Bioavailability and Bioequivalence

The first official definition of bioequivalence in Brazil was the one included in Law 9787 (February 10, 1999) (3), which introduced the concept of a generic drug product. According to this law, bioequivalence is the demonstration of pharmaceutical equivalency between two products in the same pharmaceutical dosage form, with identical qualitative and quantitative amount of active substance(s), and that show comparable bioavailability when evaluated using the same study design.

The Resolution RDC 16 (March 2, 2007), guidance for the registration of generic products, defines two drug products as bioequivalent when they are pharmaceutical equivalents and when administered at the same molar dose, under the same experimental conditions and do not show statistically significantly differences in bioavailability (15).

According to the ANVISA definition, bioequivalent drug products are pharmaceutical equivalents that, upon administration of the same molar dose in the same experimental conditions do not present significant statistical differences concerning bioavailability.

Resolution RDC 17 (March 2, 2007), guidance for the registration of similar products, defines relative bioavailability as the rate and extent of absorption of an active ingredient that reaches systemic circulation as a result of the extravascular administration of a preparation compared to those of a reference product that contains the same active ingredient (16).

Due to the innovative character of these activities, it became evident during the assessment of the bioequivalence centers by ANVISA that more training and discussion on validation of the analytical methods, statistical evaluation, handling and storage of biological samples, and so on was needed. As a consequence, ANVISA organized discussion groups including representatives from academia, industry, regulatory bodies, and so on to prepare a manual on good practices on relative bioavailability and bioequivalence and a check list for the inspections of those centers (17). The final version of the *Manual on Good Practices in Bioavailability and Bioequivalence* was published in 2002, and on May 2003, Resolution RDC 103 became official. This resolution defines the activities of the bioequivalence centers and contains a check list for the inspection carried out during the certification of the national and international bioequivalence centers. Three months later, only certified centers were authorized to carry out bioavailability/bioequivalence studies for the registration of drug products in Brazil.

Pharmaceutical Equivalency

According to ANVISA, pharmaceutical equivalents are drug products that contain the same drug substance, in the same salt form or free base, in the same amount, in the same type of dosage form, with or without the same excipients. These products should comply with the corresponding monograph in the Brazilian Pharmacopeia. In the absence of a specific monograph in the Brazilian Pharmacopeia, they should comply with the monograph in any other compendia accepted by the Brazilian health authorities, or with any other applicable quality standard. These quality standards include identity, strength, purity, potency, uniformity, disintegration, and dissolution, when applicable. In Brazil, the evaluation of pharmaceutical equivalency should be done by a center qualified by ANVISA (18). In June 2001, ANVISA implemented its quality system and posted all standard operating procedures (SOPs) and criteria for the certification of pharmaceutical equivalency centers on the agency website (www.anvisa.gov.br).

Good Manufacturing Practices
GMP was adopted in Brazil in 1995 through the *Portaria 16* from March 6, 1995. This law established the technical regulations and the check list for the verification of compliance to GMP (19). This regulation was updated in 2003 by the *Resolução* RDC 210 (August 4, 2003), based on the guidances from the World Health Organization and on the experience acquired during inspections carried out in Brazil. The new law provides more details regarding validation of processes and methods, qualification of suppliers, and manufacturing process of products that need segregation such as antibiotics, hormones, etc. (19). Among the objectives of the GMP inspection is to certify that the pharmaceutical company complies with current GMP guidances. This certification is included in *Resolução* RDC 210 from 2003. This activity is sponsored by some fees paid by the company requesting this certification. A GMP certificate is given for each line/dosage form at each pharmaceutical plant, and it is valid for one year (20).

Evolution of the Technical Regulations
The major aspects of the first regulation for generic drug products in Brazil are summarized in Table 1.

TABLE 1 Major Characteristics of the First Regulation for Generic Drugs Products in Brazil

Law	Characteristics
RDC 391 (Aug 9, 1999) Technical Guidance for the Registration of Generic Products in Brazil	√ Mandatory presubmission
	√ Production and Quality Control Technical Report
	√ Manufacturing report for three pilot batches
	√ Validation of the manufacturing process
	√ Name(s) of the supplier of the drug substance (not more than three)
	√ Description of the synthesis route, including isomers and polymorphs
	√ Specifications and validated analytical methods
	√ Pharmaceutical equivalency
	√ Certificate of GMPs and QC
	√ Stability studies according to Zone 4
	√ Bioequivalence—Quantification of the drug substance or metabolite
	- Washout period of minimum of five t^{1}_{2}
	- Number of healthy volunteers (in general 24)
	- Weight variance $\pm$ 10% of the normal value
	- Use of the Brazilian or international reference product.
	√ Prescription: In the network by Brazilian Common Nomenclature (DCB)—In private practice: brand name or DCB
	√ Dispensing: Possibility of the pharmacist replacing the brand name product by the generic version, if there are no restrictions by the private practitioner

The experience acquired during the first years of implementation of the regulations for generic products and the need for compliance to international standards were the major reasons for the several updates made in the regulations. This process is summarized below:

Law	Major Characteristics
First revision RDC 391/99 ⇓	√ Presubmission not mandatory √ Guidance for imported generic products √ The list of reference products began to be published and updated by ANVISA on the Web site
RDC 10/01	√ Biowaivers; in vitro studies √ New guidances
Second revision RDC 10/01 ⇓	√ Annexes—Guidances changed to Resolutions (RE) √ New Guidances: Manufacturing of pilot batches, scale-up, and postapproval changes—Experimental design for bioequivalence studies
RDC 84/02	√ Authorization for bioequivalence studies using truncated area under the curve (AUC) for drug substances with long half-life
Third revision RDC 84/02 ⇓ RDC135/03	√ Notification for the production of a pilot batch √ Dissolution profile comparison between the biobatch and the batch manufactured with the drug substance from a new supplier √ Explanation for the need for the quantification of metabolite(s) √ Possibility of employing a parallel experiment design √ Reduction in the number of healthy volunteers to 12 supported by a test power of not less than 80% √ Truncated AUC (72 hr) for drug substances with long half-life √ Special cases with multiple dosing √ For modified release dosage forms, bioequivalence must be demonstrated in both fed and fasted states. In the case of immediate release dosage forms fasted studies are necessary only in cases where the presence of food can influence its absorption. This information must be included in the Instructions for Patients in the insert included in the product packaging material of the reference product. √ Explanation for pharmacodynamic studies √ Criteria for the transportation of samples
Fourth revision RDC 135/03 ⇓ RDC 16/07	√ More details in the listing of formulation components √ Protocol for bioequivalence studies prepared by an evaluation center certified by ANVISA for oral contraceptives, endogenous hormones, and immunosuppressants √ Approval for the simultaneous manufacture of a generic product in more than one manufacturing site √ Six months report on adverse effects and nontherapeutic action for contraceptives, endogenous hormones, and immunosuppressants √ Report on adverse effects and nontherapeutic action for all other therapeutic products in the renewal of the registration √ Permit for registration of oral contraceptives and endogenous hormones

Law	Major Characteristics
Guidances for Pharmaceutical Equivalency and Dissolution Profiles RE 900 e 901/03 ⇓ RE 310/04	√ Portions of resolution RE 901 are converted into recommendations for carrying out dissolution tests of immediate release oral pharmaceutical dosage forms √ Comparison of dissolution profiles must use the same dissolution method as used in the pharmaceutical equivalency study. If the dissolution method is not a compendial one, the dissolution profiles must be obtained using at least three different media within the physiological pH range. √ Inclusion of the dissolution profile comparison between the reference product and the generic product. It is not mandatory to meet the f_2 acceptance criteria. √ Inclusion of the certificates of pharmaceutical equivalency and dissolution profiles study according to the form available at http:www.ANVISA.gov.br/REBLAS/certificados/index.htm √ Calculation and acceptance criteria for dissolution profiles comparison using independent model approach, calculating the difference factor (f_1) and similarity factor (f_2)
Relative bioavailability and bioequivalence RE 896/03 ⇓ RE 397/04	√ Quantification of metabolites only if it is not possible to determine the drug substance by itself or in special cases, with preapproval by ANVISA. Definition of the analyte quantified in the bioequivalence studies. √ Minimum of 12 volunteers, if the statistical power is more than 80%. In cases where it is not possible to determine the number of volunteers, a minimum of 24 volunteers can be used. √ Description of the cases where a fed state study should be done. Posting of Table 1 – Administration route on the ANVISA Web site. This list is only for immediate release and delayed-release (gastro-resistant) dosage forms. For extended-release dosage forms, fed studies are mandatory. This list is inteded to be updated regularly. √ Detailed explanation on the pharmacokinetic parameter AUC_{0-t}, where t is the time corresponding to the last concentration of the drug substance determined experimentally (above the quantitation limit). √ New text format for conclusions in bioequivalence studies. √ Alteration of the confidence interval requirement from 95% to 90% of the ratio of the geometric averages (AUC_{0-t} test/AUC_{0-t} reference and C_{max} test/C_{max} reference) for the evaluation of bioequivalence of drug products containing narrow therapeutic index drugs such as carbamazepine, valproic acid, clindamycin, etc.
Relative bioavailability and bioequivalence RE 397/04 ⇓ RE 1170/06	√ Definition of the analyte used to establish bioequivalence. Listing 2—Analyte used to establish relative bioavailability/bioequivalence √ The sampling schedule should allow the appropriate characterization of the plasma profile of the drug or its metabolite (concentration x time) considering a time equal or higher than 3 to 5 times the elimination half-life of the compounds.

(*Continued*)

Law	Major Characteristics
	√ Inclusion of transdermal products and depot dosage forms with information regarding the sampling frequency.
	√ Exclusion of the age limit of the participants in the studies, maintaining the age limit of not less than 18 years.
	√ Restrictions regarding age and gender of the volunteers. In the cases of products used by specific populations, the study should be carried out with volunteers representing these specific populations. In the case of contraceptives, the studies should be run with women of child-bearing age.
	√ Inclusion of the control of body weight variability (maximum 10%) among volunteers for oral contraceptives
	√ Inclusion of transdermal patches
	√ Inclusion of more information regarding endogenous compounds, taking into account the basal levels.

The new regulations published in 2003 for fixed combination products (FDCs), required the comparison of the bioavailability of the drug substance(s) alone and in combination, demonstrating that the combination product does not modify the bioavailability of the individual drugs. For the registration of new strengths, new dosage forms, and/or new routes of administration, within the therapeutic range, relative bioavailability studies may replace phase 2 and phase 3 studies (21,22).

At the beginning of the implementation of the new regulations for the registration of generic products in Brazil, bioequivalence studies were permitted to use an international reference product (reference products from countries/regions such as United States, Canada, or United Kingdom). This facilitated the registration of products in a more expedient way. The review of international bioequivalence dossiers was a learning experience for reviewers at ANVISA. The knowledge obtained helped them to update and modify the regulations and guidances in Brazil. On May 14, 2003, Law 10669 was published, defining the deadline of June 30, 2003, for submission of studies carried out with an international reference product.

Biopharmaceutical Classification System
Currently, the Biopharmaceutical Classification System (BCS) is not accepted by ANVISA for biowaivers because of the lack of sufficient information regarding mainly the permeability of drug substances, the high cost of permeability determination, and incomplete validation of the methodology.

DESIGN AND CONDUCT OF BIOEQUIVALENCE STUDIES FOR ORALLY ADMINISTERED DRUG PRODUCTS
Bioequivalence studies in Brazil are regulated by the Guidance for Bioavailability/Bioequivalence Studies Resolution 1170 from 2006 (23).

Study Design
An open, cross-over, randomized design is required where the volunteers receive test and reference drugs on separate occasions (periods), in a single- or multiple-dose regimen. If necessary, a parallel design can be used, for long terminal half-life drugs (longer than 24 hours).

Subjects

Number of Subjects

The number of subjects included must provide adequate statistical power for bioequivalence demonstration. Numbers of participants can be calculated from the coefficient of variation (intraindividual) and statistical power, and no less than 12 volunteers may be used. If no literature data are available the investigator can opt for the inclusion of at least 24 volunteers. The guidance for planning the statistical issues of bioavailability/bioequivalence studies [Resolution 898 from 2003 (24)] describes a method to calculate the number of volunteers based on the procedure described by Chow and Liu (25).

Selection of Subjects

Sex/Age
Subjects must be 18 years of age or older and capable of giving informed consent. They may be of either/both sexes but if males and females are included, there must be an equal distribution among the study sequences. If the drug product is to be used in a specific population (age and gender) the investigator must include volunteers of this population. As an example, bioequivalence studies of oral contraceptives must include women of child-bearing age.

Mass
The weight of the subjects should be within 15% of the normal range for males and females, taking into account their height and physical type. It also recommends the inclusion of woman within 10% of the normal range in bioequivalence studies for oral contraceptives.

Informed Consent
Informed consent must be approved by an ethics committee that has to be submitted for ANVISA evaluation together with the bioequivalence study report. The informed consent form must contain the drug name, the dose by unit, the pharmaceutical form, and the name of the manufacturer(s). Detailed information about the informed consent content is presented in Resolution 196 from 1996 (26).

Medical Screening

The Guidance for bioavailability/bioequivalence studies [Resolution 1170 from 2006 (23)] does not state any information regarding medical screening but the Guidance for protocol of bioavailability/bioequivalence studies [Resolution 894 from 2003 (27)] states that the protocol must contain the clinical evaluation of the volunteers, a list of the laboratory tests (blood, biochemistry, hepatitis B and C, HIV, beta-HCG for women, type I urine tests, and electrocardiogram).

Smoking/Drug and Alcohol Abuse

Smokers and individuals with drug abuse history should be avoided. If smokers are included, they must be identified.

Inclusion of Patients Instead of Healthy Subjects

For cytotoxic drugs, the study must include patients whose disease process is stable and who are being treated with cytotoxic medicines.

Phenotyping/Genotyping
The Brazilian regulations for bioavailability/bioequivalence studies do not discuss this topic.

Standardization of Study Conditions

Dosing Times
Multiple-dose studies should be avoided, since single-dose studies are usually more sensitive to show differences in formulations. Multiple-dose studies can be performed when it is shown that this approach can reduce the intraindividual variability of drug absorption in cases such as high variability drugs.

Food and Fluid Intake
The products should be administered with a standard liquid volume (usually 200 mL of water). For modified-release products, two bioequivalence studies must be performed, one in the fasted state and the other in the fed state. For immediate and delayed release formulations only a fasted study is requested for the majority of the drugs. There is a list on the ANVISA Web site (List 1, available at http://www.anvisa.gov.br/medicamentos/listas/lista1.htm) that establishes whether the study should be conducted in a fasted or fed state. A fed study is requested only if the drug has a significant change in bioavailability when administered with food and the reference product indicates that the drug has to be taken with food.

Concomitant Medication, Posture, and Physical Activity
Brazilian regulations for bioavailability/bioequivalence studies do not discuss this topic.

Blood/Urine Sample Collection and Times
The study can be performed by measuring drug or its metabolite concentration in the circulation (blood, plasma, or serum) or in urine if there is a justification for this choice. It is preferable to have bioequivalence studies that quantify the nonmodified drug substance. The quantification of any other entity will be accepted only when there are analytical limitations or the drug substance undergoes very fast biotransformation, with proper justification. A possible scenario is when the metabolite is active (has the same potency as the original drug or has greater activity/potency than the original drug), it is biotransformed by presystemic metabolism and it is important for the efficacy and safety of the product. In all cases, the study protocol must be approved by ANVISA. The protocol must specify the metabolite that will be quantified during the bioequivalence studies according to List 2—Analyte for Relative Bioavailability/Bioequivalence (available on the ANVISA website), and must meet the criteria established by the regulatory agency.

Duration and Frequency of Sampling
Frequency and number of samples collected have to guarantee an adequate characterization of the pharmacokinetic profile (plasma concentration vs. time). It must cover at least three times the elimination half-life of the drug and it should be sufficient to account for at least 80% of the known area under the drug concentration versus time curve (AUC). For transdermal delivery systems

and depot dosage forms, the duration and the frequency of sampling should be sufficient to characterize the absorption, distribution, and elimination of the drug. For long terminal half-life drugs (longer than 24 hours), an alternative collection time, for at least 72 hours, or a parallel design may be used.

Number of Samples
The Brazilian regulations for bioavailability/bioequivalence studies do not discuss this topic.

Characteristics to Be Investigated

Blood/Plasma/Serum Concentration Versus Time Profiles
The pharmacokinetic parameters should be calculated based on plasma drug concentration versus time profile. When dealing with endogenous compounds, the baseline level should be measured and the analysis must be done with and without basal correction. The following pharmacokinetic parameters should be presented: AUC_{0-t}, $AUC_{0-\infty}$, C_{max}, t_{max}, and the half-life, $t_{1/2}$. For multiple-dose studies, the parameters should be $AUC_{0-\tau}$, C_{max}, t_{max}, C_{min}, C_{avg}, and the fluctuation index where

AUC_{0-t}—Area under the plasma/serum/blood concentration–time curve from time zero to time t, where t is the last time point with measurable concentration for individual formulation;

$AUC_{0-\infty}$—Area under the plasma/serum/blood concentration–time curve from time zero to time infinity.

t_{max}—Time to maximum plasma concentration.

C_{max}—The maximum observed plasma concentration after dose administration.

Urinary Excretion Profiles
The Brazilian regulations for bioavailability/bioequivalence studies do not provide the pharmacokinetic parameters required for urine studies.

Pharmacodynamic Studies
Pharmacodynamic studies are indicated when it is not possible to measure the drug in plasma or in urine in an accurate way, for example, ophthalmic suspensions and local acting inhalers.

Bioanalysis
Details about performing and validating bioassays are stated in the Guidance for bioanalytical method validation [Resolution 899 from 2003 (28)].

Stock Solutions Stability
The stability of stock solutions of the drug and the internal standard should be evaluated at room temperature for at least six hours and, if the stock solutions are refrigerated or frozen for a relevant period, stability should be done in this condition as well.

Stability of Analyte in Biological Matrix:
Freeze and thaw stability: the analyte stability should be determined after three freeze and thaw cycles.

Short-term stability: the aliquots of low and high concentrations should be thawed at room temperature and kept at this temperature from 4 to 24 hours.

Long-term stability: the storage time evaluation should exceed the time between the date of first sample collection and the date of last sample analysis.

Processed sample stability: the stability of processed samples, including the resident time in the autosampler, should be determined.

Validation

Specificity

Analyses of blank samples of the appropriate biological matrix (blood, plasma, serum, urine, or other matrix) should be obtained from at least six sources (four normal, one lipemic, and one hemolyzed). Each blank sample should be tested for interference, and selectivity should be ensured at the lower limit of quantification (LLOQ).

Precision and Accuracy

Precision should be measured using a minimum of five determinations per concentration (high, middle, and low). The precision determined at each concentration level should not exceed 15% of the coefficient of variation (CV) except for the LLOQ, where it should not exceed 20% of the CV. Accuracy is determined by replicate analysis of samples containing known amounts of the analyte. Accuracy should be measured using a minimum of five determinations per concentration (high, medium, and low). The mean value should be within 15% of the actual value except at LLOQ, where it should not deviate by more than 20%.

Lower Limit of Quantification

The limit of quantification should be at least five times the response compared to blank response or analyte peak (response) should be identifiable, discrete, and reproducible with a precision of 20% and accuracy of 80% to 120%.

Limit of Detection

The limit of detection (LOD) should be at least two or three times the response compared to blank response.

Recovery

The recovery of the analyte does not need to be 100%, but the extent of recovery of an analyte and of the internal standard should be consistent, precise, and exact. Recovery experiments should be performed by comparing the analytical results for extracted samples with unextracted standards that represent 100% recovery.

Response Function

The simplest model that adequately describes the concentration–response relationship should be used. Selection of weighting and use of a complex regression equation should be justified.

Robustness

This parameter should be considered during the method development phase. If measurements are susceptible to variations in analytical conditions, the analytical conditions should be suitably controlled or a precautionary statement should be included in the procedure. Examples of typical variations are stability of analytical solutions and extraction time. In the case of liquid chromatography, examples of typical variations are

influence of variations of pH in a mobile phase,
influence of variations in mobile phase composition,
different columns (different lots and/or suppliers),
temperature, and
flow rate.

In the case of gas chromatography, examples of typical variations are

different columns (different lots and/or suppliers),
temperature, and
flow rate.

Calibration Curves

A calibration curve should be generated for each analyte in the sample. A sufficient number of standards should be used to adequately define the relationship between concentration and response. Concentrations of standards should be chosen on the basis of the concentration range expected in a particular study. A calibration curve should consist of a blank sample (matrix sample processed without internal standard), a zero sample (matrix sample processed with internal standard), and six to eight nonzero samples covering the expected range, including LLOQ. The following conditions should be met in developing a calibration curve: 20% deviation of the LLOQ from nominal concentration and 15% deviation of standards other than LLOQ from nominal concentration. At least four out of six nonzero standards should meet the above criteria, including the LLOQ and the calibration standard at the highest concentration. The exclusion of any standards should not change the model used.

Quality Controls Sample

Quality Control (QC) samples at three concentrations [one near the LLOQ (i.e., smaller or equal than $3 \times$ LLOQ), one in midrange, and one close to the high end of the range (between 75% and 90% of the higher concentration)] should be incorporated in each assay run. The number of QCs should not be less than 5% of the total number of samples in an analytical run, and in the case that the run has less than 120 samples, a minimum of six QCs. The results of the QC samples provide the basis of accepting or rejecting the run. At least four of every six QC samples should be within 15% (or 20% to LLOQ) of their respective nominal value. Two of the six QC samples may be outside the 15% (or 20% to LLOQ) of their respective nominal value, but not both at the same concentration.

Standard Operating Procedures

The following SOPs should be presented in the final report:

Bioanalytical method
Preparation, storage, and criteria of acceptance of the stock solutions, calibration
standards, QCs, dilution standards, and reference solutions
Validation of bioanalytical method
Acceptance of analytical runs

Reassay
Chromatographic analyses
Reintegration
Others.

Choice of Reference Product

The Brazilian reference products are listed on the ANVISA Web site (www.anvisa.gov.br).

Study Products and batch size

Before starting a bioequivalence study, the recommendation is to perform all in vitro tests comparing test and reference products. These tests have to be performed at a certified laboratory in Brazil. Both test and reference products have to comply with the drug monograph and the potency difference between test and reference drug content cannot be higher than 5%. The same batch product should be used for the in vitro and in vivo (bioequivalence) tests.

Data Analysis

The statistical evaluation should be done according to the Guidance for planning and accomplishment of the statistical phase of bioavailability/bioequivalence studies [Resolution 898 from 2003 (24)].

Statistical Analysis

Pharmaceutical parameters shall be obtained from the blood concentration curves of the drug versus the time, and statistically analyzed to determine the bioequivalence; in case of endogenous substances, statistical analysis should be conducted using quantified plasma concentrations with and without baseline level correction, while the decision on bioequivalence should be based on the corrected values.

The following pharmacokinetic parameters should be determined:

- AUC_{0-t}, $AUC_{0-\infty}$, C_{max} of the drug and/or metabolite, and the time to reach this peak (T_{max}) should be directly obtained, without data interpolation; elimination half-life ($t_{1/2}$) of the drug and/or metabolite should also be determined, although there is no need for statistical treatment.
- For evaluation of bioequivalence, the parameters AUC_{0-t}, C_{max}, and T_{max} should be used and in the case of multiple-dose studies, steady state after administration of the test and reference drugs must be confirmed.
- The exclusion of more than 5% of volunteers that participated in the study is not permitted until the conclusion or absence of over 10% of the blood concentration values of the drug from the administration of each drug per volunteer.

Data Presentation

a) Present a table containing individual values, averages (arithmetical and geometrical), standard deviation, and variation coefficient of all pharmacokinetic parameters related to the administration of test and reference drugs.
b) It is recommended that the AUC_{0-t} and C_{max} parameters be transformed into natural logarithms.

c) Conduct variance analysis (ANOVA) of pharmacokinetic parameters AUC_{0-t} and C_{max} transformed to evaluate sequence effects, of volunteer within the sequence, period, and treatment. In addition, present ANOVA table containing source, release level, sum of squares, mean square, F statistics, p values, and the intra- and interindividual coefficient values.

d) Build a reliability interval (RI) of 90% for the differences measured from the transformed data of the test and reference drugs, for AUC_{0-t} and C_{max} parameters. The RI antilogarithm obtained constitutes the RI of 90% for the amount of geometric means of the parameters.

The construction of this IC should be based on the mean residual square of the ANOVA obtained according to (c) above.

e) T_{max} shall be analyzed as individual difference (= test – reference), building the RI of 90%, using a nonparametric test.

f) This method based on RI is equivalent to the procedure of two corresponding unicaudal tests with the null possibility of bioinequivalence, with level significance level of 5% ($\alpha = 0.05$).

g) Validated statistical programs should be used.

h) When necessary, suitable statistical models, depending on the type of study (for instance, of multiple doses) should be adopted.

i) In case of volunteers presenting discrepant behavior in the pharmacokinetic parameters in relation to the others, their exclusion from the study should be justified. Present study results with and without the inclusion of their data.

j) Information on the software used for statistical data analysis to be included.

Acceptance Range for Pharmacokinetic Parameters

Two drugs shall be considered bioequivalent if the extreme values of the RI of 90% of the ratios of the geometric means (AUC_{0-t}test/AUC_{0-t} reference and C_{max}test/C_{max} reference) are greater than 0.8 and less than 1.25. Other RI of 90% limits for C_{max} previously established in protocol can be accepted upon scientific justification. When clinically relevant, T_{max} shall also be considered.

Single-Dose Studies

The following pharmacokinetic parameters have to be presented in the report: C_{max}, t_{max}, AUC_{0-t}, $AUC_{0-\infty}$, and $t_{1/2}$. The 90% CI for C_{max} and AUC_{0-t} has to be between 80% and 125%.

Steady-State Studies (Immediate Release Dosage Forms, Controlled/Modified Release Dosage Forms)

The following pharmacokinetic parameters have to be presented for a multiple-dose study: C_{max}, t_{max}, $AUC_{0-\tau}$, average concentration at steady-state and the fluctuation index. It has to be proven that steady state was reached. The 90% CI for C_{max} and $AUC_{0-\tau}$ (AUC during a dosage interval) has to be between 80% and 125%. No special request is made for modified release dosage forms.

Reporting of Results

The Resolution 895 from 2003 (29) describes all the information that should be presented in the bioequivalence report, which includes the test and reference data, administrative information, responsible personnel and sites, study protocol, ethics committee approval, and the specific information about each phase.

Clinical Report
The documents should include study design, randomization list, drug information, drug accountability, and all the information of the study population (demographic, laboratory tests), inclusion and exclusion criteria, time of drug administration and blood collections, fasted and fed times, dose washout period, adverse effects, and protocol deviations.

Analytical Report
Should contain bioanalytical method description and validation, reference standards, method to calculate drug concentrations, complete information about the analytical runs (date, initial and final time, volunteers analyzed, calibration standard and bias, quality controls, and bias), reanalysis (initial and reanalysis values, cause for reanalysis, criteria for choosing the value), and complete series of chromatograms from 20% of the volunteers.

Pharmacokinetic and Statistical Report
It should contain the calculation of sample size, ANOVA table for the pharmacokinetic parameters, CI for C_{max} and AUC, statistical program outputs, and the study conclusion. The individual plasma concentrations and individual pharmacokinetic parameters should be tabulated. These data can be presented in a spreadsheet and can be submitted in electronic format.

Quality Assurance
The Resolution does not say anything about quality assurance, but the study can be performed only in contract research organizations (CROs) previously certified by ANVISA.

BIOAVAILABILITY AND BIOEQUIVALENCE REQUIREMENTS

Orally Administered Drug Products Intended for Systemic Action

Solutions
For oral solutions, a bioequivalence study can be waived according to item 1.2 of the Guidance for waiver [Resolution 897 from 2003 (30)]. However, if the solution contains an excipient that can change the intestinal motility, a bioequivalence study is required.

Suspensions
A bioequivalence study is always required.

Immediate Release Products—Tablets and Capsules
For immediate release products a bioequivalence study is always required.

Other Oral Dosage Forms
If the drug is absorbed, a bioequivalence study is required.

Orally Administered Drug Products Intended for Local Action
For orally administered drug products intended for local action that are not absorbed, a bioequivalence study can be waived according to item 1.8 of the

Guidance for waiver and substitution of bioequivalence studies [Resolution 897 from 2003—oral products in which the drug substance is not absorbed by the gastrointestinal system (30)].

Parenteral Solutions and Suspensions/Emulsions

For aqueous parenteral solutions, which have the same drug, at the same concentration and excipients with the same function, a bioequivalence study can be waived according to item 1.1 of the Guidance for waiver and substitution of bioequivalence studies [Resolution 897 from 2003 (30)]. A bioequivalence study is required for parenteral suspensions whereas Brazilian regulations do not give any information about emulsions for parenteral use.

Topically Administered Drug Products

If intended for systemic action, bioequivalence study is required, including transdermal dosage forms.

Topical Products for Local Action

For topically administered drug products intended for local action, that have the same drug, at the same concentration and excipients with the same function, a bioequivalence study can be waived according to the Guidance for waiver and substitution of bioequivalence studies [Resolution 897 from 2003 (30)].

Topical Products Containing Corticosteroids

For nasal and oral inhalation products containing corticosteroids, which have the same drug, at the same concentration and excipients with the same function, a bioequivalence study can be waived only for solutions. For other pharmaceutical forms, a bioequivalence study is required. The type of study to be performed, bioavailability or bioequivalence or clinical study, must be discussed with ANVISA before undertaking a study.

Topical Products Other Than Those Containing Corticosteroids and Which are Not Simple Solutions

If intended for systemic action a bioequivalence study is required. For any topical product, such as cream, lotions, ointments, etc., with no systemic action, only in vitro studies are required.

Products Intended for Other Routes of Administration

For oral products for local use, dermal, rectal, vaginal, etc. intended to act without systemic absorption, bioequivalence studies can be waived according to item 1.6 of the Guidance for waiver and substitution of bioequivalence studies (Resolution 897 from 2003 (30). Bioequivalence studies for nasal products other than solutions are required and the guidance for nasal products and products for inhalation (Public Consult no. 22/2008) are under discussion by ANVISA.

Variations or Postregistration Amendment Requirements

In the case of postregistration, the Guidance for alterations, inclusions and notifications of postregistration [Resolution RE no 893 from 2003 (31)] should be followed. A bioequivalence study is required only when there is a significant change in the formulation that can result in different bioavailability.

WAIVERS OF IN VIVO BIOEQUIVALENCE STUDIES

Immediate Release Drug Products

Bioequivalence assessment of drug products that contain acetyl salicylic acid, or acetaminophen, or dipyrone or ibuprofen, in the solid form, are not required.

Different Strength Dosage Forms

For immediate release products with different doses, in the same pharmaceutical dosage form and with proportional formulations, manufactured by the same producer, at the same *locale*, a bioequivalence study should be performed with the higher dosage strength and the other strengths can be waived if the dissolution profiles are comparable according to the Guidance for assay dissolution for solid oral pharmaceutical forms of immediate liberation [Resolution 310 from 2004 (32)]. It also states that the waiver is acceptable if there is linearity in the drug's pharmacokinetics.

BCS Class Exemptions

Brazilian regulations for bioavailability/bioequivalence studies do not discuss this issue.

Controlled/Modified Release Dosage Forms

Capsules Containing Beads—Lower Strengths

Brazilian regulations for bioavailability/bioequivalence studies do not discuss this type of dosage form.

Tablets—Lower Strength

Tablets with different strengths/doses, in the same pharmaceutical dosage form and with proportional formulations, same release mechanism, manufactured by the same producer, at the same site, a bioequivalence study should be performed with the higher dosage strength and a bioequivalence study for the other strengths can be waived if the dissolution profiles are comparable among all strengths according to the Guidance for assay dissolution for solid oral pharmaceutical forms [Resolution 310 from 2004 (32)]. For modified release dosage forms, a comparative dissolution study using three different dissolution media (such as pH 1.2; pH 4.5, and pH 6.8 media) should be performed. In addition, the dissolution profile should be comparable among test and reference dosage forms for all dosage strengths.

REFERENCES

1. Brazil. Health Minister. National Health Surveillance Agency. MS/GM Ordinace no. 3916 of October 30, 1998. Approves the National Drug Policy. Brasília, DF; 1998. http://e-legis.bvs/leisref/public/showAct.php?id=751. Accessed June 16, 2009.
2. Brazil. Law no. 9782 of January 26, 1999. Defines the National Health Surveillance System, establishing the National Health Surveillance Agency, among other provisions. Brasília, DF; 1999. http://e-legis.bvs/leisref/public/showAct.php?id=182. Accessed June 16, 2009.
3. Brazil. Law no. 9787 of February 10, 1999. Amending the Act no 6360, to September 23, 1976, which deals with health surveillance, provides the generic product, provides for the use of generic names in pharmaceutical and other provisions.

Brasília, DF; 1999. http://e-legis.bvs/leisref/public/showAct.php?id=245. Accessed June 16, 2009.

4. Brazil. Health Minister. National Health Surveillance Agency. Resolution no. 391 of August 9, 1999. Technical regulation for generic drugs. Brasília, DF; 1999. http://e-legis.bvs/leisref/public/showAct.php?id=251. Accessed June 16, 2009.

5. Storpirtis S, Oliveira PG, Rodrigues D,et al. Biopharmaceutical considerations in the manufacturing of generic products: factors affecting dissolution and absorption of drugs. Rev Bras Cienc Farm 1999; 35(1):1–16.

6. Vernengo M. Technical elements of a generic drug policy: an agenda of essential medicines and technology for health care. Geneve, Switzerland: PAHO/WHO, 1993.

7. Brazil. Decree no. 793 of April 5, 1993. Amendment to the Decree No. 74,170s of 10 June 1974 and 79094 of January 5, 1977, which govern, respectively, Laws paragraphs 5991 of January 17, 1973 and 6360, from September 23 1976, among other provisions. Brasília, DF; 1993. http://e-legis.bvs/leisref/public/showAct.php?id=513&word. Accessed June 16, 2009.

8. Brazil. Law no. 9279 of May 14, 1996. Regulates rights and obligations relating to industrial property. Brasília, DF; 1996. http://e-legis.bvs/leisref/public/showAct.php?id=5565&word. Accessed June 16, 2009.

9. Dias HP. Flagrant judicial system of health. Brasília: ANVISA, 2004.

10. Brazil. Decree no. 3675 of November 28, 2000. Provides for special measures relating to the registration of generic drugs, as mentioned in art. 4 of Law no 9787 of 10 February 1999. Brasília, DF; 2000. http://e-legis.bvs/leisref/public/showAct.php?id=236. Accessed June 16, 2009.

11. Brazil. Decree no. 3841 of June 11, 2001. Gives new wording to devices of Decree no. 3675 of 28 November 2000, which provides for special measures relating to the registration of generic drugs, as mentioned in art. 4 of Law no 9787 of 10 February 1999. Brasília, DF; 2001. http://e-legis.bvs/leisref/public/showAct.php?id=513&word. Accessed June 16, 2009.

12. Brazil. Health Minister. National Health Surveillance Agency. Market survey on the price of medicines.(Monitoramento de mercado reverte aumento dos preços de medicamentos). Boletim Informativo da Anvisa, Brasília, 2002; (16):4–5. http://www.anvisa.gov.br/divulga/public/boletim/16_02.pdf. Accessed June 16, 2009.

13. Pró Genéricos—Brazilian Association of Industries of Generic Drugs. World market for generic. Sao Paulo: Pró Genéricos; 2008. Available at: http://www.progenericos.org.br/mercado.shtml. Accessed June 16, 2009.

14. Brazil. Health Minister. National Health Surveillance Agency. RDC Resolution no. 135 of May 29, 2003. Approves Technical Regulation for Generic Drugs. Brasília, DF; 2003. http://www.anvisa.gov.br/eng/legis/resol/135_03_rdc_e.htm. Accessed June 16, 2009.

15. Brazil. Health Minister. National Health Surveillance Agency. RDC Resolution no. 16 of March 2, 2007. Approves the technical regulation for generic drugs, annex I. Accompanying this regulation to annex II, entitled "face sheet of the registration process and post-registration of generic drugs. Brasília, DF; 2007. http://e-legis.anvisa.gov.br/leisref/public/showAct.php?id = 25960&word = Accessed June 16, 2009.

16. Brazil. Health Minister. National Health Surveillance Agency. RDC Resolution no. 17 of March 2, 2007. Approves the technical regulation for registration of similar drug. Brasília, DF; 2007. http://e-legis.anvisa.gov.br/leisref/public/showAct.php?id = 26132&word = Accessed June 16, 2009.

17. National Health Surveillance Agency. Manual of good practices in bioavailability and bioequivalence. Brasília: ANVISA, 2002.

18. Brazil. Health Minister. National Health Surveillance Agency. Resolution no. 41 of April 28, 2000. ANVISA (National Health Surveillance Agency) sets the minimum criteria for acceptance of units that perform testing of pharmaceutical equivalence, bioavailability and bioequivalence of drugs. Brasília, DF; 2000. http://e-legis.bvs/leisref/public/showAct.php?id=2114. Accessed June 16, 2009.

19. Brazil. Health Minister. National Agency for Sanitary Surveillance. SVS/MS Ordinance no. 16 of March 6, 1995. Approves the regulation on appropriate practices

for the manufacture and quality inspection of drugs. Brasília, DF; 1995. http://
e-legis.bvs/leisref/public/showAct.php?id=5355. Accessed June 16, 2009.

20. Brazil. Health Minister. National Health Surveillance Agency. RDC Resolution no.
210 of August 4, 2003. Determines all establishments manufacturers of drugs, compliance with the guidelines established in the technical regulation of the practice for the manufacture of drugs. Brasília, DF; 2003. http://e-legis.anvisa.gov.br/
leisref/public/showAct.php?id = 22321&word = Accessed June 16, 2009.

21. Brazil. Health Minister. National Health Surveillance Agency. RDC Resolution no.
134 of May 29, 2003. Available on the suitability of drugs already registered Brasília,
DF; 2003. http://e-legis.bvs/leisref/public/showAct.php?id=7904. Accessed June
16, 2009.

22. Brazil. Health Minister. National Health Surveillance Agency. RDC Resolution no. 136
of May 29, 2003. Provides for the registration of new drug. Brasília, DF; 2003. http://e-
legis.bvs/leisref/public/showAct.php?id=7914. Accessed June 16, 2009.

23. Brazil. Health Minister. National Health Surveillance Agency. RE Resolution no.
1170 of April 19, 2006. Determines the publication of guidelines for evidence
on bioavailability/bioequivalence of drugs. Brasília, DF; 2006. http://e-legis.bvs/
leisref/public/showAct.php?id=21746&word. Accessed June 16, 2009.

24. Brazil. Health Minister. National Health Surveillance Agency. RE Resolution no. 898
of May 29, 2003. Guide to planning and performing the step of statistical studies on
the bioavailability/bioequivalence. Brasília, DF; 2003. http://e-legis.anvisa.gov.br/
leisref/public/showAct.php?id = 2489&word = Accessed June 16, 2009.

25. Chow SC, Liu, JP. Design and Analysis of Bioavailability and Bioequivalence Studies.
New York: Marcel Dekker, 2000.

26. Health Minister. National Health Surveillance Agency CNS Resolution no. 196
of 10 October 10, 1996. Requirements for conduct of clinical research using
health products for humans. Brasília, DF; 1996. http://e-legis.anvisa.gov.br/leisref/
public/showAct.php. Accessed June 16, 2009.

27. Brazil. Health Minister. National Health Surveillance Agency. RE Resolution no.
894 of May 29, 2003. Determines the protocol for publication of the guide and
technical report of the bioequivalence study. Brasília, DF; 2003. http://e-legis.bvs/
leisref/public/showAct.php?id=1914. Accessed June 16, 2009.

28. Brazil. Health Minister. National Health Surveillance Agency. RE Resolution no.
899 of May 29, 2003. Determines the publication of guidelines for validation of
analytical and bioanalytical methods, is repealed RE Resolution no 475 of March
19, 2002. Brasília, DF; 2003. http://e-legis.bvs/leisref/public/showAct.php?id=5745.
Accessed June 16, 2009.

29. Brazil. Health Minister. National Health Surveillance Agency. RE Resolution no.
895 of May 29, 2003. Determines the publication of "Guidelines for preparation of
technical report on the study of bioavailability/bioequivalence". Brasília, DF; 2003.
Accessed June 16, 2009.

30. Brazil. Health Minister. National Health Surveillance Agency. RE Resolution no. 897
of May 29, 2003. Determines the publication of guidelines for relief and replacement of bioequivalence studies. Brasília, DF; 2003. http://e-legis.bvs/leisref/public/
showAct.php?id=1775. Accessed June 16, 2009.

31. Brazil. Health Minister. National Health Surveillance Agency. RE Resolution no. 893
of May 29, 2003. Determines the publication of the guide to make changes, addi-
tions and reporting post-registration of drugs. Brasília, DF; 2003. http://e-legis.bvs/
leisref/public/showAct.php?id=1909. Accessed June 16, 2009.

32. Brazil. Health Minister. National Health Surveillance Agency. RE Resolution no. 310
of September 1, 2004. Determines the publication of the guide for the study and
reporting of pharmaceutical equivalence and profile of dissolution. Brasília, DF; 2004.
http://e-legis.bvs/leisref/public/showAct.php?id=12431. Accessed June 16, 2009.

Canada

Iain J. McGilveray

McGilveray Pharmacon Inc., Ottawa, Ontario, Canada

INTRODUCTION

As a result of 1969 amendments to the Patent Act (1) (compulsory licensing), Canadian regulators were the first to apply pharmacokinetics (PKs) to safety and efficacy risk assessment of generic drug products. However, formal guidelines developed by an Expert Advisory Committee (EAC), currently referred to as the Scientific Advisory Committee (SAC), were not published until the 1990s. Currently, guidelines are being updated and new SAC initiatives such as Guidances for nonproportional PK drugs and for drug products requiring fed studies have been published in draft form for stakeholder comments and consultation while others, such as guidance for critical dose drugs, have been finalized.

The Canadian Food and Drugs Act and consolidated Regulations in Division 8 (2) require that for new drugs (which also means drug products), the manufacturer must file a New Drug Submission (NDS) that must provide evidence of safety, efficacy, and consistency of quality. (This requirement includes all drugs submitted after 1962, but for some older drugs, it may be applied if presented for new claims or in a new dosage form or strength.) However, for second and subsequent entry new drugs (generic drug products), an abbreviated NDS (ANDS), the regulation was revised in 1995 as **C.08.002.1.**, noting that a manufacturer may file an abbreviated new drug submission (ANDS) for the generic product (i.e., "the new drug"), where, in comparison with a Canadian reference product,

a. the new drug is the pharmaceutical equivalent of the Canadian reference product;
b. the new drug is bioequivalent with the Canadian reference product, based on the pharmaceutical and, where the Minister considers it necessary, bioavailability characteristics;
c. the route of administration of the new drug is the same as that of the Canadian reference product;
d. the conditions of use for the new drug fall within the conditions of use for the Canadian reference product.

In this regulation "pharmaceutical equivalent" means a drug product that, in comparison with another drug product, contains identical amounts of the identical medicinal ingredients, in comparable dosage forms, but does not necessarily contain the same nonmedicinal ingredients; and "Canadian Reference Product C.08.001.1." (http://www.canlii.org/ca/regu/crc870/secc.08.001.1.html) means:

a. a drug in respect of which a notice of compliance is issued pursuant to Section C.08.004 and which is marketed in Canada by the innovator of the drug,

b. a drug, acceptable to the minister, that can be used for the purpose of demon-
    strating bioequivalence on the basis of pharmaceutical and, where applica-
    ble, bioavailability characteristics, where a drug in respect of which a notice
    of compliance has been issued pursuant to Section C.08.004 cannot be used
    for that purpose because it is no longer marketed in Canada, or

c. a drug, acceptable to the minister, which can be used for the purpose of
    demonstrating bioequivalence on the basis of pharmaceutical and, where
    applicable, bioavailability characteristics, in comparison to a drug referred
    to in paragraph *a*

In general, this means an innovator product, but there is a policy (3) for
exceptions when the innovator has withdrawn from the Canadian market. This
regulation was developed after application of internal policies and decisions by
reviewers of applications over about 30 years. These ad hoc policies resulted after
amendments to the Canadian Patent Act in June 1969 (1), which made provision
for compulsory licensing and facilitated registration of generic drug products in
Canada. ["In a *compulsory license*, a government forces the holder of a patent,
copyright or other exclusive right to grant use to the state or others. Usually, the
holder does receive some royalties, either set by law or determined through some
form of arbitration" (4). In Canada, this royalty was set at 4%.]

The Patent Act amendment in Canada, facilitated growth of a substantial
generic market (the first in a developed country), although the amendment was
attenuated in 1988 when the patent life was extended to international norms. Following the legal changes, introduction of generic products led to a need for assuring the safety and efficacy of such products and a research program into comparative bioavailability was undertaken in the 1970s. In fact, Health Canada was
the first jurisdiction to apply bioequivalence (BE) to safety and efficacy review of
new drug products (5).

Many of the internal decisions for generic product approval involved an
Expert Advisory Committee on Bioavailability first formed in 1971, which was
at that time largely a review committee for early BE studies carried out in Health
Canada laboratories. However, the EAC not only provided decisions from review
of study results for generic studies, but also was the first to define bioequivalence from comparative bioavailability and to provide an early standard that
such products should have a bioavailability of at least 80% relative to a "reference formulation" according to a statistically sound design (6).

The major work of the EAC toward the end of the 1980s was the production of three reports A, B, and C that were the basis of two guidelines (7,8) and
several policies that provide the foundation of most of this chapter. There is
also draft guidance, "Preparation of Comparative Bioavailability Information for
Drug Submissions in the CTD Format" (9), which provides a list of the requirements to be met and integrates unique Canadian bioequivalence requirements
with the ICH harmonized approach (Common Technical document, CTD) for
drug product registration, to which Canada is a signatory.

However, it must be understood that Canada is governed as a Confederation of Provinces. Thus, the regulations and guidelines for BE are federal, leading to a Notice of Compliance (NOC) to sponsors for marketing in Canada and
while their application leads to a Declaration of Bioequivalence for specific products, the Provinces in the Canadian Confederation have the responsibility for

health care and the licensing of health professions. Thus, provincial formulary committees consider interchangeability of products and may not accept the federal decision of BE (which also evaluates quality) to be sufficient to list a particular product. Also, each province has different formulary rules, usually permitting product substitution of prescribed brand name by pharmacists and also governing prices. Of course, before being approved federally with a Notice of Compliance, the generic product must comply with the patent laws. This has been problematic in some cases, as slight changes in innovator products have gained extensions and often delayed introduction of generic products into the market.

Most of the Canadian requirements for bioavailability and particularly BE are, as previously mentioned, provided in two guidances, Part A for immediate-release or conventional oral products (7) and Part B for modified-release (MR) products (8). Both the guidances, Part A and B, state that "this guidance document deals with drugs that have *uncomplicated Characteristics.*"

Part A provides examples of drugs that may require methodology and standards to be modified for certain drugs (products)—for example, those with one or more of the following characteristics:

a. MR dosage forms.
b. Complicated or variable PKs, for example,
   • nonlinear kinetics;
   • substantial first-pass effect (greater than 40%);
   • variable kinetics owing to different genetic phenotypes;
   • stereochemical effects such as in vivo inversion of configuration;
   • an effective half-life of more than 24 hours.
c. (An important) time of onset of effect or rate of absorption.
d. High toxicity or a narrow therapeutic range.
e. Little or no absorption with activity exerted locally in the gastrointestinal tract.
f. A drug measurement methodology insufficiently sensitive or reliable to determine blood concentrations to at least three terminal half-lives.
g. Combination products.
h. Biologicals.

Guidances for these types of drugs were never finalized, but report C of the EAC (10) suggested modifications for some of the above and, more recently, some specific guidances or policies have been developed for these exceptions, as will be discussed later in this chapter.

DESIGN AND CONDUCT OF BIOEQUIVALENCE STUDIES FOR ORALLY ADMINISTERED DRUG PRODUCTS

Part A (7) is akin to the FDA general guidance and lays out the scope and conditions for both bioavailability and bioequivalence (comparative bioavailability) studies.

Study Design

This is described under Section 4 of the guidance, "Study Design and Environment," which is introduced with the statement "The design of a bioavailability study should minimize variability that is not attributable to the drug *per se* and should eliminate bias as much as possible. The guidances in this section serve for the usual case. *Other designs may be permissible after consultation with Health Canada (HC) before the study is initiated."*

Section 4.3 states that "The basic design to be used is a two-period cross-over, in which each subject is given the test and reference formulations at different times in a blinded manner). In cases where more than two formulations are under study, or are studied under different conditions, each volunteer should receive all treatments in a restricted randomized design. However, when the number of treatments results in a study that is longer than a month, a balanced incomplete block design may be considered." The latter type of design is rarely used.

Number and Selection of Subjects or Patients and Phenotyping/Genotyping

Section 3 of the guidance deals with the selection of subjects for a study.

There is considerable advice given in Part A, Section 3.3, on the calculation of study size.

"The number of subjects to be used in the cross-over study should be estimated by considering the standards that must be passed (discussed later) and the drug products being compared. The probability that a study of a given size will pass the standards depends on the expected mean difference between the test and reference formulations of both AUC_T (or last) and C_{max}, and the anticipated intrasubject coefficient of variation (CV) of both AUC_T and C_{max}. For drugs with uncomplicated characteristics, the intrasubject CV is generally less than 20%; however, as a result of sampling, or if the study is poorly run, the intra-subject CV can be higher."

"The minimum number of subjects is 12, but a larger number is often required." Probability diagrams are included and calculations are suggested to estimate the appropriate number of subjects.

Section 3.2 notes that drugs with uncomplicated characteristics can usually be tested in normal, healthy volunteers. Subjects of both genders are acceptable and investigators should ensure that female volunteers are not pregnant or likely to become pregnant during the study. Confirmation should be obtained by urine tests just before the first and last doses of the study.

"In some instances, studies may be required to ascertain bioavailability in patients or subjects with special characteristics—for example, for drugs to be used in the treatment of conditions accompanied by altered absorption or distribution, or for drugs to be used in special age groups such as children or the elderly." This is relatively rare for BE, except for some critical dose drugs discussed later. However, women subjects would be preferred for testing oral contraceptives and products for treatment of menopause.

It is also noted that an important objective in the selection of subjects is to reduce the intrasubject variability in PKs that may be attributable to certain characteristics of the subject. *"Subjects should be assigned in such a way that the study design is balanced for any factors that are suspected to contribute to variability."* A range of subject age from 18 to 55 years of age is specified. As this is an older guidance, the height/weight ratio suggested is based on insurance tables and "should be within 15% of the normal range." However common body mass index (BMI) limitations are accepted.

Section 3.2.c deals with health checks and since adoption of the ICH good clinical practice (GCP) guidance (11) the Health Canada guidance has been superseded, although in general the statement "The health of the volunteers must be determined by the supervising physician through a medical examination and review of results of routine tests of liver, kidney, and hematological functions" covers many of the GCP expectations. Psychological evaluation of candidate subjects is also required to exclude patients unlikely to comply with study restrictions or unlikely to complete the study. Also, when the drug has a cardiac effect, an electrocardiogram should be included in the study documentation. Of course, as in the GCP guidance (11), full documentation of aberrant laboratory values should be rechecked and a summary must be presented along with the physician's opinion. While there is no actual direct mention of informed consent from volunteers, the requirement for ethical review board approval (now according to clinical trial certification (12) with advice for BE applications (13) and citation of good clinical practice guidelines (now ICH) are very clear on documentation of consent, as well as responsibilities of investigators and suitability of clinical, laboratory and analytical facilities involved in the study. For details see also "CTD preparation" document (13).

Nonsmoking subjects are preferred, but if smokers are included they must be identified, with discussion on any likely effect on the study included in the study report. Volunteers should not take any other drug, including alcoholic beverages and over-the-counter (OTC) drugs, for an appropriate interval before, as well as during, the study (Section 4.1). In the event of emergency, the use of any drug must be reported (dose and time of administration). It is expected that alcohol and other recreational drug abuse would be detected during health checks. There is no direct mention of genotyping or phenotyping in the older guidance. However for subject safety, GCP requires that poor metabolizers be protected from adverse effects.

Standardization of Study Conditions

Section 4.1 in Part A states, "Every effort should be made to standardize the study conditions in every phase of the study—for example; exercise, diet, smoking, and alcohol use." Section 4.4 details the expectations in standardizing administration of food and fluids, viz:

> The administration of food and fluid should be controlled carefully. Normally, subjects should fast for 10 hours before drug administration. A fast means that no food or solids are to be consumed, although alcohol-free and xanthine-free clear fluids are permissible the night prior to the study. On the morning of the study, up to 250 mL of water may be permitted up to two hours before drug administration. The dose should be taken with water of a standard volume (e.g., 150 mL) and at a standard temperature. Two hours after drug administration, 250 mL of xanthine-free fluids are permitted. Four hours after drug administration, a standard meal may be taken. All meals should be standardized and repeated on each study day.

"Some drugs are given with food to reduce gastrointestinal side effects. Studies of such drugs should include studies with standard meals" refer to "Guidance for industry. Bioequivalence Requirements: Comparative Bioavailability Studies Conducted in the Fed State" (14). This suggests that for uncomplicated drugs, as defined in Part A, BE should be demonstrated in a single-dose study under fasting conditions. The exception would be if there is a documented serious safety risk to subjects from single-dose administration of the drug or drug product in the absence of food.

For complicated drugs contained in immediate-release dosage forms (critical dose drugs and nonlinear drugs) and for drugs in MR dosage forms, BE should generally be demonstrated under both fasted and fed conditions. While there an example of a test meal is provided, sponsors must be able to justify the choice of meal in a fed BE study and relate the specific components and timing of food administration.

Section 4.5 discusses posture and physical activity during the study. "For most drugs, subjects should not be allowed to recline until at least two hours after drug ingestion. Physical activity and posture should be standardized as much as possible to limit effects on gastrointestinal blood flow and motility. The same pattern of posture and activity should be maintained for each study day."

Section 4.6 considers the interval between doses. "The interval between study days should be long enough to permit elimination of essentially the entire previous dose from the body. The interval should be the same for all subjects and, to account for variability in elimination rate between subjects, normally should be not less than 10 times the mean terminal half-life of the drug. Normally, the interval between study days should not exceed three to four weeks. Furthermore, the drugs must be administered at approximately the same time on each study day and, where possible, the same day of the week." Long half-life drugs are considered in Report C (10) and in a separate guidance discussed later in the chapter.

Blood/Urine Sample Collection and Times

Section 4.7 of Part A discusses sampling times of blood and urine, "The duration of blood or urine sampling in a study should be sufficient to account for at least 80% of the known AUC to infinite time (AUCI). This period is usually at least three times the terminal half-life of the drug. To permit calculation of the relevant pharmacokinetic parameters, from 12 to 18 samples should be collected per subject per dose. An account of inter-subject variability should be used in the placement and number of samples. The exact times at which the samples are taken must be recorded and spaced such that the following information can be estimated accurately:

a) Peak concentration of the drug in the blood (C_{max})
b) The area under the concentration time curve (AUC) is at least 80 percent of the known AUCI, and
c) The terminal disposition rate constant of the drug.

There may be considerable inaccuracies in the estimates of the terminal disposition rate constant if the constant is estimated from linear regression using only a few points. To reduce these inaccuracies it is preferable that four or more points be determined during the terminal log-linear phase of the curve."

Long half-life drugs are considered in Report C (10) and in a separate guidance discussed later, but blood in these cases need only be collected to 72 hour.

If urine is used as the biological sampling fluid (see below), then sufficient samples must be obtained to permit an estimate of the rate and extent of renal excretion.

Characteristics To Be Investigated

Blood/Plasma/Serum Drug Concentration Versus Time Profiles
Section 4.8 of the guidance A discusses sampling of blood or urine. "Under normal circumstances, blood should be the biological fluid sampled to measure the concentrations of the drug. In most cases the drug may be measured in serum or plasma; however, in some cases, whole blood may be more appropriate for analysis. If the concentrations in blood are too minute to be detected and a substantial amount (>40 percent) of the drug is eliminated unchanged in the urine, then the urine may serve as the biological fluid to be sampled. When urine is collected, the volume of each sample must be measured immediately after collection and included in the report. Urine should be collected over no less than three times the terminal elimination half-life. For a 24-hour study, sampling times of 0 to 2, 2 to 4, 4 to 8, 8 to 12, and 12 to 24 hours are usually appropriate. Quantitative creatinine determinations on each urine sample are also required. Sometimes the concentration of drug in a fluid other than blood or urine may correlate better with effect. Nevertheless, the drug must first be absorbed prior to distribution to the other fluids such as the cerebrospinal fluid, bronchial secretions, and so on. Thus, for bioavailability estimations, blood is still to be sampled and assayed."

Examples of drugs for which urinary excretion profiles may be used for BE are potassium chloride and biphosphonates. Pharmacodynamic studies are rarely used for BE, but report C has a section on this (10). It is a very general advice. Of interest is the suggestion on appropriate standards:

> The requirements of a pharmacodynamic study should be comparable to those of standard bioavailability or bioequivalence studies, including measures of the magnitude, onset and duration of response. For approval under such circumstances, criteria similar to those defined for bioavailability and bioequivalence studies that use drug concentration measurements must be derived; e.g., AUC of measured pharmacodynamic response and maximum response. In addition, similar standards should be met in these criteria to establish bioavailability and bioequivalence.

In fact such studies are drug specific and usually wider standards are necessary. The human skin blanching assay (HSBA) for topical corticosteroid products and the change in pulmonary function tests with albuterol are the known applications and are noted later.

Bioanalysis
Section 6 of the guideline, Part A, entitled "Measurement Methodology" describes the attributes of such methods for parent drug, and when appropriate, metabolites in the matrix and the validation procedures required in reports to establish and maintain selectivity, range, precision, and accuracy. There is a draft guidance on metabolites (15), which essentially notes that parent drug should be measured for BE, even for prodrugs, except if the sensitivity does not permit the performance standards to be met to obtain an adequate profile. The 1992 guidance refers to the report (16) of a bioanalytical methods validation workshop (the first of its kind), which laid out internationally accepted practices for this area but has been superseded by the report of a subsequent (second) conference (17). The report describes the requirements for method development validation and then partial validations required for changes in processes, methodology, or

analytical equipment, for example, different mass spectrometers with the final validation being the application to the study samples. Documentation requires listing of standard operation procedures (SOPs) for sample storage, processing, standard and quality control sample preparation, and evaluation of stability (for analyte and internal standard) under various conditions, such as benchtop as well as long-term frozen samples. A Canadian difference compared to international practice requirements, given in the Part A guidance, was that there was to be reanalysis of 15% randomized replicate (or incurred) samples for each study, but this was revoked in 2005 (18). However, the during the May 2006 third bioanalytical methods validation conference discussed the need for validating methods by using incurred as well as the spiked samples normally evaluated for precision and accuracy. The "15% randomized replicate" request was an attempt to demonstrate reproducibility in practice. However, it was difficult to apply as other jurisdictions did not require incurred sample stability in that era. In fact determination of the need (or not) for incurred sample investigation may be expected in the original bioanalytical method validation report rather than the report for each study (Eric Ormsby, TPD, personal communication, May 2008).

Choice of Reference Product—C.08.001.1.

"Canadian Reference Product" (http://www.canlii.org/ca/regu/crc870/secc. 08.001.1.html) has previously been defined

An associated policy (3) (1995) expands on the paragraph c intent. While this recognizes that globalization of the industry leads to innovators marketing the same formulation in many countries and even having one plant manufacture a product for the entire world, the interpretation is very narrow. As well as requiring considerable documentation of the provenance of the reference product from another country, the following criteria are to be met by the medicinal ingredient and the drug product.

The medicinal ingredient exhibits an aqueous solubility of more than 1% and is "uncomplicated" as defined in the Drugs Directorate Guideline, Part A (as noted earlier).

In addition, the drug product contains a single medicinal ingredient in the same quantity as the innovator product marketed in Canada and is the same as the drug product marketed in Canada with respect to color, shape, size, weight, type of coating (e.g., uncoated, film-coated, or sugar-coated), and must demonstrate individual and mean values of the dissolution profiles comparable to the product marketed in Canada. (Usually this means meeting the FDA f_2 dissolution criteria.) The dissolution profiles should be determined in at least three media within the physiological range (pH 1–7.5), for example, water, 0.1 N HCl, and pharmacopoeial buffer media at pH 4.5, 6.5, and 7.5. One dissolution medium should be that described in the USP or BP monograph, if one exists. The percentage of drug content released should be measured at a number of suitably spaced time points, for example, at 10, 20, and 30 minutes, and continued to achieve virtually complete dissolution.

If *any* of the above conditions are not met, the manufacturer must demonstrate the equivalence of the second-entry product to the innovator's product marketed in Canada by the appropriate BE study or studies.

The result of these requirements and criteria is that non-Canadian references for BE studies for generic products are rarely accepted. For changes in innovator products requiring proof of BE between an older and revised (changed) product, much of the information is available from the company's files.

Study Products and Batch Size
The Part A document (Section 5.2) calls for "the bioequivalence batch to be representative of proposed production batches, comparable in size and produced using the same type of equipment and procedures proposed for the formulation to be marketed."

Since the time that the guidance was written, commonly, lots used in stability studies according to ICH (19) are tested and these are primary batches that should be of the same formulation and packaged in the same container closure system as proposed for marketing and are at least pilot scale batches.

Data Analysis
Section 7 of guideline, Part A details the PK and statistical analysis expected.

Statistical Analysis
The requirements for "analysis of data" are described in Section 7.0 of the guidance. The analysis of variance (ANOVA) tables submitted with the study documentation (report) should include the appropriate statistical tests of all effects in the model. The output from ANOVAs appropriate to the study design and execution must be expressed with enough significant figures to permit further calculations.

> *The analyses should include all data for **all** subjects on measured data. Supplementary analyses may also be carried out with selected points or subjects initially excluded from the analyses. Such exclusions must be justified. It is rarely acceptable to exclude more than 5 percent of the subjects or more than 10 percent of the data for a single subject-formulation combination.*

In fact, rejection of outliers is rarely accepted and remains an issue to be resolved, despite several advisory committee discussions. In the impending revision of Canadian guidelines it will be proposed (20) that values from subjects identified as extreme with an acceptable outlier test can be rejected, if after retesting the subjects with both formulations, the retested values are found to be within the normally expected range.

An ANOVA should be carried out on the time to peak concentration, t_{max} and terminal slope (λ) data, and on the logarithmically transformed AUC_T, AUC_I, and C_{max} data. The analysis and results for each parameter should be reported on a separate page as detailed in the guidance (see Section 8, "Sample Analysis for a Comparative Bioavailability Study"). The reported results must include

a. means and CVs (across subjects) for each product;
b. the ANOVA, containing source, degrees of freedom, sum of squares, mean square, F and p values and the derived intra- and intersubject CVs;
c. AUC_T and C_{max} ratios for test versus reference products;
d. the 90% confidence interval (CI) about the mean AUC_T;
e. estimates of measured content for each formulation being compared, and a separate table showing points c and d above, corrected for measured content.

Appropriate tabulation of PK parameters is required, as well as individual and mean plasma (or urinary excretion) profiles graphed both in linear and semilogarithmic scales.

The Canadian guideline requires BE results to be calculated (and standards should be met) with and without correction for drug content, although this is under consideration (observed data may be accepted without the correction if test and reference assays for labeled content are within 5% of each other) (20).

Add-On Studies

There is an allowance for add-on studies. The latter thus permits two or more studies to be pooled if certain requirements are met, viz

a. the same protocol must be used for all studies. Specifically, this means that the same analytical method is to be used, the blood samples drawn at the same time, and the same lots of the same formulations used.
b. two consistency tests must be done on the studies to ensure that pooling is meaningful. The first test is the test of equality of the residual mean squares and the second test is the formulation by study interaction.

Add-on and sequential designs are currently under consideration for a revised guidance (20). Consistency tests would not be mandatory, but an adjustment (Bonferroni correction—see footnote d) will be required to keep the error rate at 5% to protect the α. Sequential designs should be considered when variation and expected mean differences are uncertain, but use of this approach should be stated a priori in the clinical study protocol.

Acceptance Ranges for Pharmacokinetic Parameters

Single-Dose Studies

Health Canada requires two PK metrics to be measured for uncomplicated drugs [see Guidance Part A (8), Section 7.1]: the area under the drug (or metabolite) concentration in plasma (or serum, or whole blood) versus time curve from zero (time of dose) to the time of last quantifiable concentration. (AUC_T), often known as AUC_{last}, and C_{max}, the observed maximum or peak concentration of drug (or metabolite) in plasma (or serum, or whole blood). As already discussed these concentrations are usually for parent drug. The area under the drug concentration versus time curve from zero (time of dose) extrapolated to infinity (AUC_I) should also be estimated and one of the acceptance criteria is that the AUC_T is at least 80% of the known AUC_I, but otherwise AUC_I is not usually applied as a standard by the TPD.

Also, for a drug with a terminal elimination half-life greater than 24 hours, BE standards in comparative bioavailability studies will be applied to AUC0–72 hours rather than to AUC_T and an estimate of AUC_I is not required (21). Alternate designs such as parallel studies could be considered for long half-life drugs.

Part A has no description for urinary standards, but the cumulative urinary excretion (A_e) and the maximum rate of urinary excretion are reflective of the plasma AUC_T and C_{max}, respectively and the identical acceptance limits for plasma and urine metrics are invoked.

Part A, Section 7.1 states the Standards for Bioequivalence:

For drugs with uncomplicated characteristics, the following standards, obtained in single dose cross-over comparative bioavailability studies, determine BE as follows:

a. The 90% CI of the relative mean AUC_T of the test to reference product should be within 80% to 125%.
b. The relative mean measured C_{max} of the test to reference product should be between 80 and 125%.

These standards must be met on log-transformed parameters calculated from the measured data, and data corrected for measured drug content [percent potency of label claim, but, as noted above, this is under consideration (20)]. Thus, unlike FDA and European requirements, only the geometric mean ratio (point estimate) is required for C_{max}, but the AUC_T acceptance ranges are identical [currently for both measured and corrected for drug content data, but under consideration for application to measured data only (20)].

There is a guidance for critical dose drugs (22), which includes drugs that commonly exhibit adverse effects and limit the therapeutic use to doses close to those required for the therapeutic effect (e.g., "narrow therapeutic range drugs") or drugs for which the therapeutic use may result in dose- or concentration-dependent adverse effects, which are persistent, irreversible, or slowly-reversible, or life threatening (e.g., "highly toxic drugs"). The guidance provides the following acceptance ranges:

The 90% CI of the relative mean AUC_T of the test to reference formulation should be within **90.0% to 112.0%**; the relevant AUC or appropriate AUC (as described in Guidelines A and B) are to be determined.

1. The 90% CI of the relative mean measured C_{max} of the test to reference formulation should be between **80.0% and 125.0%**.
2. These requirements are to be met in both the **fasted and fed states**.
3. These standards should be met on log transformed parameters calculated from the measured data and from data corrected for measured drug content (percent potency of label claim). The latter is also currently under consideration for amendment (20).

The list of critical dose drugs is described in an appendix and includes cyclosporine, digoxin, flecainide, lithium, phenytoin, sirolimus, tacrolimus, theophylline, and warfarin.

There is also a Guidance for Bioequivalence Requirements for Drugs for Which an Early Time of Onset or Rapid Rate of Absorption is Important (rapid onset drugs) (23). These are defined as drugs for which the time of onset of effect is important because of therapeutic or toxic effects; for example, an analgesic for rapid relief of pain. In addition to the 90% CI criterion for AUC_T, from Part A above, the following standards are recommended for this group.

1. The 90% CI of the relative mean measured C_{max} of the test to reference formulation should be between 80% and 125%.
2. The relative mean $AUC_{Reftmax}$ of the test to reference formulation should be within 80% to 125%, where $AUC_{Reftmax}$ for a test product is defined as the area under the curve to the time of the maximum concentration of the

reference product, calculated for each study subject. To date this standard has only been applied to two drugs, oral gel formulations of ibuprofen, and tablet formulations of sumatriptan.

Another update of report C was the Notice to Industry: Bioequivalence Requirements for Combination Drug Products (24). The purpose is to state BE requirements specific to combination drug products. Further work is being done on requirements for fixed dose combinations, particularly in New Drug Submissions.

For studies of combinations, the PK parameters to be reported and assessed according to acceptance criteria, are those which would normally be required of each drug if it were in a formulation as a single entity, as described in current TPD guidelines (Parts A and B) and policy statements (which are given earlier).

There is a draft policy on Bioequivalence Requirements: Drugs Exhibiting Non-Linear Pharmacokinetics (25), which states that a drug will be considered to exhibit nonlinear PK if this is indicated in the peer-reviewed scientific literature or the approved labeling for the drug. However, the drug may be treated in the same way as those exhibiting linear PKs, if evidence is provided to show that dose-normalized AUC deviate (increase or decrease) by less than 25% over the labeled dose range for the proposed indication.

Drugs that exhibit nonlinear PK characteristics with single or multiple doses of approved strengths should meet standards for BE as outlined in the TPD Guideline on the *Conduct and Analysis of Bioavailability and Bioequivalence Studies: Part A* or *Part B*, as applicable.

These requirements should be met in single-dose studies in both the *fasted and fed* states for all nonlinear drugs, with the following exceptions:

a. If nonlinearity occurs after the drug enters the systemic circulation unless there is evidence that a product exhibits a food effect;
b. If a condition (fasted or fed) for product ingestion is contraindicated, that condition may be waived in a BE trial.

This draft policy is also under consideration (20) and it is proposed that if the product monograph for the reference clearly states the type of nonlinearity, then the drug is considered to be nonlinear (non–dose-proportional) and the appropriate dose should be used in the study or studies. A fed study would **not** be required.

Steady-State Studies
The major group of drugs with different criteria are MR products as described in the guideline, Part B (9). This guidance defines the general acceptance criteria for the required multiple-dose, steady-state studies, which are also applicable to multiple-dose studies that may be required for some nonlinear drugs and with the narrower limits previously described, for critical dose drugs. However, the application of steady-state studies to extended-release products is also being reconsidered (20) in harmony with the FDA general guidance. Steady-state studies do not seem to provide any better comparison of the two formulations. The general acceptance criteria for BE from steady-state studies are described in Section 4.3 of the guideline, Part B, and are based on the geometric mean ratio of test to reference and the standards described below must be met for parameters

calculated from the observed concentrations, as well as those corrected for measured drug content.

AUC: The 90% CI of the relative mean AUC_τ (generally known as AUC_{ss}: AUC at steady state for the dosage interval) of the test to reference formulation should be within 80% to 125%.

C_{max}: The relative mean measured C_{max} at steady state of the test to reference formulation should be within 80% to 125%.

C_{min}: The relative mean measured C_{min} at steady state of the test to reference formulation should not be less than 80%.

Reporting of Results

As Health Canada applies the ICH guidances and the CTD to submission review, submissions are required to follow the subsections of this document. There is an additional advice in the draft guidance for industry "Preparation of Comparative Bioavailability Information for Drug Submissions" in the CTD Format, Health Canada, May 2004 (9). Thus the clinical report, analytical report, and PK and statistical reports would be included in CTD Module 5 Section 5.3.1, Biopharmaceutic studies, with reports according to the ICH good clinical practice guideline, including "adverse events." In addition, the Part A and other TPD guidance standards and tabulations would be expected. Quality information must be included in the quality overall summary and details in CTD Module 2, Quality.

BIOAVAILABILITY AND BIOEQUIVALENCE REQUIREMENTS

Possible changes to Canadian guidances are under consideration (20) but the formal consultation process has just begun and final changes may take some time to be promulgated.

Orally Administered Drug Products Intended for Systemic Action

The Part A (7) and Part B (8) (MR) guidances include bioavailability and bioequivalence requirements, but Part A in particular has little information on bioavailability studies which are expected for the characterization of new drugs. Absolute bioavailability is described as "Comparison of the AUC values following oral *versus* intravenous administration of an equivalent dose of the same active ingredient provides an estimate of *absolute bioavailability* for most drugs." As noted earlier, Report C (10) and the draft policy on Bioequivalence Requirements: Drugs Exhibiting Non-Linear Pharmacokinetics (25) comment on testing for dose proportionality, with the latter recommending a limit of ±25% of dose-normalized AUC deviation as the definition of linearity. However, there is no information on the statistics or model to be applied in this calculation. As noted previously, this guidance is under review (20). Other jurisdictions, such as FDA and EMEA also mention nonlinear kinetics, but with few details on the determination thereof.

Solutions

There is no specific guidance for PK studies of solutions, including bioavailability of solutions of new drugs. However, during development, basic PK studies in support of general safety and efficacy would be expected. To integrate with ICH guidelines, there is a relatively new guidance on Pharmaceutical Quality of Aqueous Solutions (26), which replaced a previous policy on *Waiver of Comparative Bioavailability Studies for Oral Solutions*. While the guidance mainly states the expectations concerning testing parameters that should be considered during the pharmaceutical development of these new drug products (NDS), it is also intended to provide guidance on the quality of aqueous solutions for generic products and certain changes to drug products contained in NDS and ANDSs. To support the request for a waiver of the requirement to demonstrate in vivo BE for solutions other than parenteral generic aqueous solutions (e.g., oral, dermatological, ophthalmic, otic), the nonmedicinal ingredients in the formulation of the test product, when compared to the reference product, should be qualitatively *the same* and quantitatively *essentially the same*. For the purposes of this document, *essentially the same* would be interpreted as the amount (or concentration) of each excipient in the test product to be within ±10% of the amount (or concentration) of each excipient in the reference product. Differences beyond these criteria should be scientifically justified and the potential impact on the safety and efficacy of the drug product should be discussed. However, those excipients that could potentially modify the absorption of the drug substance should be qualitatively *the same* and quantitatively *essentially the same* (as defined earlier). These would include those excipients that could enhance absorption (e.g., polysorbate 80, polyethylene glycol, ethanol) and those that could inhibit absorption (e.g., sorbitol, manitol).

In addition, to support the request for a BE waiver for aqueous solutions, the results of a study comparing the physicochemical properties of the test product against the reference product should be provided, showing their essential similarity.

Suspensions
Bioequivalence studies are expected for oral suspensions for systemic effects according to the 1995 Drugs Directorate Guidelines, Preparation of drug identification number submissions (27). Demonstration of bioequivalence is generally required and oral suspensions, powders/granules for oral suspension, intended for systemic effect containing prescription drugs are affected. The standards for AUC_T and C_{max} given in the guideline, Part A (7), are identical to those listed below for immediate-release conventional formulations of uncomplicated drugs. For complicated drugs, such as critical dose drugs, the acceptance ranges are as described in "Data Analysis" section of this chapter.

Immediate-Release Products—Tablets and Capsules
For immediate-release products (conventional formulations) for systemic effects, BE standards are given in "Acceptance Ranges for Pharmacokinetic Parameters" section of this chapter. In the guideline Part A, the general standard for uncomplicated drugs is given, with an example calculation.

AUC: The 90% CI of the geometric mean ratio of AUC_T of the test to reference product should be within 80% to 125%.

C_{max}: The geometric mean ratio of C_{max} of the test to reference product should be between 80% and 125%.

As previously mentioned, this standard differs from most other countries, but acknowledges that the C_{max} metric is more variable than AUC and for uncomplicated drugs, this part measure of rate is not particularly robust. As noted earlier in "Acceptance Ranges for Pharmacokinetic Parameters" section, tighter standards are invoked for some complicated drugs, such as critical dose drugs.

Other Oral Dosage Forms

MR Dosage Forms (Previously Defined as MR—Page 13, 2.9.2.1)
Guideline, Part B (8) for MR dosage forms notes that a separate guidance and standards are required for each of the circumstances in which an MR formulation might be developed. These circumstances fall into three groups:

- **Group I**—This group includes original MR dosage forms when the drug is not marketed in either a conventional-release or MR dosage form.
- **Group II**—This group includes first market-entry MR dosage forms that are developed after a conventional-release or a different kind of MR formulation is marketed.
- **Group III**—This group includes second and subsequent market-entry (SME) MR dosage forms that are developed to be bioequivalent to marketed MR dosage forms.

This guidance mainly addresses Groups II and III. If the MR formulation is the original market entry of the chemical substance (i.e., group I), selected PK parameters must be determined as part of demonstrating the product's efficacy and safety.

For the assessment of first market-entry MR formulations (group I), studies must be performed using single-dose administration, both in subjects who have fasted and in subjects who have eaten a meal standardized to challenge the formulation. The following PK parameters should be calculated from the concentrations in plasma (or blood or serum): AUC_X (dose interval), AUC_T, AUC_I, C_{max}, T_{max}, and λ, the terminal slope.

For formulations that are likely to accumulate (i.e., $AUC_X/AUC_I < 0.8$), demonstration of safety requires that steady-state studies be performed in addition to single-dose studies. The following PK parameters should be calculated from the concentrations in plasma (or blood or serum):

- AUC_τ (steady-state dosage interval)
- C_{max}
- T_{max}
- C_{pd} (predose observed concentration)—The concentration observed immediately before administering a dose at or near steady state. This is not always the same as C_{min}
- C_{min}
- Fluctuation (The range of steady-state concentrations divided by the average concentration in percentage) $(C_{max} - C_{min})/(AUC_\tau/\tau) \times 100$

Where the AUC_X/AUC_I ratio cannot be reliably determined, accumulation must be assumed to occur.

For group I products no standards are provided since the objectives are to define the PK characteristics of the drug as background to the demonstration of safety and efficacy for the NDS.

For group II, first market-entry MR dosage forms, the same single- and multiple-dose PK studies as already described for group I products are required. The studies should be generally pursued in the context of demonstrating the efficacy and safety of the recently developed drug product, and comparable bioavailability should be demonstrated against an appropriate reference formulation. The single-dose study should be a comparison between a single dose of the first market-entry MR formulation and the doses of the innovator's conventional formulation that the MR formulation is intended to replace. (The doses of the conventional formulation are administered according to the conventional dosing regimen.) When identical doses of conventional and MR formulations cannot be administered, a proportionality correction must be made for the calculation of relevant parameters. Unless there are subject safety concerns, both fasted and fed studies should be completed. In addition for group II products, steady-state studies are required for formulations used at a dose interval likely to lead to accumulation ($AUC_X / AUC_I < 0.8$ for the MR product). The comparison should be made between the first market-entry MR formulation and equivalent doses of the conventional formulation over the dosing interval (claimed for the MR product) at steady state. Generally, steady-state studies should be performed under fasting conditions unless subject safety might be compromised, in which situation a fed study may be applied. In this case, manufacturers should consult with Health Canada before undertaking a study.

The *bioavailability* standards below are suggested for group II products. It should be noted that usually these types of products are supported by extensive clinical trials and such information can justify results of comparative bioavailability versus immediate-release reference that do not meet these standards (usually lower). The fed study is both an attempt to check for dose-dumping and to provide labeling information.

Acceptance criteria suggested for group II products:

- AUC—The geometric mean ratio AUC of the MR formulation to the conventional formulation should be between 80% and 125% in the fasting state.

  The AUC may be evaluated by determining AUC_T, provided that AUC_T obtained by the linear trapezoidal rule is at least 80% of the extrapolated AUC_I (i.e., $AUC_T / AUC_I \geq 0.80$).
- C_{max}—The geometric mean ratio of C_{max} of a single dose of the MR formulation to the conventional formulation should not exceed 125% in the fasting state.

  In some cases, the intended use of the MR formulation may call for a modification of the stated C_{max} criterion. In such cases, before undertaking the study, manufacturers should consult with Health Canada.

  Suggested criteria for steady-state studies:

- AUC_τ—The relative mean AUC_τ (geometric mean ratio) at steady state of the MR formulation to the conventional formulation should be between 80% and 125%.
- C_{max}—The relative mean measured C_{max} (geometric mean ratio) at steady state of the MR formulation to the largest peak concentration of the conventional formulation should not exceed 125%.

If the MR formulation is a second or SME product (group III), developed while marketed MR product(s) are already available, then BE studies must be performed in healthy subjects, using an appropriate Canadian reference product. Single-dose comparative studies, both in the fasted and in the fed state should be completed and if there is accumulation ($AUC_X / AUC_I < 0.8$ for the MR product) then steady-state comparative studies are required, usually only in the fasting state.

The BE criteria for group III MR products are given below for single-dose studies. The standards described later must be met for parameters calculated from the observed concentrations, as well as those corrected for measured drug content. [The latter is currently under consideration (20).]

- AUC—The 90% CI of the geometric mean ratio AUC^a of the test to reference formulation should be within 80% to 125% in the fasting state and after the administration of an appropriate meal at a specified time before taking the drug.
- C_{max}—The geometric mean ratio C_{max} of the test to reference formulation should be between 80% and 125% in the fasting state and after the administration of a standardized meal to challenge the formulation.

For steady-state studies, the BE standards for group III MR products are as follows:

AUC_τ—The 90% CI of the geometric mean ratio AUC_τ at steady state of the test to reference formulation should be within 80% to 125%.

C_{max}—The geometric mean ratio C_{max} at steady state of the test to reference formulation should be within 80% to 125%.

C_{min}—The geometric mean ratio C_{min} at steady state of the test to reference formulation should not be less than 80%.

Modified-Release, Delayed-Release, or Enteric-Coated

For the special MR case of delayed release, the guideline, Part B, states that comparative bioavailability can usually be demonstrated using the AUC and C_{max} requirements for uncomplicated drug formulations, provided that the only difference between the enteric-coated formulation and the corresponding immediate release product is a time shift in the concentration–time curve (i.e., no other modification of release occurs). Studies must be carried out using both subjects that have fasted and those that have eaten an appropriate meal at a specified time before taking the drug. The reference product for group III is to be the innovator's enteric-coated product (or the market leader's enteric-coated product if there is no recognized innovator). The general AUC and C_{max} standards above for immediate-release products are applied to both fasted and fed studies, with both the observed and drug-content corrected results [the latter under consideration (20)]. For complicated drugs, the appropriate acceptance ranges, as described in "Acceptance Ranges for Pharmacokinetic Parameters" section, will be applied.

---

[a] AUC may be evaluated by determining AUC_T, provided that AUC_T obtained by the linear trapezoidal rule is at least 80% of the extrapolated AUC_I (i.e., $AUC_T / AUC_I \geq 0.80$).

Orally Administered Drug Products Intended for Local Action

The TPD does not have a class guidance for this type of product. Each drug would be dealt with on a case by case basis. In most cases, unless there was a pharmacodynamic surrogate, some type of comparative clinical trial would be required.

Parenteral Solutions and Suspensions/Emulsions

There is a 1990 policy "submissions for generic parenteral drugs" (28), which lists requirements for the four defined categories of parenteral products;

Category I products:

a. Water-soluble powders for reconstitution with no nonmedicinal ingredients.
b. Aqueous solutions, no nonmedicinal ingredients other than the vehicle; non-aqueous single solvent solutions, other than oil preparations, no nonmedicinal ingredients other than the vehicle.

Category II: Products that include lyophilized powders, buffered powders, aqueous solutions, with nonmedicinal ingredients, nonaqueous solutions, other than oil preparations, with nonmedicinal ingredients.

Category III products: Products include oil soluble preparations involving single oil.

Category IV: Products include special products, such as suspensions, emulsions, preparations involving cosolvent systems, MR preparations, special classes of drugs and biological products.

Essentially, except for category I products, for which a biowaiver may be given as described by the ICH guideline (based on pharmaceutical equivalence; see "Solutions" section earlier in the chapter), comparative bioavailability studies would be expected and the standards would be those of Part A guidance.

Topically Administered Drug Products

Topical Products for Local Action

There is a policy on "submissions for generic topical drugs" (29) dealing particularly with dermal–topical drug products, that is, those applied on the skin for the purposes of achieving localized effects, for example, antiacne preparations.

Two categories for such products are defined here:

*   Category I products of simple formulation, such as solutions containing the drug substance in which the solvent does not include nonmedicinal ingredients that may affect the penetration/absorption of the drug through the skin and Category II products of complex formulation, such as emulsions, suspensions, ointments, pastes, foams, gels, sprays, and medical adhesive systems. Category I products may not require clinical trials. In place of these trials, extensive testing of physicochemical parameters in comparison to the innovator's product may be accepted as providing sufficient evidence of the safety and efficacy of the generic product.
*   Category II are products of complex formulation, such as emulsions, suspensions, ointments, pastes, foams, gels, sprays, and medical adhesive systems. Because of the complexity of Category II product formulation, extensive testing of physicochemical parameters will not suffice to demonstrate safety and efficacy. Direct evidence of safety and efficacy through clinical trials or via surrogate models must be provided for these products.

In view of the considerations that may apply to these products, a written opinion on special requirements for individual products will be provided on request, upon submission of chemistry and manufacturing data, and proposed labeling. In general, well-controlled clinical trials are required to establish safety and efficacy of the generic preparation.

The third category mentioned in this guidance are transdermal drug products discussed later in the chapter.

Topical Products Containing Topical Corticosteroids.
The policy quoted above (29) states that "the Vasoconstrictor assay which was developed by McKenzie and Stoughton, has proved to be the most useful method to compare topical corticosteroid products" and it seems to be accepted in lieu of clinical trials for this type of product. (This is now more appropriately called the HSBA.) There is a *caveat* remarked in the policy, "However, problems may be faced in applying this method for BE testing since there is great variation in subject response and it is critical that it be carried out by experienced investigators."

Topical Products Other Than Those Containing Topical Corticosteroids and Which are Not Simple Solutions
As noted in "Topical Products for Local Action" section, usually clinical trials for equivalence would be required, but a written opinion can be obtained for an individual product upon request.

Topical Products for Systemic Action
The policy for submissions for generic topical drugs (25) defines three types of product, but only deals with the first of these. For BE of transdermal systems (patches) for which the active is absorbed through the skin and is transported by blood to sites of action, comparison of plasma levels is possible. BE of these products is assessed by single-and multiple-dose studies against the Canadian reference transdermal patch and the standards for group III MR (ER) products described in "Topical Products for Local Action" section are applied.

There is a special policy on submissions for topical nonsteroidal anti-inflammatory drugs (topical NSAIDs) (30), which have not been approved for marketing in Canada, due to concern about drug sensitization. The policy outlined in this document was developed because TPD identified a risk of potential sensitization to these products. The policy concentrates on handling this concern and gives some guidance in the conduct of clinical studies (mainly by skin-patch sensitivity tests) to help ultimately in assessing the risks versus benefits of each product on its own merits. However the one topical NSAID (diclofenac), which was accepted for marketing in Canada was on the basis of an approved review of a full NDS.

In the consultation with stakeholders, the following response is of interest

> While the policy is intended for topical NSAIDs, the principles outlined therein with respect to the safety issue of sensitization could be applied to any topically administered product that has shown potential for sensitization.

Products Intended for Other Routes of Administration

These include oral products for local use, nasal, inhalation, dermal, rectal, vaginal, etc. intended to act without systemic absorption where bioavailability measurement of the active in a biological fluid is not applicable. Submission guidelines indicate that clinical trials would be required for equivalence to be determined between a Canadian reference and a generic product in this category with the exception of short-acting β_2-agonists (albuterol) and corticosteroids, for which there are specific guidances that are as follows:

Guidance to establish "Equivalence or Relative Potency of Safety and Efficacy of a Second Entry Short-Acting Beta$_2$-Agonist Metered Dose Inhaler (MDI)" (31) and Draft Guidance Document: Submission Requirements for Subsequent Market Entry Inhaled Corticosteroid Products for Use in the Treatment of Asthma (32)

The β_2-agonist metered dose inhaler (MDI) guidance (31) requires a safety study and two pharmacodynamic studies in patients. Details of study design for both the bronchodilator and bronchoprotection study are given. The bronchodilator study endpoint of test versus reference is measured by the magnitude and duration of increase in FEV_1 (forced expiratory volume in one second) and the acceptance standard set is that the 90% CIs for relative potency for maximum FEV_1 and the area under the FEV_1 curve (AUFC) must be contained entirely within 80% to 125%. Cardiovascular and other adverse effects with the test product must not be significantly greater than that seen with the reference product.

For the bronchoprotection study the relative potency is measured by the magnitude of protection against methacholine airway constriction [PC_{20} methacholine (provocation concentration of a bronchoconstrictor agonist causing a 20% fall in FEV_1)] for the test to reference product. The acceptance standard is that the 90% CI for the relative potency for test to reference PC_{20} methacholine must be contained entirely within 80% to 125%. Also, cardiovascular and other adverse effects with the test product must not be significantly greater than that seen with the reference product.

For inhaled corticosteroids, there is a draft guidance (32) and again pharmacodynamic and PK tests are recommended. For the therapeutic equivalence study an adequately designed, well controlled, double-blinded, randomized study is recommended in which asthmatic patients are randomized into three parallel arms: Canadian reference product (R), SME product (T), and placebo (formulation placebo). This will use sputum eosinophils as inflammatory markerr to see inflammatory and clinically significant improvement and/or to allow time for the endpoints to plateau. Prebronchodilator FEV_1 also should be estimated. Very recently (June 2009, Ormsby CSPS presentation (20), it was noted that the eosinophil measure would be the primary endpoint and FEV_1, as well as blood nitric oxide would be supportive. The 80% to 125% standard would only apply to the primary metric, but the FEV_1, as well as blood nitric oxide results should go in the same direction (see footnote c). In both of the above guidances (31,32), if the drug is detectable in blood plasma after dosing, systemic exposure must be shown to be comparable between the test (T) and the reference (R) products. This will be with the usual type of PK exposure data measured in BE studies, following a single-dose study at the upper limit of the dosing range. The BE standards for AUC_T and C_{max} are identical to those in the guideline, Part A (see "Immediate-Release Products—Tablets and Capsules" section).

There is also a Draft Guidance Document: Submission Requirements for Subsequent Market Entry Steroid Nasal Products for Use in the Treatment of Allergic Rhinitis (33)

It is noted that drug sponsors may be exempt from the requirement to conduct comparative bioavailability, pharmacodynamic, and/or clinical studies for SME drug products formulated as simple solutions, since in vitro studies may provide sufficient information to support a proposal of equivalence to the Canadian reference product. However, if the biowaiver justification for simple solutions is not accepted as well as for more complex formulations, such as suspensions and emulsions, determination of therapeutic equivalence with the Canadian reference product will be required. This requires direct evidence of comparative safety and efficacy, through well designed comparative clinical trials with appropriate outcome measures, which must be provided to demonstrate equivalence with the Canadian reference product. For most products, this would require using the same clinical endpoint(s) with which the Canadian reference product was granted a NOC. However, in certain instances alternate clinical endpoints may be acceptable where advances in clinical medicine support that these are commonly used and accepted for the therapeutic indication(s) of interest at the time of application.

The clinical studies required to support equivalence of the SME product with the Canadian reference product should also include a PK study for exposure. The latter may be waived should blood or plasma levels be too low to allow reliable analytical measurements. In such cases of lack of concentration measurement, the equivalence for systemic exposure may be carried out by measuring the steroid side effects on the hypothalamic–pituitary–adrenal axis (HPA).

The following clinical study is required when using patients with seasonal allergic rhinitis:

An adequately designed, well-controlled, double-blinded, randomized study in which patients with seasonal allergic rhinitis (SAR) are randomized into three parallel arms: Canadian reference product (R), SME product (T), and placebo (formulation placebo).

An appropriate primary efficacy endpoint is the change from baseline in the Total Nasal Symptom Score (TNSS) for the entire double-blind treatment period (2–3 weeks). For therapeutic equivalence the criteria are the following

An absolute difference in TNSS mean change from baseline of at least one unit on a 12-point scale should be demonstrated between test and placebo, using all available data points. A change of less than one unit would be considered insignificant on the 12-point TNSS scale. A percentage difference would be too variable, dependent on both the baseline value and allergy season. The weighted average of all scores' (e.g., days 1–21) change from baseline for the SME product (T) and CR product (R) must be significantly different from the placebo. Test and reference should not be significantly different from one another. Test values can be better but not worse than those of the reference product. To demonstrate the BE of the test product compared with the reference product, the 90% CI of the T/R ratio mean change of the TNSS from baseline (based on log transformed data) must be within 80% to 125%.

If a PK exposure study (single-dose cross-over) is possible, the BE standards for AUC_T and C_{max} are identical to those in guideline Part A (see "Immediate-Release Products—Tablets and Capsules" section).

Variations or Postregistration Amendment Requirements

There is a Health Canada policy on changes to marketed new drug products (34), which defines four levels of change and the required documentation.

Level 1, documented by a supplemental new drug submission (S/NDS), includes changes in the dosage form or strength of the drug product, (and/or) formulation, (and or) method of manufacture, (and/or) equipment, (and/or) process control of the drug product that would usually require supporting clinical or BE data.

Level 2, documented by notifiable change (Notice of Intention to Change) may be a change of site, (and/or) method of manufacture or formulation and would probably require BE to be demonstrated, except for cases of a waiver noted elsewhere in the document.

Level 3, documented by notice of change, includes changes in packaging materials, specifications and analytical methods and would not require BE studies. Level 4 are any other changes not listed in Levels 1 to 3 and which may be made without notification, but manufacturers are expected to maintain a list of Level 4 changes.

There is a Draft Guidance Document, Post-Notice of Compliance Changes: Quality Document, issued in February, 2007 (35). It provides an Appendix 1 that has examples intended to assist with the classification of changes made to the quality information. The information summarized in the tables provides recommendations for

a. the conditions to be fulfilled for a given change to be classified as a either a Level, 1, 2, 3, or 4 change (above).
b. the supporting data for a given change, either to be submitted to Health Canada and/or maintained by the sponsor. Where applicable, the corresponding modules of the ICH CTD for the supporting data have been identified in brackets.
c. the reporting category (e.g., supplement, notifiable change, or annual notification).

Examples of changes in drug products that may require supporting clinical or comparative bioavailability data or a request for a waiver of in vivo studies include the following:

- Addition of a dosage form or strength.
- Change in the composition of a solution dosage form.
- Change in the composition of an immediate release solid oral dosage form containing a single drug substance.
- Change in the composition (qualitative or quantitative) in the release controlling agent of a MR solid oral dosage form.

WAIVERS OF IN VIVO BIOEQUIVALENCE STUDIES

Immediate Release Drug Products

Different Strength Dosage Forms
There is a policy (**C.08.001.1.**) on BE of proportional solid oral dosage formulations (33) which states; "With some exceptions, it is generally accepted that

when a product is marketed in more than one strength, if the formulation of each strength contains the **same** ingredients in the **same** proportion (i.e. the formulations are proportional), the results of a single comparative bioavailability study can be extrapolated to all strengths in the series. Extrapolation becomes more difficult, however, when the proportion of ingredients changes among the strengths or when there are pre- or post-marketing formulation changes. In general if different strengths are proportional in formulation, or have only 'minor' differences in the proportion of ingredients, a comparative bioavailability study is required for only one strength (preferably the highest)." Differences in proportion are considered to be "minor" when no strength within a range differs from a studied strength, by more than the percentages listed in the policy for various classes of excipients (e.g., <5% for filler, <0.25 for magnesium stearate as lubricant). The total additive effect of all excipient changes should not be more than 5%. Changes in coatings that are not designed to play a role in the drug release mechanism are also generally concluded to be "minor." In all instances, if comparative bioavailability data are not provided for each formulation, the sponsor must provide scientific justification for a waiver of this requirement. This justification will include dissolution using a validated QC method. In this regard, a validated method is one that has been demonstrated to be sensitive to changes in formulation and manufacturing, including the physicochemical attributes of formulation ingredients, as documented in method development studies. In the absence of a validated method, the comparative dissolution profiles should be determined in three media, as described in "Choice of Reference Product—C.08.001.1." section. Media should be selected to emphasize possible differences between the products, for example, a medium in which the dissolution rate is relatively slow (e.g., pH of the medium close to the pKa value of the drug) may offer some advantages. The percentage of drug content released should be measured at a number of suitably spaced time points, for example, at 10, 20, and 30 minutes, and continued to achieve virtually complete dissolution. At least six dosage units of each should be tested.

The policy applies only to drugs with uncomplicated characteristics as defined in the guideline, Part A. Comparative bioavailability studies would likely be required for all strengths of critical dose drugs for generic (or formulation change) approval [see, Guideline Part A, Section 5.1 and may be considered on a case-by-case basis, Ref. (36).]

Biopharmaceutics Classification System (BCS)
Health Canada has no formal guideline accepting the BCS, however sponsors can argue for waivers for different strengths and minor changes, on a scientific basis, including BCS principles. Use of BCS is implied in the Draft Guidance Document Post-Notice of Compliance Changes: Quality Document (35).

Controlled/Modified-Release Dosage Forms
The policy on BE of proportional solid oral dosage formulations (36) also applies to (possible waiver) MR dosage forms, as defined in the guideline, Part B, except for ingredients that affect the release of drug from the formulation. Where differences exist in the proportion of ingredients that may affect the release, comparative bioavailability studies are required for each different formulation. In particular, this possibility of a BE waiver would apply to capsules containing beads,

with the BE study of higher strength allowing a waiver of an in vivo study for lower strengths, if the appropriate quality attributes, including dissolution profiles are documented and appropriate.

CONCLUSIONS

While the general Canadian guidelines for BE have many similarities to the FDA and European guidances, there remain differences that can cause difficulties in submitting second entry products (generics) to TPD. The Canadian reference or comparator situation in general still requires studies to be repeated if a foreign reference has been used in an ANDS (3). This is in distinction to an NDS when the company can provide full details of global strategy. The nonlinear draft guidance (25) is different from other jurisdictions and often results in an additional fed study being required for Canadian approval. The correction for potency is still required (but note below). As yet, for biowaivers, the Canadian regulators have not embraced the BCS, or at least made clear the tolerances for acceptance using the approach. Clearly for oral dosage forms for systemic use, the guidelines are clear and comprehensive. Nonoral, locally acting drugs (such as dermal or inhaled routes), with few exceptions require clinical trials. The exceptions are when some relevant pharmacodynamic measures can be applied, such as with corticosteroids [dermal (30) and inhaled (32)].

As has been noted, within the Health Canada, Therapeutic Products Directorate, changes to the general guideline A[b] are under consideration (20), although, the draft for consideration has not been released at the time of writing this chapter (June 2009). The objective is a revision of the guideline Part A (7) to remove ambiguities. It will also include Guideline Part B (8) for MR products and attempt to integrate report C classes (10) as well as modifications of other special guidances[c] or policies on BE.

With the *caveat* that these are not yet final, the main points for change are provided below:

- As noted in several sections of this chapter , the potency correction requirement will be removed, provided the certificates of analysis of test and reference products demonstrate that the drug content assay results are within 5% of each other.

---

[b] Guidance 1 will combine current "Conduct and analysis" guidelines Parts A (1992) and B (1996)—clarifies some positions and adds some new approaches. Guidance 2—defines classes of drugs and their required studies and standards (Report C drugs, 1994).

[c] Guidance 3—data requirements and standards for inhaled corticosteroid products used in the treatment of asthma. Use of steroid-naive patients with mild to moderate asthma, and 3% eosinophil count and FEV1>60%. Three-treatment (reference, test, and placebo) parallel study design for three weeks by using the lowest dose of the Canadian Reference Product and eosinophil recommended as primary outcome measure (also FeNO, FEV1). Secondary measurements (FEV1, asthma symptoms) where both test and reference must be 50% over placebo (% change from baseline). Test versus Reference 90% CI between 80% and 125% and systemic absorption test to reference must meet usual BE standards. Guidance 4—data requirements and standards for nasal steroid products.

- Add-on studies will continue to be allowed (with a 12 subject minimum) but the requirement for consistency tests will be removed and there will be an adjustment to keep the error rate at 5%.[d]
- Sequential designs can be accepted when the variation and expected mean differences are uncertain.[e,f] Clearly, the design would be a priori and appropriate statistical aspects would need to be included.
- Despite the existence of a draft guidance on the use of metabolite data (15), it is still unclear when metabolite data can be used to determine BE. However, whenever possible the drug administered should be measured and used for BE decisions. If this is not possible, then measurements of a metabolite (preferably primary) could be applied to BE decisions. The use of metabolite must be justified and stated in the study protocol. Metabolite results can be applied as a covariate.
- The draft policy on nonlinear PK drugs (25) was never finalized, it states that nonlinearity could be determined from the literature and that for nonlinear products, fed, as well as fasted BE studies must be run. The proposal is that if the reference (innovator) product monograph states clearly the type of non-linearity, the appropriate dose should be used for BE comparison and a fed study will not be required.
- Although there is guidance on food studies (14) it is not entirely clear. Fed (and fasted) studies will continue to be required for MR products. However, the reference (innovator) product monograph information on food effect (or lack thereof) will be used to decide when a fed study will be required (instead of or in addition to the fasted study).
- The requirement for steady-state studies for extended-release product BE comparisons will be removed.
- Currently the statistical analysis (7,8) recommends least squares analysis using ANOVA with fixed effects (e.g., PROC GLM); the proposal is to apply mixed-effects methodology (e.g., PROC MIXED). The mixed-effects model analyses all types of cross-over designs and handles missing values, such as when a subject misses one period.[g]
- There is consideration for dealing with extreme values indicated as "subjects which give statistically significantly larger differences that other subjects but

---

[d] For "add-ons"—consideration being given to use a Bonferroni adjustment to keep error rate at 5% and remove consistency tests.

[e] Sequential design being considered when variation and expected mean differences are uncertain, viz, expected CV and mean difference to be stated, total N determined from table, declaration of $n < N$ where interim analysis will be performed; 93% CI to be applied as penalty for stopping early and BE accepted. N.B. All Must Be Defined in Study Protocol.

[f] Adaptive designs, similar to sequential design being considered to allow re-estimation of sample size based on observed CV of the new study, viz, estimation of expected CV and mean difference and determination of sample size N; $n < N$ to be defined where interim analysis will be performed; reestimation of sample size required based on observed CV. N.B. Must Be All Defined in Study Protocol.

[g] More flexibility to be provided in analysis using mixed effects methodology (e.g., Proc mixed in SAS) since mixed effects models analyze all types of cross-over designs (replicate) and handles missing values, that is, when subject misses one period; per-protocol analysis not intent-to-treat analysis being considered.

still give well-characterised profiles i.e., not caused by poor placement of sampling times." Currently (7) (but very rarely) 5% of subjects can be removed with justification. The proposals include application of nonparametric 90% CIs or allowing removal of identified extreme values from an acceptable outlier test after retesting the subject(s) identified with both test and reference formulations. The reanalysis with new values would indicate if the outliers can be removed. If the subject continues to show up as an outlier, there is a problem.[h]

- Incurred sample analysis will not be required for BE studies, but in bioanalytical method development the "reproducibility of samples must be shown to be adequate for the intended purpose."

  Some other areas to be considered are the following:
- Highly variable drugs and drug products (HVDP)[i]
- Definition of an "acceptable plasma profile"
- Dealing with endogenous substances, for example, adjusting for baseline
- How best to analyze urinary excretion data when concentrations in blood or its components are too low to be quantified for BE assessment.

REFERENCES

1. Bill C-102 An Act to Amend the Patent Act, the Trademarks Act and the Food and Drugs Act. http://www. curc.clc-ctc.ca/lang. Accessed September 2008.
2. Department of Justice Canada, http://laws.justice.gc.ca/en/F-27/C.R.C.-c.870/234078.html. Current to February 21, 2006.
3. Health Canada, Drugs and Health Products Drugs Directorate Policy regarding the use of a non-Canadian Reference Product under the provisions of Section C.08.002.1(c) of the Food and Drug Regulations. December 1995. http://www.hc-sc.gc.ca/dhp-mps/prodpharma/applic-demande/pol/crp_prc_pol-eng.php. Accessed September 2008.
4. http://en.wikipedia.org/wiki/Compulsory_license. Accessed September 2008.
5. McGilveray IJ. Bioequivalence: A canadian regulatory perspective. In: Welling PG, Tse FLS, Dighe SV, eds. Pharmaceutical Bioequivalence. Drugs and the Pharmaceutical Sciences, Vol 48. New York: Marcel Dekker, Inc., 1991:381–418.
6. Morrison AB, Cook D, Casselman WGB. Clinical Equivalency: A health protection branch perspective. Can Med Assoc J 1973; 109:800–808.

---

[h] Consideration being given to define an outlier as a value which is 3 standard deviations away from mean; does not overlap with other observations; has a studentized residual greater than at the 0.02 level. Two approaches to deal with outlier, viz, construction of a nonparametric 90% CI or two allowance of removal of identified subjects' data after retesting subjects with both formulations and new results not identified as outliers using same approach. These considerations are based on the premise that cause of extreme values unlikely to be formulation related.

[i] HVDPs defined as drugs where intrasubject CV of AUC and/or C_{max} greater than 30%. Recommendations being considered, viz: outlier analysis to be done; adjustment of bioequivalence interval (BI) as a function of the intrasubject CV in a stepwise manner as CV increases, for example, if observed CV from study is between 30% and 39% BI will be 77% to 130% and likewise if observed CV from study is between 60% and 69% BI will be 61 to 160%, which will allow sample sizes to be significantly reduced while retaining 80% power and means at 100%. Intention to provide consistent decision-making for same Canadian Reference Product from different sponsors.

7. Guidance for Industry. Conduct and Analysis of Bioavailability and Bioequivalence Studies—Part A: Oral Dosage Formulations Used for Systemic Effects. Health Canada, 1992. http://www.hc-sc.gc.ca/dhp-mps/alt_formats/hpfb-dgpsa/pdf/prodpharma/bio-a_e.pdf. Accessed September 2008.
8. Guidance for Industry. Conduct and Analysis of Bioavailability and Bioequivalence Studies–Part B: Oral Modified Release Formulations, Health Canada, 1996. http://www.hc-sc.gc.ca/dhp-mps/prodpharma/applic-demande/guide-ld/bio/bio-b-eng.php. Accessed September 2008.
9. Draft guidance for industry Preparation of Comparative Bioavailability Information for Drug Submissions in the CTD Format. Health Canada, May 2004. http://www.hc-sc.gc.ca/dhp-mps/prodpharma/applic-demande/guide-ld/ctd/draft_ebauche_ctdbe-eng.php. Accessed September 2008.
10. Expert Advisory Committee on Bioavailability, Health Protection Branch, December 1992. Report C: Report on bioavailability of oral dosage formulations, not in modified-release form, of drugs used for systemic effects, having complicated or variable pharmacokinetics. Health Canada, 1992. http://www.hc-sc.gc.ca/dhp-mps/prodpharma/applic-demande/guide-ld/bio/biorepc_biorapc-eng.php. Accessed September 2008.
11. International Conference on Harmonisation (ICH) of technical requirements for registration of pharmaceuticals for human use ICH harmonised tripartite guideline for Good Clinical Practice, E6(R1) Current Step 4 version,10 June 1996 (including the Post Step 4 corrections). http://www.ich.org/LOB/media/MEDIA482.pdf. Accessed September 2008.
12. Guidance for clinical trial sponsors: Clinical Trial Applications Health Canada, June 2003. http://www.hc-sc.gc.ca/dhp-mps/prodpharma/applic-demande/guide-ld/clini/ctdcta_ctddec-eng.php. Accessed September 2008.
13. Draft guidance CTAs for Comparative Bioavailability Studies. Health Canada, 2001. http://www.hc-sc.gc.ca/dhp-mps/prodpharma/applic-demande/guide-ld/bio/ctabio_decbio-eng.php. Accessed September 2008.
14. Guidance for industry. Bioequivalence Requirements: Comparative Bioavailability Studies Conducted in the Fed State, Health Canada, June 2005. http://www.hc-sc.gc.ca/dhp-mps/prodpharma/applic-demande/guide-ld/bio/fedstate_sujetsjeun-eng.php. Accessed September 2008.
15. Draft guidance for industry Use of Metabolite Data in Comparative Bioavailability Studies, Health Canada, May 2004. http://www.hc-sc.gc.ca/dhp-mps/prodpharma/applic-demande/guide-ld/bio/draft_ebauche_metabolites-eng.php. Accessed September 2008.
16. Shah VP, Midha KK, Dighe SV, et al. Analytical methods validation: Bioavailability, bioequivalence and pharmacokinetic studies (Conference report), Pharm Res 1992: 9(4):588–592.
17. Shah VP, Midha KK, Findlay JWA, et al Workshop/Conference Report, bioanalytical method validation—a revisit with a decade of progress. Pharm Res 2000; 17:1551–1557.
18. Notice to industry: Removal of Requirement for 15% Random Replicate Samples, Health Canada, 2003. http://www.hc-sc.gc.ca/dhp-mps/prodpharma/applic-demande/guide-ld/bio/15rep_e.html. Accessed September 2008.
19. Guidance for Industry. Stability Testing of New Drug Substances and Products ICH Topic Q1A(R2), Health Canada. http://www.hc-sc.gc.ca/dhp-mps/prodpharma/applic-demande/guide-ld/ich/qual/q1a(r2)-eng.php. Accessed September 2008.
20. (i)Presentation by Eric Ormsby (Health Canada) to the Canadian Society of Pharmaceutical Scientists Annual Meeting, Banff, Alberta, May 24, 2008. (ii) Presentation by Eric Ormsby (Health Canada) to the Canadian Society of Pharmaceutical Scientists Annual Meeting, Toronto, Ontario, June 6, 2009.
21. Notice to industry: Bioequivalence Requirements for Long Half-life Drugs. Health Canada, June 2005. http://www.hc-sc.gc.ca/dhp-mps/prodpharma/applic-demande/guide-ld/bio/notice_longhalflife_avis_longuedemivie-eng.php. Accessed September 2008.

22. Guidance for industry, Bioequivalence Requirements: Critical Dose Drugs, Health Canada, May 2006. http://www.hc-sc.gc.ca/dhp-mps/prodpharma/applic-demande/guide-ld/bio/critical_dose_critique-eng.php. Accessed September 2008.

23. Notice to industry. Bioequivalence Requirements for Drugs for Which an Early Time of Onset or Rapid Rate of Absorption Is Important (rapid onset drugs). Health Canada, June 2005. http://www.hc-sc.gc.ca/dhp-mps/prodpharma/applic-demande/guide-ld/bio/notice_rapidonset_avis_apparitionrapide-eng.php. Accessed September 2008.

24. Notice to industry: Bioequivalence Requirements for Combination Drug Products, Health Canada, June 2005. http://www.hc-sc.gc.ca/dhp-mps/prodpharma/applic-demande/guide-ld/bio/notice_avis_comb-eng.php. Accessed September 2008.

25. Draft policy on Bioequivalence Requirements: Drugs Exhibiting Non-Linear Pharmacokinetics, Health Canada, July 2003. http://www.hc-sc.gc.ca/dhp-mps/prodpharma/applic-demande/pol/nonlin_pol-eng.php. Accessed September 2008.

26. Guidance for industry, Pharmaceutical Quality of Aqueous Solutions. Health Canada, February 2005. http://www.hc-sc.gc.ca/dhp-mps/prodpharma/applic-demande/guide-ld/chem/aqueous_aqueuses-eng.php. Accessed September 2008.

27. Drugs Directorate Guidelines, Preparation of drug identification number submissions, Health Canada, 1995. http://www.hc-sc.gc.ca/dhp-mps/prodpharma/applic-demande/guide-ld/din/pre_din_ind-eng.php. Accessed September 2008.

28. Policy issue "Submissions for generic parenteral drugs", Health Canada, 1990. http://www.hc-sc.gc.ca/dhp-mps/prodpharma/applic-demande/pol/gen_subm_pres_pol-eng.php. Accessed September 2008.

29. Policy issue Submissions for generic topical drugs, Health Canada, 1990. http://www.hc-sc.gc.ca/dhp-mps/prodpharma/applic-demande/pol/gener_pol-eng.php. Accessed September 2008.

30. Policy on Submissions for Topical Non-Steroidal Anti-inflammatory Drugs (Topical NSAID's). http://www.hc-sc.gc.ca/dhp-mps/prodpharma/applic-demande/pol/topnsaids_ainstop_pol-eng.php. Accessed September 2008.

31. Guidance to establish Equivalence or Relative Potency of Safety and Efficacy of a Second Entry Short-Acting Beta2-Agonist Metered Dose Inhaler (MDI). http://www.hc-sc.gc.ca/dhp-mps/prodpharma/applic-demande/guide-ld/inhal-aerosol/mdi_bad-eng.php. Accessed September 2008.

32. Draft Guidance Document: Submission Requirements for Subsequent Market Entry Inhaled Corticosteroid Products for Use in the Treatment of Asthma, May 2007. http://www.hc-sc.gc.ca/dhp-mps/prodpharma/applic-demande/guide-ld/inhal_corticost-eng.php. Accessed September 2008.

33. Draft guidance document Submission Requirements for Subsequent Market Entry Steroid Nasal Products for Use in the Treatment of AllergicRhinitis. http://www.hc-sc.gc.ca/dhp-mps/prodpharma/applic-demande/guide-ld/nas_rhin-eng.php. Accessed September 2008.

34. Health Canada Policy Issues From the Drugs Directorate Changes to marketed new drug products, April 1994. http://www.hc-sc.gc.ca/dhp-mps/prodpharma/applic-demande/pol/changmar_nd_dn_pol-eng.php. Accessed September 2008.

35. 5 Post-Notice of Compliance Changes: Quality Document, July 2006. http://www.hc-sc.gc.ca/dhp-mps/prodpharma/activit/proj/post-noc-apres-ac/noc_postnotice_ac_apreavis-eng.php. Accessed September 2008.

36. Health Canada Policy issue from the drugs directorate Bioequivalence of Proportional Formulations: Solid Oral Dosage Forms, March 1996. http://www.hc-sc.gc.ca/dhp-mps/prodpharma/applic-demande/pol/bioprop_pol-eng.php. Accessed September 2008.

5 The European Union

Roger K. Verbeeck
School of Pharmacy, Catholic University of Louvain, Brussels, Belgium; and Faculty of Pharmacy, Rhodes University, Grahamstown, South Africa

Joelle Warlin
Federal Agency for Medicines and Health Products, Brussels, Belgium

INTRODUCTION

The generic drug product market is projected to grow from US $15 billion in 2004 to US $27 billion in 2009 in the United States, and from US $9 billion to US $14 billion in Western Europe (1). Moreover, the growth opportunities for generic drug products in the near future are significant with an estimated US $100 billion worth of branded pharmaceutical products to go off patent by 2010 (1). The substantial growth of the world generics drug market has been driven by a number of factors, but in particular the need to contain public health care spending, including the expenditure on drug products. In response to the important growth of the generic pharmaceutical industry during the last 10 to 15 years, regulatory agencies in countries all over the world, such as the Food and Drug Administration (FDA) in the United States, Canada's Health Products and Food Branch (HPFB), and the European Medicines Agency (EMEA) in the European Union (EU), have established requirements which must be met by a generic drug product to receive marketing authorization (2,3).

The EU offers four routes for the registration of generic drug products: (*i*) a national procedure, (*ii*) a mutual recognition procedure (MRP), (*iii*) a decentralized procedure (DCP), and (*iv*) a centralized procedure (CP) (4). The national procedure may lead to marketing authorization of the generic drug product in the concerned member state. This national procedure is still being used, but is strictly limited to medicinal products that are not authorized in more than one member state. The MRP is based on the principle of mutual recognition of national authorizations and, therefore, provides for the extension of marketing authorizations granted to one member state, the so-called reference member state (RMS), to one or more member states identified by the applicant. Since November 2005, the applicant may make use of the DCP and submit an application to each of the member states where it is intended to obtain a marketing authorization and choose one of them as the RMS. The RMS prepares a draft assessment report and collects all comments received from the concerned member states that are forwarded to the applicant. Further steps are managed by the RMS to reach a consensus and to finalize the procedure.

Since 2004, it is possible to apply for marketing authorization of medicinal products in the EU by using the CP. According to this procedure, a single application is introduced by the applicant and is subject to a single evaluation. The scientific evaluation of this latter type of application is carried out within the Committee for Medicinal Products for Human Use (CHMP) of the EMEA and is valid throughout the EU and confers the same rights and obligations in each of the member states.[a]

Although the requirements for the approval of generic drug products may still differ among countries, one or more comparative bioavailability studies showing bioequivalence (BE) between the generic drug product and a reference product usually constitute an important part of the information requested by the regulatory agencies of most countries for marketing authorization. In addition, as for all medicinal products, the applicant must demonstrate that the manufacturing process leads to a generic product of sufficient and reproducible quality which will be maintained for the entire duration of its shelf-life.

The concept of BE and the methodology to assess BE have evolved over the past several decades. The first "European" BE guidelines were published in 1991 by the Commission of the European Communities in an attempt to harmonize the registration of generic drug products in the various member states of the European Community (EC), which has been called the EU following the Treaty of Maastricht in 1993 (5). Until the publication of this first Note for Guidance related to BE assessment, generic drug products were registered by the national authorities of the member states. In those days, the registration dossiers were not comprehensive and the assessment was based according to principles published in the scientific literature, FDA guidelines, and the first European guidelines on pharmacokinetic studies in man (6). In 1995 the EMEA, a decentralized body of the EU with headquarters in London, was established. Its main responsibility is the protection and promotion of public and animal health through the evaluation and supervision of medicines for human and veterinary use. In 2001, the EMEA Committee of Proprietary Medicinal Products (CPMP) published the current version of the Note for Guidance on the Investigation of Bioavailability and Bioequivalence (7).

This chapter provides a short overview of the EMEA Guidelines on Bioavailability and Bioequivalence studies for generic drug products and includes comments on a few of the controversial issues regarding these guidelines. Two main Notes for Guidance, prepared by the CHMP of the EMEA, are currently operational: (*i*) the Note for Guidance on the Investigation of Bioavailability and Bioequivalence that came into effect in January 2002, and (*ii*) the Note for Guidance on Modified Release Oral and Transdermal Dosage Forms: Section II (Pharmacokinetic and Clinical Evaluation), which came into operation in January 2000 (7,8). The complete text of these guidelines can be consulted and downloaded from the EMEA website (http://www.emea.europa.eu). The objective of these guidelines is to define, for medicinal products with a systemic effect, when in vivo BE studies are necessary and to formulate requirements for their design, conduct, and evaluation. After the current Note for Guidance on

---

[a] Norway, Iceland and Liechtenstein form the European Economic Area (EEA) with the 25 member states of the EU. These countries have, through the EEA agreement, adopted the complete EU acquis on medicinal products and are consequently parties to the EU procedures. Where in this text reference is made to member states of the EU this should be read to include Norway, Iceland and Liechtenstein.

the Investigation of Bioavailability and Bioequivalence came into effect (January 2002), it appeared that some harmonization regarding the interpretation of critical parts of the guideline was needed. As a result, a Questions & Answers document was published in 2006 by the CHMP Efficacy Working Party (EWP), which clarifies some of the critical parts of this EMEA guidance (9). A revised version of the current BE guidelines for oral, immediate release drug products with systemic action has been in preparation for some time by the CPMP efficacy working party on pharmacokinetics (EWP-PK) of the EMEA. The draft version of this revision of the BE guidelines, entitled Guideline on the Investigation of Bioequivalence, was made publicly available in August 2008 on the EMEA website and a modified version will probably come into effect in 2010 (10).

The application for marketing authorization of a generic drug product, the so-called "generic" application, is an abridged application because the applicant is neither required to provide the results of pharmacological or toxicological tests nor the results of clinical trials if it can be demonstrated that the medicinal product is essentially similar to a product that has been authorized within the community (i.e., the member states of the EU plus Norway, Iceland, and Liechtenstein) for not less than 6 to 10 years and is marketed in the member state for which the application is made (4). According to the EMEA Note for Guidance on the Investigation of Bioavailability and Bioequivalence (7):

> A medicinal product is essentially similar to an original product where it satisfies the criteria of having the same qualitative and quantitative composition in terms of active substances, of having the same pharmaceutical form, and of being bioequivalent unless it is apparent in the light of scientific knowledge that it differs from the original product as regards safety and efficacy.

The Note for Guidance further explains (7)

> By extension, it is generally considered that for immediate release products the concept of essential similarity also applies to different oral forms (tablets and capsules) with the same active substance.

As pointed out in the EMEA Note for Guidance on the Investigation of Bioavailability and Bioequivalence, demonstration of BE is generally the most appropriate method of substantiating therapeutic equivalence between medicinal products, but for pharmaceutical alternatives containing a different salt or ester of the active substance, additional safety data may be needed in some cases (7,11).

ORAL IMMEDIATE RELEASE DOSAGE FORMS WITH SYSTEMIC ACTION

BE studies are clinical studies involving human subjects and, therefore, must follow regulations on good clinical practice (GCP). The design, conduct, and evaluation of BE studies for oral immediate release dosage forms intended to act, following absorption of the active moiety into the systemic circulation are described in the Note for Guidance on the Investigation of Bioavailability and Bioequivalence (7). In what follows, a brief description of the important aspects of BE studies for oral immediate release dosage forms as laid out in this guidance will be presented together with some critical comments and comparisons with the BE guidelines of other countries such as Canada and the United States. Where necessary, reference will be made to the new revised EMEA Guideline on the Investigation of Bioequivalence but it should be kept in mind that, at this stage, it is only a draft version (10).

Study Design

For many drugs a large intersubject variability in pharmacokinetic parameters, such as the extent of absorption (F), the apparent volume of distribution (V), and plasma clearance (CL), is generally observed. The intrasubject variability usually is substantially smaller than the between-subject or intersubject variability and, therefore, a cross-over design is generally recommended for BE studies (12,13). The EMEA Note for Guidance on the Investigation of Bioequivalence and Bioavailability is clear in this regard and recommends a two-period, two-sequence cross-over design, with random allocation of the subjects to each sequence, when comparing the bioavailability of two medicinal products (7). Other study designs may be acceptable, such as a parallel study design for long half-life substances and replicate designs for substances with highly variable pharmacokinetics (14). In the case of a cross-over design, treatments should be separated by a sufficiently long washout period (usually at least five times the terminal plasma half-life of the active drug substance or its metabolites) to ensure that all of the drug and/or its metabolite(s) has been cleared from the body prior to the time of the subsequent administration.

The number of subjects required for a BE study should ideally be estimated at the design stage and is determined by (*i*) the error variance (σ^2) of the primary BE metrics to be studied, (*ii*) the significance level (α), (*iii*) the expected deviation, with respect to the primary BE metrics, between the two formulations which is considered compatible with BE (e.g., $\pm$ 20% for AUC), and (*iv*) the required statistical power (15,16). An estimate of the error variance can be obtained from the published literature, a previous BE study, or by undertaking a pilot study. Nomograms of the number of subjects required for various ratios of the expected means for test and reference products and various intrasubject coefficients of variation have been published by Diletti et al. (15,17). The guidance document of Canada's HPFB allows an add-on study (stated a priori in the protocol) when the results from the first study fail to reach the required power, under the condition that appropriate statistical tests validate the analysis of the combined data (18). The draft version of the revised EMEA Guideline on the Investigation of Bioequivalence includes a recommendation regarding under which circumstances a sequential design (a so-called two-stage approach) may be used (10).

For oral immediate release dosage forms the EMEA guidance favors a study where a single dose is taken on an empty stomach, that is, following an overnight fast, with a fixed volume of fluid (at least 150 mL). However, if it is recommended in the summary of product characteristics (SPC) that the reference medicinal product should be taken with a meal, the BE study should be carried out under fed conditions if the recommendation of food intake has any pharmacokinetic implications such as a higher bioavailability (9). All subsequent meals and drinks as well as other test conditions (e.g., posture during the first few hours following intake of the medicinal products, physical activity, etc.) should be standardized to minimize the variability in the bioavailability metrics unrelated to a possible difference in the formulations. For the same reason of minimising variability, BE studies are recommended to be carried out in healthy volunteers, of either sex, between 18 and 55 years old having a normal body weight based on body mass index, preferably nonsmokers, and without a history of alcohol or drug abuse. For an active substance known to be subject to major genetic polymorphism in its metabolic elimination, phenotyping/genotyping "should be considered" according to the EMEA Note for Guidance on the Investigation of Bioavailability and Bioequivalence when using a parallel design (7). Phenotyping/genotyping "may be considered" as well for crossover BE studies for safety or pharmacokinetic reasons (7). Indeed, plasma concentrations of an active substance that is a substrate for an enzyme showing genetic polymorphism may be much higher and half-lives much longer in poor metabolizers, thus necessitating longer sampling schedules compared to extensive metabolizers (19).

Although, in general, a single-dose study will suffice to show that a generic drug product is bioequivalent to an approved reference product, according to the EMEA Note for Guidance on the Investigation of Bioavailability and Bioequivalence there are situations in which steady-state studies may be required or can be considered (7). These situations may include BE studies for active substances undergoing dose- or time-dependent kinetics, or for active substances with high intraindividual variability for which it may be difficult or even impossible to demonstrate BE in a reasonably sized single-dose study. In addition, steady-state studies can be considered when problems of analytical sensitivity preclude sufficiently precise measurement of analyte plasma concentrations after single-dose administration. According to the revised EMEA Guideline on the Investigation of Bioequivalence, a multiple-dose study as an alternative to a single-dose study may also be acceptable if problems of sensitivity of the analytical method preclude sufficiently precise plasma concentration measurements after single-dose administration. However, if possible, C_{max} should be determined as a measure of peak exposure following administration of the first dose of the multiple-dose study. AUC, a measure of extent of exposure, should be determined at steady state. Moreover, in a multiple-dose BE study the administration scheme should preferably follow the highest usual dosage recommendation (10).

Post hoc exclusion of outliers based on pharmacokinetic or statistical reasons alone is not accepted (7,9). Nonstatistical reasons to exclude the data of a particular subject from the final statistical analysis should have been prospectively defined in the protocol, or according the EMEA Questions & Answers document on the bioavailability and BE guideline: "... at the very least, established before reviewing the data." Acceptable explanations to exclude pharmacokinetic data or to exclude a subject from the final statistical analysis would be protocol violations such as vomiting, diarrhoea, analytical failure, etc.

Reference and Test Products

In a BE study which is carried out as part of an application for marketing authorization of a generic medicinal product, the bioavailability of the generic product (test) is compared to the bioavailability of an innovator medicinal product (reference). The batches of the test and reference product used in the BE study are called the "biobatches." The requirements for the test product used in the BE study are clearly spelled out in the EU guidance. The test product should usually originate from a batch of at least 1/10 of production scale or

100,000 units, whichever is greater, unless otherwise justified. As far as the reference product is concerned, the EMEA guidance specifies that the choice of the reference product should be justified by the applicant. The reference product should normally be the innovator, a medicinal product authorized on the basis of a full dossier. When the innovator is no longer on the market, the product that is the market leader may be used as the reference product provided that it has been authorized for marketing and its efficacy, safety, and quality have been fully established and documented. In the case of a MRP, application for marketing authorization to numerous member states based on a BE study with a reference product from one member state, that is, the RMS, can be made. In general, the qualitative and quantitative composition of the reference product is the same in all the member states of the EU. However, if the reference products marketed in the various member states slightly differ in terms of qualitative/quantitative composition of the excipients as well as the manufacturing process, extrapolation of the results of the BE study carried out with the reference medicinal product marketed in one particular member state to BE claims in comparison to reference products marketed in the other member states is not always straightforward. Comparative in vitro dissolution profiles between the reference product used in the BE study and the one registered and marketed in the member state where marketing authorization is requested may be asked by the assessors. The in vitro dissolution method should be discriminating and in accordance with the pharmacopoeial requirements. The in vitro dissolution profiles may be compared by calculating an f_2 similarity factor. An f_2 value between 50 and 100 suggests that the two dissolution profiles are similar. Alternative methods to prove similarity of dissolution profiles are accepted as long as they are justified. In cases where more than 85% of the drug is dissolved within 15 minutes, dissolution profiles are considered to be similar without further mathematical evaluation. In cases where the composition and/or the manufacturing process of the reference product used in the BE study compared to the reference product registered and marketed in the member state(s) where marketing authorization is requested differ to such an extent that the bioavailability may be affected, a BE study with the latter may be requested. The EMEA Note for Guidance on the Investigation of Bioavailability and Bioequivalence, however, is not very helpful in this regard:

> Concerned Member States may request information from the first Member State on the reference product, namely on the composition, manufacturing process and finished product specification. Where additional bioequivalence studies are required, they should be carried out using the product registered in the concerned Member State as the reference product.

Indeed, no indication whatsoever is given as to the nature and/or importance of the differences in composition and manufacturing process between reference products marketed in the various member states that would necessitate a new BE study. Perhaps a series of guidelines such as those issued by the FDA in the case of scale-up and postapproval changes (SUPAC) would be helpful to decide when an additional BE study between the generic product versus the reference medicinal product registered in a particular concerned member state would be required (20). Although the EMEA Note for Guidance on the Investigation of Bioavailability and Bioequivalence suggests that in vitro dissolution studies can be used as "bioequivalence surrogate inference" to demonstrate similarity between the reference products from different member states, they cannot replace for most active substances an in vivo BE study unless an in vitro/in vivo correlation (IVIVC) has been demonstrated (21,22).

BE Metrics

The area under the plasma (serum, blood) concentration of the parent compound versus time curve (AUC_t, AUC_∞) generally serves as a measure of the extent of absorption. T_{max} and the corresponding maximum plasma concentration, C_{max}, may serve as characteristics of the rate of absorption. However, it should be emphasized that C_{max} is not a pure measure of absorption rate but is confounded with the extent of absorption (23). Urine excretion data may also be used to determine the extent of absorption provided elimination is predominantly renal as intact drug substance and is dose proportional (7,9). In the revised EMEA Guideline on the Investigation of Bioequivalence the condition that "elimination is dose-linear and is predominantly renal as intact drug" is no longer mentioned (10). However, the use of urinary data has to be carefully justified when used to estimate the rate of absorption (9,10,24).

AUC_∞, the area under the curve extrapolated to infinity, can only be reliably measured if the terminal plasma half-life can be accurately determined, which is not always the case. AUC_t, the area under the curve from the time of administration to the last measurable plasma concentration at time t, is therefore considered to be the most reliable measure of the extent of absorption provided that it covers at least 80% of AUC_∞. Literature data support the notion that BE assessment for long half-life drugs is not adversely affected by using truncated AUC (25–29). Blood sampling time in this case should be sufficiently long to ensure completion of gastrointestinal transit of the drug product (approximately two to three days) and consequently the absorption process of the drug substance. The Canadian HPFB guidelines, for example, accept AUC_{0-72}, the AUC from time 0 to 72 hours following administration, as a measure of extent of absorption for drug substances with a half-life of more than 12 hours (30).

Moiety To Be Measured: Parent Drug Versus Metabolite(s)

In most cases the evaluation of BE should be based on the measurement of plasma concentrations of the parent compound. The rationale for this approach is that the concentration–time profile of the parent drug is more sensitive to changes in formulation performance than that of the metabolite, which includes the processes of metabolite formation, distribution and elimination. However, according to the EMEA Note for Guidance on the Investigation of Bioavailability and Bioequivalence: "In some situations, however, measurements of an active or inactive metabolite may be necessary instead of the parent compound." A clear consensus on the role of metabolites for the assessment of BE has not yet been achieved within the scientific community (31–33). This is reflected in the different views expressed in the current national and international regulatory guidelines concerning the role of the measurement of metabolites in BE assessment. According to the EMEA Note for Guidance on the Investigation of Bioavailability and Bioequivalence, measurement of metabolites is required to assess BE between two medicinal products in the following cases: (*i*) if the concentration of the parent compound is too low to be accurately measured and (*ii*) if the parent compound is unstable in the biological matrix or its half-life is too short (7). The same guidelines further state:

> In particular if metabolites significantly contribute to the net activity of an active substance and the pharmacokinetic system is non-linear, it is necessary to measure both parent drug and active metabolite plasma concentrations and evaluate them separately.

The most recent version of the general BE guidance from the FDA requests that only the parent compound should be measured to assess BE (34). Only when a metabolite is formed as a result of gut wall or other presystemic metabolism and the metabolite contributes to safety and efficacy is the metabolite measured to provide supportive evidence. In all other instances only the parent compound is measured for BE. According to the guidelines of Canada's HPFB, the determination of BE is based on measurement of the active ingredient, or its metabolite, or both, as a function of time (18). They further specify that normally measurement of the parent compound is sufficient but in some cases measurement of the metabolite could be required. For example, when a prodrug is administered, the active metabolite should be measured. The Questions & Answers document on the Bioavailability and Bioequivalence Guideline which were recently formulated by the CHMP Efficacy Working Party of the EMEA, also deals with the issue when metabolite data should be used to establish BE (9). This document stipulates that metabolite data can only be used if the applicant presents state-of-the-art evidence that measurements of plasma concentrations of the parent compound are unreliable. In addition, it is pointed out that C_{max} of the metabolite is less sensitive to differences in the rate of absorption than C_{max} of the parent compound:

> Therefore, when the rate of absorption is considered of clinical importance, bioequivalence should, if possible, be determined for C_{max} of the parent compound if necessary following administration of a higher dose. (9)

In their excellent review of the topic of measurement of metabolites for BE assessment, Jackson et al. conclude that the parent compound is the entity most sensitive to formulation changes (31,33). The continuing belief by some that activity is important and should be considered for BE assessment is the major reason for most of the controversy regarding metabolite measurement. Another argument for using metabolites in BE assessment is that metabolite concentrations are generally associated with a lower intrasubject variability and consequently their use allows a decrease in the number of subjects required to establish BE. However, analysis of both parent drug and metabolite to assess BE is problematic since it would decrease Type I error (consumer risk) and increase Type II error (producer risk) (33).

Calculation of Confidence Interval and Acceptance Limits

Estimation of BE is based on the "two one-sided tests" procedure (35) in which the 90% confidence interval (CI) around the geometric mean ratio of the test and reference values of an appropriate bioavailability measure, such as AUC or C_{max}, is required to fall within preset BE limits. One of the important objectives of BE testing is to assure that two medicinal products containing the same active substance are interchangeable in any individual patient. For this reason, the "two one-sided tests" procedure is based on the intrasubject variability that is commonly estimated from the mean square error (MSE), also called the residual mean square, of an analysis of variance in which the fixed effects are typically formulation, period, sequence, and subject nested within sequence. The intrasubject variability can be estimated from the MSE by calculating the CV_{anova} (ANOVA coefficient of variation) as follows:

$$CV_{anova}(\%) = \left(\sqrt{e^{MSE} - 1} \right) 100$$

The width of the 90% CI depends on the magnitude of MSE and the number of subjects in the BE study. Active substances whose AUC and C_{max} show a high intrasubject variability have high values for CV_{anova} (>30%) and are called highly variable drugs (HVDs). The larger the CV_{anova}, the higher the number of subjects required to give adequate statistical power (16,17).

The usual acceptance limit for the 90% CI around the geometric mean ratio for AUC and C_{max}, that is, 0.80 to 1.25 (or 80–125%), is based on a consensus amongst clinical experts that a difference of ±20% in plasma concentrations of the active substance following administration of two different medicinal products would have no clinical significance for most drugs (36). Since measures derived from plasma concentrations such as AUC and C_{max} are log-normally distributed, this ±20% translates into an asymmetric acceptance limit, for example, 0.80 to 1.25. The EMEA Note for Guidance suggests that in specific cases of active substances with a narrow therapeutic index (NTI) the acceptance interval may need to be tightened but does not give more specific information (7). The draft version of the EMEA revised Guideline on the Investigation of Bioequivalence adds that "... the need for narrowing the acceptance interval for both AUC and C_{max} or for AUC only should be determined on a case by case basis." (10)

The HPFB of Canada has issued Guidance for Industry on the bioequivalence requirements for critical dose drugs (37). According to this guidance, "critical dose drugs" are defined as those drugs for which comparatively small differences in dose or concentration lead to dose- and concentration-dependent, serious therapeutic failures and/or serious adverse drug reactions. For these "critical dose drugs," the 90% CI of the relative mean AUC of the test to reference formulation should lie within 90% to 112%, according to Canada's HPFB guidance. In addition, the 90% CI of the relative mean C_{max} of the test to reference formulation for these "critical dose drugs" should be between 80% and 125%. For "uncomplicated" drugs, Canada's HPFB requires the point estimate of C_{max} to simply lie between 80% and 125%. These requirements for "critical dose drugs" are to be met in both the fasted and fed states. In an appendix to this HPFB guidance a list of 9 "critical dose drugs" is given (37). The FDA Guidance for Industry on Bioavailability and Bioequivalence Studies for Orally Administered Drug Products recommends that the usual BE limit of 80% to 125% for non-NTI drugs remain unchanged for the bioavailability measures (AUC and C_{max}) for NTI drug substances unless otherwise indicated by a specific guidance (34).

On the other hand, wider BE limits for the 90% CI may be acceptable for C_{max} (in certain cases) and for AUC (in rare cases) according to the EMEA Note for Guidance (7). Indeed, when the intrasubject variability in AUC and C_{max} is high the estimated 90% CI is wide and it is very difficult to be entirely located within the usually accepted BE limits of 0.80 to 1.25. Among the methods proposed during recent years in the scientific literature to evaluate the BE of these highly variable drugs and drug products, scaled average BE and expanding the usual BE limits to, for example, 0.75 to 1.33, were recently shown to be sensitive to differences between means and, consequently, highly effective for assessing the equivalence of average kinetic responses (38–40). Recently, a commentary was published in which the authors proposed to adjust the BE limits for highly variable drugs/drug products by scaling to the intrasubject variability of the reference product in the study (41). The recommendation for the use of reference scaling is based on the general concept that reference variability should be used as an index for setting the public standard expressed in the BE limit. The use of the reference-scaling approach necessitates a study design that evaluates the reference variability via replicate administration of the reference product to each subject.

BE studies using a replicate design, for example a three-period or four-period study, have certain advantages over the classical two-period design. They allow the comparison of the intrasubject variance and the evaluation of the subject-by-formulation interaction. Information on these variances associated with the test and reference formulations allows assessment of the pharmaceutical quality of a new test product compared to the pharmaceutical quality of the marketed innovator product. At the moment, none of the major health authorities (EMEA, FDA, HPFB), however, provide clear recommendations on how to assess BE of highly variable drugs or drug products.

The draft version of the revised EMEA Guideline on the Investigation of Bioequivalence is clear in this respect. According to this guideline, it is acceptable to widen the 90% acceptance range of C_{max}, but not AUC, from 0.80–1.25 to 0.75–1.33 under the following conditions: (*i*) the 0.75 to 1.33 acceptance range has been prospectively defined in the study protocol, (*ii*) it has been prospectively justified that widening of the acceptance criteria for C_{max} does not affect clinical efficacy or safety, and (*iii*) the BE study is of a replicate design where it has been demonstrated that the intrasubject variability for C_{max} of the reference compound in the study is >30% (10).

Exemptions from In Vivo BE Studies (Biowaivers)

The biopharmaceutics classification system (BCS) provides a scientific framework for classifying active substances based on their aqueous solubility and intestinal permeability (21,42,43). When combined with the in vitro dissolution characteristics of the drug product, the BCS takes into account the major factors, that is, solubility and intestinal permeability, which are fundamental in controlling the rate and extent of oral drug absorption from immediate release solid oral dosage forms. In August 2000, the FDA issued a guidance for industry on waivers of in vivo bioavailability and BE studies for immediate release solid oral dosage forms (44). This guidance recommends that applicants may request biowaivers for highly soluble and highly permeable drug substances (BCS class I) in immediate release solid oral dosage forms provided that they exhibit rapid in vitro dissolution rates and a few other conditions are met. The methods for determining solubility, permeability, and in vitro dissolution are described in this FDA biowaiver guidance as well as the approaches recommended for classifying drug substances according to the BCS.

The EMEA Note for Guidance on the Investigation of Bioavailability and Bioequivalence under certain conditions also allows exemptions from in vivo BE studies for oral immediate release dosage forms with systemic action (7). Although this exemption from in vivo BE studies is based on similar considerations as those described in the FDA Guidance for Industry (44), the EMEA Note for Guidance on the Investigation of Bioavailability and Bioequivalence is much less detailed in the description of the criteria on which a biowaiver may be granted. Biowaivers are still rarely used in the EU probably due to uncertainties by both the pharmaceutical companies and the regulatory authorities regarding the application of the biowaiver principles. An example of a biowaiver, accepted by the German regulatory authority, that is, the Bundesinstitut für Arzneimittel und Medizinprodukte in Bonn, has been described in the scientific literature for 80 and 160 mg immediate release tablets containing sotalol hydrochloride, a BCS class I substance (45). To reach an optimal and harmonized application based on biowaiver principles, the draft version of the revised EMEA Guideline on the Investigation of Bioequivalence addresses the issue of BCS-based biowaivers in much more detail than the current EMEA Note for Guidance (10,46). According to this revised EMEA Guideline on the Investigation of Bioequivalence, BCS-based biowaivers will be considered not only for BCS class I drug substances (high solubility, high permeability), but also for BCS class III substances (high solubility, low permeability), as has been proposed in multiple scientific commentaries (47–49). In the latter case, special attention will have to be paid to the excipients since it is known that the absorption of BCS class III substances is more susceptible to transporter-mediated excipient–drug interactions (50,51).

Formulation Changes and Variations

Information to document BE following reformulation of an approved generic (or innovator) drug product or following a modification in its manufacturing process or manufacturing equipment used is obviously required. Volume 2 of the publication "The rules governing medicinal products in the European Union" contains a list of regulatory guidelines related to procedural and regulatory requirements such as renewal procedures, dossier requirements for variation notifications, summary of product characteristics, package information, readability of the label, and package leaflet requirements (51). Since 2003, new categories of variations, that is, notifications type IA and type IB, have been introduced in the EU (52). Type IA variations are "minor" variations, for example, a change in the name and/or address of the marketing authorization holder, a change in the name of the active substance or its ATC (anatomical therapeutic chemical) code, which do not require a new in vivo BE study. Examples of type IB notifications are a minor change in the manufacturing process of the active substance, a minor change in the manufacturing process of the finished product, replacement of an excipient with a comparable excipient. For some type IB notifications, a justification for not submitting a new BE study and/or comparative dissolution data must be provided. Type II variations constitute "major" changes and an in vivo BE study is required unless a biowaiver can be granted on the basis of in vitro dissolution tests (BCS-based biowaiver, in vitro–in vivo correlation).

Unlike the FDA, which has a specific guidance on SUPAC, the EMEA Note for Guidance on the Investigation of Bioavailability and Bioequivalence only has a small paragraph on variations:

> If a product has been reformulated from the formulation initially approved or the manufacturing method has been modified by the manufacturer in ways that could be considered to impact on the bioavailability, a bioequivalence study is required, unless otherwise justified. Any justification presented should be based upon general considerations . . ., or on whether an acceptable in vivo/in vitro correlation has been established. (7).

From a regulator's and sponsor's point of view it would be desirable to have clear and more detailed guidelines on this important issue, such as the SUPAC guidelines of the FDA, to guarantee the continuing quality of a generic drug product even during the postapproval period.

Bioequivalence of Chiral Drugs

Attempts have been made in the scientific literature to examine the stereochemical aspects of BE and several examples demonstrate that BE between two medicinal products containing a mixture of stereoisomers based on nonstereospecific assays alone may not be extended to the pharmacologically relevant stereoisomer(s) (53–55). The results of these studies suggest that stereospecific assays are necessary for at least some chiral drugs. However, at this time no consensus has been reached regarding the conditions whereby BE of medicinal products containing a mixture of stereoisomers should be assessed (56). According to the current EMEA Note for Guidance on the Investigation of Bioavailability and Bioequivalence:

> . . . bioequivalence studies supporting applications for essentially similar medicinal products containing chiral active substances should be based upon enantiomeric bio-analytical methods unless (1) both products contain the same stable single enantiomer; (2) both products contain the racemate and both enantiomers show linear pharmacokinetics.

The draft version of the revised EMEA Guideline on the Investigation of Bioequivalence provides much clearer recommendations on this issue (10).

Locally Applied Drug Products

According to the EMEA Note for Guidance on the Investigation of Bioavailability and Bioequivalence, for drug products for local use (oral, nasal, ocular, dermal, rectal, vaginal, inhalation, etc.) intended to act without systemic absorption the approach to assess BE on the basis of systemic concentrations of the active substance is not applicable and pharmacodynamic or comparative clinical studies are in principle required (7,57).

The EMEA is currently working on a detailed guideline describing the requirements for clinical documentation for abridged applications for orally inhaled formulations and variations/extensions to a marketing authorization with respect to demonstrating therapeutic equivalence between two inhaled products for use in the management and treatment of asthma and chronic obstructive pulmonary disease (58).

As far as the assessment of therapeutic equivalence between topical corticosteroid products is concerned, the current EMEA guidance in question has been in operation since 1987 (59). More recently, a Questions & Answers document was released by the EMEA dealing more specifically with the vaso-constriction (human skin blanching) assay that may reduce the need for data from clinical trials when assessing therapeutic equivalence between topical cor-ticosteroid products (60). This document refers to the FDA guidance for industry for a detailed description of how to perform this vasoconstriction assay (61).

MODIFIED RELEASE ORAL AND TRANSDERMAL DOSAGE FORMS
In January 2000, the EMEA Note for Guidance on Modified Release Oral and Transdermal Dosage Forms: Section II (Pharmacokinetic and Clinical Evaluation) came into effect (8). The primary purpose of this guidance was

> to define the studies necessary to investigate the properties and effects of the new delivery system in man and to set out general principles for design-ing, conducting and evaluating such studies.

Although the guidance only deals with oral modified release formulations and transdermal dosage forms, most recommendations are also applicable to implants and intramuscular/subcutaneous depot formulations. Paragraph 5 of this document specifically deals with applications for modified release dosage forms essentially similar to a marketed modified release form, that is, so-called generic applications. A distinction is made between prolonged release oral for-mulations, delayed release oral formulations and transdermal drug delivery sys-tems (TDDS).

Prolonged Release Oral Formulations
Whereas BE for oral immediate release dosage forms with systemic action is established on the basis of a single-dose study usually carried out in the fasting state, the EMEA Note for Guidance on Modified Release Oral and Transdermal Dosage Forms recommends that assessment of BE of prolonged release oral for-mulations should be based on single- *and* multiple-dose studies. Typically, single- and multiple-dose studies are carried out with the test and reference formulation following an overnight fast. In addition, a single-dose study has to be carried out with both test and reference formulation administered after a predefined high-fat meal. The effect of this high-fat meal on the in vivo bioavailability should be comparable for both preparations. The conditions to apply for a biowaiver in case the application concerns multiple strengths are different for single unit and multiple unit prolonged release oral formulations (8). It is interesting to note that the FDA guidance recommends only single-dose studies (a fasting study and a food-effect study) for modified release products submitted as Abbreviated New Drug Applications (ANDA) (34). Their argument is that single-dose studies are more sensitive to assess BE between two drug products. Canada's HPFB guide-lines for BE assessment on oral modified release formulations also recommend single-dose BE studies under fasting and fed conditions. In addition, for formu-lations that are likely to lead to accumulation of the active substance in plasma, the HPFB also recommends a BE study at steady state, that is, after multiple-dose administration (62).

According to the EMEA Note for Guidance on Modified Release Oral and Transdermal Dosage Forms: "Assessment of bioequivalence will be based on AUC_τ, C_{max}, and C_{min} applying similar statistical procedures as for the imme-diate release formulations." However, in the case of prolonged release formula-tions, which at steady state may show relatively flat plasma concentration–time curves often with multiple peaks, C_{max} is of limited value to characterize the rate of absorption. Therefore, other measures such as the half-value duration, peak-trough fluctuation (PTF), and percent swing may be useful alternatives (63,64). Any widening of the usual 0.80 to 1.25 acceptance criterion should be prospectively established in the study protocol and should be clinically justified.

Delayed Release Oral Formulations
An enteric-coated formulation, the most common example of a delayed release formulation, is designed to protect the active substance from the acid environment of the stomach or to protect the stomach from the active substance. The EMEA Note for Guidance on Modified Release Oral and Transdermal Dosage Forms only specifies for this particular case of a modified release oral dosage form that (*i*) BE is assessed using the same main characteristics and statistical procedures as for immediate release oral formulations and (*ii*) postprandial bioequivalence studies are necessary. It is not clear, though, whether only a food-effect study has to be carried out, or whether in addition a fasting study is recommended (8).

Transdermal Drug Delivery Systems
A TDDS or transdermal patch is defined as a flexible pharmaceutical preparation of varying size containing one or more active substances to be applied on the intact skin to provide a slow delivery of the active substance(s) into the systemic circulation. Transdermal patches are often highly variable drug products and consequently BE studies with replicate designs are recommended by the EMEA Note for Guidance on Modified Release Oral and Transdermal Dosage Forms (8). A replicate study is required if the systemic bioavailability of TDDS with different release mechanisms, for example, reservoir versus matrix, is compared because this design allows the assessment of the subject-by-formulation interaction. In general, the BE of TDDS should be assessed after single-dose and multiple-dose administration. When the application for marketing authorization concerns multiple strengths of a TDDS, BE studies can be performed on the highest strength only provided certain conditions are met such as (*i*) the strength is proportional to the effective surface area of the TDDS, and (*ii*) an acceptable in vitro release test exists. Finally, test product and reference product should demonstrate the same (or less) degree of local irritation, phototoxicity, sensitization and systemic adverse events, and a similar degree of adhesiveness to the skin. Although the EMEA guidance does not further elaborate on this last point, the FDA has published a guidance for industry specifically treating skin irritation and sensitization testing of generic transdermal drug products (65).

FIXED COMBINATION DRUG PRODUCTS
For fixed combination drug products, in vivo BE should be evaluated for each individual active substance. The study design and BE assessment methodology and criteria are the same as those applied to oral immediate release formulations. The reference product used in the BE study should be the originator fixed combination product (7).

A Questions & Answers document was released by the EMEA in 2005 regarding the clinical development of fixed combinations of drugs belonging to different therapeutic classes in the field of cardiovascular treatment and prevention (66). This "guideline" discusses what is required in case a new combination product is developed of active substances as substitution therapy for patients adequately controlled with the same active substances given concurrently at the same dose level and dosing interval but as separate single-substance drug products. In this particular case, only BE should be demonstrated between the already existing single-substance drug products and the fixed combination drug product according to the recommendations described in the EMEA Note for Guidance on the Investigation of Bioavailability and Bioequivalence (7). The possibility that the active substances may interact pharmacokinetically should be documented.

CONCLUSION: TOWARD GLOBAL HARMONIZATION?

The generic drug approval process has evolved over the past 30 years and regulatory agencies in a number of western countries have now established stringent requirements for the design, performance and evaluation of BE studies to protect the consumer of being exposed to drug products of inferior quality. Although the current BE guidelines and recommendations of the major regional and national health authorities show a fair degree of consistency, a number of outstanding BE issues and concerns remain to be resolved. The most obvious of these controversial issues, such as the BE acceptance limits for NTI drugs and HVDs, the role of metabolites in BE assessment, the use of stereospecific bioanalytical assays to determine BE of chiral drugs, the choice of the reference product, conditions to grant biowaivers, are not dealt with in the same way by the various guidelines. For example, the World Health Organization (WHO), which is not a regulatory body but publishes technical reports and guidelines that are recommendations to national authorities especially in developing countries, not only allows biowaivers for BCS class I substances but also allows biowaivers under certain circumstances for class II and class III substances (67). At this moment, the FDA and the EMEA do not allow biowaivers for BCS class II substances (68). This creates confusion that in turn leads to suspicion by health care providers and patients, especially since many national authorities give these WHO reports regulatory status. All stakeholders in the development and registration of new drug products must balance the need for scientific rigor in assuring BA/BE (and hence product quality toward consistent therapeutic outcomes) with the time and expense of conducting in vivo BE studies, and the overall impact on product costs and timely availability to patients. Ideally these guidelines should be the same worldwide to ensure that patients all over the world can benefit from affordable and safe medicinal products.

Global harmonization should therefore be the next logical step in the continuing process to improve the BE guidelines as a means to guarantee safe and efficacious drug products for the consumer in all parts of the world. Global harmonization efforts by the International Conference on Harmonization (ICH) and the WHO should be stepped up in collaboration with the regulatory agencies of the western world as more nations throughout the world have come to rely on low-cost, good-quality multisource (generic) pharmaceutical products as means of providing lower health care costs without sacrificing important public health goals. However, as already pointed out, a consensus on a number of BE issues has not even been reached at this point in time among international regulatory agencies. In addition, differing levels of commitment and resources by the various countries and regions constitute another formidable barrier that has to be overcome to harmonize BE approaches to ensure development of optimally performing and affordable drug products for use by health practitioners and patients in he global community.

NOTE ADDED IN PROOF

The following important Questions & Answers document was published after completion of the manuscript: "Positions on specific questions addressed to the EWP therapeutic subgroup on Pharmacokinetics", EMEA/618604/2008 Rev. 1, London, 23 July 2009. http://www.emea.europa.eu/pdfs/human/ewp/61860408en.pdf (accessed on November 9, 2009).

REFERENCES

1. Tempest B. India: A global strategic asset for developed world market businesses. J Generic Med 2006; 4:37–42.
2. Nation RL, Sansom LN. Bioequivalence requirements for generic products. Pharmac Ther 1994; 62:41–55.
3. Chen M-L, Shah V, Patnaik R, et al. Bioavailability and bioequivalence: An FDA regulatory overview. Pharm Res 2001; 18:1645–1650.
4. EC Notice to Applicants, volume 2A: Procedures for marketing authorisation. The Rules Governing Medicinal Products in the European Union, Chapter 1: Marketing Authorisation, November 2005. http://ec.europa.eu/enterprise/pharmaceuticals/eudralex/ vol-2/a/vol2a_chap1_2005–11.pdf. Accessed May 2009.
5. CPMP Note for Guidance on the Investigation of Bioavailability and Bioequivalence, Brussels, December 1991.
6. Pharmacokinetic studies in man, 1987. http://www.emea.europa.eu/pdfs/human/ewp/3cc3aen.pdf. Accessed May 2009.
7. EMEA Note for Guidance on the Investigation of Bioavailability and Bioequivalence. London: CPMP/EWP/QWP/EMEA, July 2001. http://www.emea.europa.eu/pdfs/human/qwp/140198enfin.pdf. Accessed May 2009.
8. EMEA Note for Guidance on Modified Release Oral and Transdermal Dosage Forms: Section II (Pharmacokinetic and Clinical Evaluation). London: CPMP/EWP/280/96, July 1999. http://www.emea.europa.eu/pdfs/human/ewp/028096en.pdf. Accessed May 2009.
9. Questions & Answers on the Bioavailability and Bioequivalence Guideline. EMEA, London: EMEA/CHMP/EWP/40326/2006, July 26, 2006. http://www.emea.europa.eu/pdfs/human/ewp/4032606en.pdf. Accessed May 2009.
10. EMEA Guideline on the Investigation of Bioequivalence, EMEA. London: CPMP/EWP/QWP/1401/98, July 24, 2008. Rev. 1. http://www.emea.europa.eu/pdfs/human/qwp/140198enrev1.pdf. Accessed May 2009.
11. Verbeeck RK, Kanfer I, Walker RB. Generic substitution—the use of medicinal products containing different salts and implications for safety and efficacy. Eur J Pharm Sci 2006; 28:1–6.
12. Schuirmann DJ. Design of bioavailability/bioequivalence studies. Drug Inf J 1990; 24:315–323.
13. Pidgen AW. Statistical aspects of bioequivalence—A review. Xenobiotica 1992; 22:881–893.
14. Bolton S, Bon C. Statistical considerations—Alternative designs and approaches for bioequivalence assessments. In: Kanfer I, Shargel L, eds. Generic Product Development—Bioequivalence Issues. New York: Informa Healthcare, 2008:123–141.
15. Diletti E, Hauschke D, Steinijans VW. Sample size determination for bioequivalence assessment by means of confidence intervals. Int J Clin Pharmacol Ther Toxicol 1992; 30(suppl 1):S51–S58.

16. Chow S-C, Wang H. On sample size calculation in bioequivalence trials. J Pharmacokinet Pharmacodyn 2001; 28:155–169.
17. Diletti E, Hauschke D, Steinijans VW. Sample size determination: Extended tables for the multiplicative model and bioequivalence ranges of 0.9 to 1.11 and 0.7 to 1.43. Int J Clin Pharmacol Ther Toxicol 1992; 30(suppl 1):S59–S62.
18. HPFB Guidance for Industry. Conduct and Analysis of Bioavailability and Bioequivalence Studies—Part A: Oral Dosage Formulations used for Systemic Effects. Ottawa, 1992. http://www.hc-sc.gc.ca.
19. Olsson B, Szamosi J. Multiple dose pharmacokinetics of a new once daily extended release tolterodine formulation versus immediate release tolterodine. Clin Pharmacokinet 2001; 40:227–235.
20. FDA Guidance for Industry: Immediate Release Solid Oral Dosage Forms. Scale-up and Postapproval Changes, Chemistry, Manufacturing, and Controls, In Vitro Dissolution Testing, and In Vivo Bioequivalence Documentation. Washington, DC: CDER/FDA, November 1995. http://www.fda.gov/cder/guidance/index.htm. Accessed May 2009.
21. Amidon GL, Lennernas H, Shah VP, et al. A theoretical basis for a biopharmaceutic drug classification: The correlation of in vitro drug product dissolution and in vivo bioavailability. Pharm Res 1995; 12:413–420.
22. Polli JE. In vitro-in vivo relationship of several "immediate" release tablets containing a low permeability drug. Adv Exp Med Biol 1997; 423:191–198.
23. Hauschke D, Steinijans V, Pigeot I. Metrics to characterize concentration-time profiles in single- and multiple-dose bioequivalence studies. In: Hauschke D, Steinijans V, Pigeot I, eds. Bioequivalence Studies in Drug Development—Methods and Applications. Chichester, West Sussex, UK: Wiley, 2007:17–36.
24. Shah SA, Rathod IS, Savale SS, et al. Determination of bioequivalence of lomefloxacin tablets using urinary excretion data. J Pharm Biomed Anal 2002; 30:1319–1329.
25. Midha KK, Hubbard JW, Rawson M, et al. The application of partial areas in assessment of rate and extent of absorption in bioequivalence studies of conventional release products: Experimental evidence. Eur J Pharm Sci 1994; 2:351–363.
26. Endrenyi L, Tothfalusi L. Truncated AUC evaluates efectively the bioequivalence of drugs with long half-lives. Int J Clin Pharmacol Ther 1997; 35:142–150.
27. Gaudreault J, Potvin D, Lavigne J, et al. Truncated area under the curve as a measure of relative extent of bioavailability evaluation using experimental data and Monte Carlo simulation. Pharm Res 1998; 15:1621–1629.
28. Sathe P, Venitz J, Lesko L. Evaluation of truncated areas in the assessment of bioequivalence of immediate release formulations of drugs with long half-lives and of Cmax with different dissolution rates. Pharm Res 1999; 16:939–943.
29. Chen M-L, Lesko L, Williams RL. Measures of exposure versus measures of rate and extent of absorption. Clin Pharmacokinet 2001; 40:565–572.
30. HPFB Guidance for Industry: Conduct and Analysis of Bioavailability and Bioequivalence Studies—Part C: Report on Bioavailability of Oral Dosage Formulations, not in Modified Release Form, of Drugs used for Systemic Effects, having Complicated or Variable Pharmacokinetics. Ottawa, 1992. http://www.hc-sc.gc.ca.
31. Jackson AJ, Robbie G, Marroum P. Metabolites and bioequivalence: Past en present. Clin Pharmacokinet 2004; 43:655–672.
32. Midha KK, Rawson MJ, Hubbard JW. The role of metabolites in bioequivalence. Pharm Res 2004; 21:1331–1344.
33. Jackson A. Role of metabolites in bioequivalence assessment. In: Kanfer I, Shargel L, eds. Generic Product Development—Bioequivalence Issues. New York: Informa Healthcare, 2008:171–183.
34. FDA Guidance for Industry: Bioavailability and Bioequivalence Studies for Orally Administered Drug Products—General Considerations. Washington, DC: CDER/FDA, March 2003. http://www.fda.gov/cder/guidance/index.htm. Accessed May 2009.

35. Schuirmann DJ. A comparison of the two one-sided tests procedure and the power approach for assessing the equivalence of average bioavailability. J Pharmacokinet Biopharm 1987; 15:657–680.
36. Williams RL. Bioequivalence and therapeutic equivalence. In: Welling PG, Tse FLS, Dighe SV, eds. Pharmaceutical Bioequivalence. New York: Marcel Dekker Inc., 1991:1–15.
37. HPFB Guidance for Industry. Bioequivalence requirements: Critical dose drugs, Ottawa, June 2006. http://www.hc-sc.gc.ca/hpfb-dgpsa/tpd-dpt. Accessed May 2009.
38. Shah VP, Yacobi A, Barr WH, et al. Evaluation of orally administered highly variable drugs and drug formulations. Pharm Res 1996; 13:1590–1594.
39. Midha KK, Rawson MJ, Hubbard JW. Individual and average bioequivalence of highly variable drugs and drug products. J Pharm Sci 1997; 86:1193–1197.
40. Tothfalusi L, Endrenyi L, Midha KK, et al. Evaluation of the bioequivalence of highly-variable drugs and drug products. Pharm Res 2001; 18:728–733.
41. Haidar SH, Davit B, Chen M-L, et al. Bioequivalence approaches for highly variable drugs and drug products. Pharm Res 2008; 25:237–241.
42. Yu LX, Amidon GL, Polli JE, et al. Biopharmaceutics classification system: The scientific basis for biowaiver extensions. Pharm Res 2002; 19:921–925.
43. Polli JE, Yu LX, Cook JA, et al. Summary workshop report: Biopharmaceutics classification system—implementation challenges and extension opportunities. J Pharm Sci 2004; 93:1375–1381.
44. FDA Guidance for Industry. Waiver of In Vivo Bioavailability and Bioequivalence Studies for Immediate Release Solid Oral Dosage Forms Based on a Biopharmaceutics Classification System. Washington, DC: CDER/FDA, August 2000. http://www.fda.gov/cder/guidance/index.htm. Accessed May 2009.
45. Alt A, Potthast H, Moessinger J, et al. Biopharmaceutical characterization of sotalol-containing oral immediate release drug products. Eur J Pharm Biopharm 2004; 58:145–150.
46. EMEA Concept Paper on BCS-Biowaiver. EMEA/CHMP/EWP/213035/2007, May, 24, 2007. http://www.emea.europa.eu/pdfs/human/ewp/21303507en.pdf. Accessed May 2009.
47. Blume HH, Schug BS. The biopharmaceutics classification system (BCS): Class III drugs—better candidates for BA/BE waiver? Eur J Pharm Sci 1999; 9:117–121.
48. Polli JE, Yu LX, Cook JA, et al. Summary workshop report: Biopharmaceutics Classification System—implementation challenges and extension opportunities. J Pharm Sci 2004; 93:1375–1381.
49. Jantratid E, Prakongpan S, Amidon GL, et al. Feasibility of biowaiver extension to biopharmaceutics classification system III drug products—cimetidine. Clin Pharmacokinet 2006; 45:385–399.
50. Wu C-Y, Benet LZ. Predicting drug disposition via application of BCS: Transport/absorption/elimination interplay and development of a biopharmaceutics drug disposition classification system. Pharm Res 2005; 22:11–23.
51. The Rules Govering Medicinal Products in the European Union, Volume 2: Notice to Applicants and Regulatory Guidelines for Medicinal Products for Human Use. http://ec.europa.eu/enterprise/pharmaceuticals/eudralex/vol2_en.htm. Accessed May 2009.
52. EC Notice to Applicants, volume 2C: Regulatory Guidelines—Guideline on Dossier Requirements for Type IA and IB Notifications, July 2006. http://ec.europa.eu/enterprise/pharmaceuticals/eudralex.
53. Karim A. Enantioselective assays in comparatve bioavailability studies of racemic drug formulations: Nice to know or need to know? J Clin Pharmacol 1996; 36:490–499.
54. Mehvar R, Jamali F. Bioequivalence of chiral drugs: Stereospecific versus non-stereospecific methods. Clin Pharmacokinet 1997; 33:122–141.

55. Boni JP, Korth-Bradley JM, Richards LS, et al. Chiral Bioequivalence: Effect of absorption rate on racemic etodolac. Clin Pharmacokinet 2000; 39:459–469.
56. Mehvar R, Jamali F. Implications of chirality for assessment of bioequivalence. In: Kanfer I, Shargel L, eds. Generic Product Development—Bioequivalence Issues. New York: Informa Healthcare, 2008:185–206.
57. EMEA Note for Guidance on the Clinical Requirements for Locally Applied, Locally Acting Products Containing Known Constituents. London: CPMP/EWP/239/95, 1995. http://www.emea.europa.eu/pdfs/human/ewp/023995en.pdf. Accessed May 2009.
58. EMEA Draft Guideline on the Requirements for Clinical Documentation for Orally Inhaled Products (OIP) including the Requirements for Demonstration of Therapeutic Equivalence between two Inhaled Products for Use in the Treatment of Asthma and Chronic Obstructive Pulmonary Disease (COPD). London: CPMP/EWP/4151/00 Rev. 1, October 18, 2007. http://www.emea.europa.eu/pdfs/human/ewp/415100enrev1.pdf. Accessed May 2009.
59. Clinical Investigation of Corticosteroids Intended for Use on the Skin, August 1987. http://www.emea.europa.eu/pfds/human/ewp/3cc26aen.pdf. Accessed May 2009.
60. EMEA Questions and Answer on Guideline Title: Clinical Investigation of Corticosteroids Intended for Use on the Skin. CHMP/EWP/21441/2006, November 16, 2006. http://www.emea.europa.eu/pdfs/human/ewp/2144106en.pdf. Accessed May 2009.
61. FDA Guidance for Industry. Topical Dermatologic Corticosteroids—In Vivo Bioequivalence. Washington, June 2, 1995. http://www.fda.gov/ohrms/dockets/dockets/04p0206/04p-0206-ref0001-08-FDA-Guidance-for-Industry-06–1995-vol3.pdf. Accessed May 2009.
62. HPFB Guidance for Industry. Conduct and Analysis of Bioavailability and Bioequivalence Studies—Part B: Oral Modified Release Formulations. Ottawa, 1996. http://www.hc-sc.gc.ca. Accessed May 2009.
63. Steinijans VW, Sauter R, Hauschke D, et al. Metrics to characterize concentration-time profiles in single- and multiple-dose bioequivalence studies. Drug Inf J 1995; 29:981–987.
64. Schultz H-U, Steinijans VW. Striving for standards in bioequivalence assessment: A review. Int J Clin Pharmacol Ther Toxicol 1992; 30(suppl 1):S1–S6.
65. FDA Guidance for Industry. Skin Irritation and Sensitization Testing of Generic Transdermal Drug Products. Washington, DC: CDER/FDA, December 1999. http://www.fda.gov/cder/guidance/index.htm. Accessed May 2009.
66. EMEA Questions & Answers document on the clinical development of fixed combinations of drugs belonging to different therapeutic classes in the field of cardiovascular treatment and prevention. London: EMEA/CHMP/EWP/191583/2005, June 2005. http://www.emea.europa.eu/pdfs/human/ewp/19158305en.pdf. Accessed May 2009.
67. Multisource (generic) pharmaceutical products: Guidelines on registration requirements to establish interchangeability. WHO Technical Report Series No. 937, 2006.
68. Gupta E, Barends DM, Yamashita E, et al. Review of global regulations concerning biowaivers for immediate release solid oral dosage forms. Eur J Pharm Sci 2006; 29:315–324.

6 India

Subhash C. Mandal
Directorate of Drugs Control, "NALANDA," Fartabad, Amtola, Kolkata, India

S. Ravisankar
GVK Biosciences Pvt. Ltd., Ameerpet, Hyderabad, India

INTRODUCTION

Although the concept of bioavailability (BA) and bioequivalence (BE) has been developed over the decades and such considerations are in practice in different research centers in academic institutions and industries in India for quite some time, BE testing was not made mandatory by Drug Regulatory Agencies, until the 1980s. The introduction of dissolution tests for seven products, chlorpromazine, digitoxin, digoxin, lithium carbonate, quinidine, tetracycline, and tolbutamide in the Indian Pharmacopoeia in 1985 (1), initiated the process of framing legislation for regulatory requirements of BE studies. After this phase (three years), BE studies became mandatory for all new drugs introduced on the markets in India, by incorporating Schedule Y of the Drugs and Cosmetics Act in 1988, followed by subsequent amendments of Schedule Y in 1989 and 2005 (2). The term "new drug" in India is defined under 122E of the Rule and includes both brand and generic.

India has a large number of BA and BE centers, mostly owned and operated by contract research organizations (CROs), largely in the private sector. More than 950 new drugs were approved between 1988 and July 2007 in India (3), following successful completion of necessary trials including BE and BA studies.

Patent Status on Pharmaceuticals

India switched over from a system of granting "process patents," to "product patents" from January 1, 2005 (4,5). The salient features of the product patent system are as follows:

1. Product patents have been extended to pharmaceuticals, which was not previously permitted as per the Patent Act of 1970.
2. Patent protection has been extended to 20 years from 7 years.
3. The Indian Act denied granting of a patent retrospective to the mailing date of the submission. Thus Indian companies that had been producing and selling patented medicines could not be subjected to patent infringement claims. The Act also provided that even after a patent had been granted, Indian companies would be able to continue production subject to payment of reasonable royalties to the patent holders. Subsection 2 of Section 5 of the act specifies the term "reasonable," but no explanation is given.

4. A most significant feature of the Act is that it provides protection from secondary patenting of the same chemical/pharmaceutical molecule. It forbids patenting of "salts, esters, polymorphs, metabolites, pure forms, particle size, isomers, mixtures of isomers, complexes, combinations and other derivatives of known substances unless they differ significantly in properties with regard to efficacy."

5. The Act also provides for pre- and postpatent objection of patents and describes 11 areas where one can raise objections on the granting of a patent. This includes objection to a patent based upon existing knowledge in the public domain (prior art). This holds good for domestic inventions and also for imported materials and according to the following "That if the invention so far as claimed in any claim of the complete specification was publicly known or publicly used in India before the priority date of the claim" [Section 25 (d)] such an application is not patentable.

6. The Act has now made a provision under Section 107 A (b), for the import of patented commodities from any part of the world, where it is cheaper, even though it is patented in India. This is known as parallel import. For this purpose, it will also not be required to obtain any authorization from the patentee. The Act simply says that "who is duly authorized under law to produce and sell or distribute the product" will become the source for Indian importers.

7. One of the most important areas of the Act is its provisions for compulsory licensing. The act clearly directs that a "'Compulsory' license is granted with a predominant purpose of supply in the Indian market and that the licensee may also export the patented product." The license shall also be granted to remedy a practice determined after judicial or administrative process to be anticompetitive. This particular clause may be carefully used to control exorbitantly high prices of patented products.

In the greater interest of a country, the compulsory license process empowers and allows a domestic company to produce a particular medicine if the patent holder company does not produce or supply the medicine. In contrast to this the Indian Act has designed the provisions of compulsory licensing, in a manner that is more suitable to the needs and traits of the Indian industry. Section 92A(1) of the Act states that a "Compulsory license will be available for manufacture and export of patented pharmaceutical products to any country having insufficient or no manufacturing capacity in the pharmaceutical sector for the concerned product to address public heath problems, provided a compulsory license has been granted by such country or such country has, by notification or otherwise, allowed importation of the patented pharmaceutical products from India." In other words, a country simply needs to announce, by notification, the need for importing any patented medicine from India. Such products can then be ordered from any Indian company for manufacturing and exporting if a compulsory license was granted to that company. The other important section of the Act is that a compulsory license may be requested on the grounds that the establishment or development of commercial activities in India is prejudiced. For such purpose, applicants have to make efforts to obtain a license from the patent holder on reasonable terms and conditions and when such efforts have not been successful within six months, they will be granted a compulsory license.

Generic Medicines

In India, generic medicines are those that are labeled with their generic names. There is no separate law for registering generic medicines. Drugs and drug products are classified as either (*i*) "new drugs" or (*ii*) drugs other than new drugs. However, in general "generic drugs" mean those that are no longer subject to patent protection and are being marketed by their generic name. To understand the legislation the concept of new drugs should be clear.

According to Section 122E of Drugs and Cosmetics Act new drugs are defined as the following:

(a) A drug, as defined in the Act including bulk drug substance that has not been used in the country to any significant extent (no clarification provided by the Act—it is the prerogative of the licensing authority) under the conditions prescribed, recommended or suggested in the labeling thereof and has not been recognized as effective and safe by the licensing authority under Rule 21 for the purposes claimed; provided that the limited use, if any, has been with the permission of the licensing authority.

(b) A drug already approved by the licensing authority mentioned in Rule 21 for certain claims, which are now proposed to be marketed with modified or new claims, namely, indications, dosage, dosage form (including sustained release dosage form), and route of administration.

(c) A fixed dose combination of two or more drugs, individually approved earlier for certain claims, which are now proposed to be combined for the first time in a fixed ratio, or if the ratio of ingredients in an already-marketed combination is proposed to be changed, with certain claims, viz., indications, dosage, dosage form (including sustained release dosage form) and route of administration.

It also explained that all vaccines shall be new drugs unless certified otherwise by the licensing authority under Rule 21 and a new drug shall continue to be considered as a new drug for a period of four years from the date of its first approval or its inclusion in the Indian Pharmacopoeia, whichever is earlier (6).

All new drugs are required to comply with the provisions and requirements of Schedule Y for registration in India. However some relaxation has been granted for fixed dose combinations (FDC) to be registered in India (vide Annexure VI of Schedule Y). For this purpose fixed dose combinations are categorized into the following four groups:

(a) The first group of FDC includes those in which one or more of the active ingredients are a new drug. Such FDC are treated in the same way as any other new drug, for both clinical trials and marketing permission.

(b) The second group of FDC includes those in which the active ingredients are already approved and marketed individually and are combined for the first time, for a particular claim and where the ingredients are likely to have significant interaction of a pharmacodynamic or pharmacokinetic nature For permission to carry out clinical trials with such FDCs, a summary of available pharmacological, toxicological, and clinical data on the individual ingredients should be submitted, along with the rationale for combining them in the proposed ratio. In addition, acute toxicity data (LD 50) and pharmacological data should also be submitted on the individual ingredients as well as their combination in the proposed ratio. If clinical trials have been carried out with the FDC in other countries, reports of such trials should be submitted. If the FDC is marketed abroad, the regulatory trials performed should be stated.

For marketing permission, the reports of clinical trials carried out with the FDC in India should be submitted. The nature of the trials depends on the claims to be made and the data already available.

(c) The third group of FDCs includes those that have already been marketed, but in which it is proposed either to change the ratio of active ingredients or to make a new therapeutic claim. For such FDCs, the therapeutic rationale should be submitted to obtain permission for clinical trials and the reports of trials should be submitted to obtain marketing permission. The nature of the trials will depend on the claims to be made and the data already available.

(d) The fourth group of FDCs includes those whose individual active ingredients have been widely used for a particular indication for years, their concomitant use is often necessary and no claim is proposed to be made other than convenience. Also, it must be a stable acceptable dosage form and whose ingredients are unlikely to have significant interaction of a pharmacodynamic or pharmacokinetic nature.

No additional animal or human data are generally required for those FDCs, and marketing permission may be granted if the FDC has an acceptable rationale.

If any drug does not fall under the ambit of a new drug, licenses could be issued by the state licensing authority. India has a two-tier regulatory system. One is Central Drugs Standard Control Organization (CDSCO) under the Government of India having certain powers vested in them and each state has its own drug regulatory system having certain powers. CDSCO is authorized to approve new drugs as defined under Rule 122E but the state drugs control agency cannot grant new drug licenses. Therefore it is clear that a BA/BE study is needed for manufacturing and import of new drugs, where the relevant product is compared against the innovator product.

Several different provisions and sections of the Drugs and Cosmetics Rules have been modified and amended over the years. A new provision has been introduced for the registration of manufacturing premises of foreign drug manufacturers and for drug products, before their import to India. These amendments have been in force since January 1, 2003. The notification also introduced additional provisions, some of which are as follows:

1. An increase in import license fees.
2. Increased validity period of license.
3. Deletion of the clause of exemption relating to the requirement of import license for bulk drug for actual users.
4. A minimum requirement of 60% of retained shelf life for imported drug.
5. Conditions for the importation of small amounts of new drug products by Government hospitals for treating their own patients.

Under the existing regulations, foreign manufacturers must apply for registration certificates for both their manufacturing premises and for individual drugs to be imported. The applications for these registrations can be filed by the authorized agents of foreign firms in India, along with the documents required to be accompanied by the registration application, as specified in the relevant amendments. The registration certificates are valid for three years from the date from which they were issued.

A fee of US $1500 is charged for the registration of overseas manufacturer's premises and a fee of US $1000 is charged for each individual drug. The rules now provide for inspection of the premises of a foreign manufacturer by Indian Drug Authorities, whenever so required. In such cases, an additional fee of US $5000 is charged. The rules also provide for payment of testing charges by registration holders. The foreign manufacturer or his authorized agent in India is liable to report any change in the manufacturing and testing process of a drug. However, no registration certificate is required for an inactive bulk substance used as a pharmaceutical aid for the manufacture of the drug formulation. Registration may be suspended or cancelled in the event of any violation of the conditions of registration. The new registration and import license scheme also covers diagnostic kits, viz. HIV I and II, hepatitis B surface antigen, and blood group detection reagents.

According to these rules, an import license is required for all types of drugs. An import license application submitted on Form 10 is granted after the completion of the registration of overseas manufacturers and their specific drug product(s) to be imported. The import licensee for specific drugs is valid for three years from the date on which these were granted. The import license fee has been fixed at Rs. 1000 for a single drug and at the rate of Rs. 100 for any additional drug. The fee for "import licenses for test and analysis" of a drug has been kept at Rs. 100 for a single drug and at the rate of Rs. 50 for each additional drug. Exemption from import licenses for the import of bulk drugs by the formulators for actual use under Schedule D has been deleted. A provision has been made that only drugs with a minimum of 60% of retained shelf life will be allowed to be imported in the country, for example, if a drug product has a shelf life of 30 months, it will not be allowed to be imported into India more than 12 months after manufacture.

Separate provision has been made to enable the Government hospitals to import small quantities of essential new drugs for the treatment of their own patients. The fee for such import licenses has been kept at Rs. 100 for a single drug and at the rate of Rs. 50 for each additional drug.

The following guidelines describe the requirements of an application for permission to import and/or manufacture new drugs for sale or to conduct clinical trials as per the Drugs and Cosmetics Rules (2):

GUIDELINES FOR THE APPLICATION TO IMPORT AND/OR MANUFACTURE NEW DRUGS FOR SALE OR TO CONDUCT CLINICAL TRIALS

(1) Application for permission to import or manufacture new drugs for sale or to undertake clinical trials shall be made on Form 44 accompanied with the following data in accordance with the appendices:
   (i) Chemical and pharmaceutical information as prescribed in item 2 of Appendix I.
   (ii) Animal pharmacology data as prescribed in item 3 of Appendix I and Appendix IV.

(a) Specific pharmacological actions as prescribed in item 3.2 of Appendix I, and demonstrating therapeutic potential for humans shall be described according to the animal models and species used. Wherever possible, dose–response relationships and ED50 shall be submitted. Special studies conducted to elucidate mode of action shall also be described (Appendix IV).

(b) General pharmacological actions as prescribed in item 3.3 of Appendix I and item 1.2 of Appendix IV.

(c) Pharmacokinetic data related to the absorption, distribution, metabolism, and excretion of the test substance as prescribed in item 3.5 of Appendix I. Wherever possible, the drug effects shall be correlated to the plasma drug concentrations.

(iii) Animal toxicology data as prescribed in item 4 of Appendix I and Appendix III.

(iv) Human Clinical Pharmacology Data as prescribed in items 5, 6, and 7 of Appendix I and as stated below

(a) For new drug substances discovered in India, clinical trials are required to be carried out in India from Phase I and data should be submitted as required under items 1, 2, 3, 4, 5 (data, if any, from other countries), and 9 of Appendix I.

(b) For new drug substances discovered in countries other than India, Phase I data as required under items 1, 2, 3, 4, 5 (data from other countries), and 9 of Appendix I should be submitted along with the application. After submission of Phase I data generated outside India to the licensing authority, permission may be granted to repeat Phase I trials and/or to conduct Phase II trials and subsequently Phase III trials concurrently with other global trials for that drug. Phase III trials are required to be conducted in India before permission to market the drug in India is granted.

(c) The data required will depend upon the purpose of the new drug application. The number of study subjects and sites to be involved in the conduct of a trial will depend upon the nature and objective of the study. Permission to carry out these trials shall generally be given in stages, considering the data emerging from earlier phase(s).

(d) Application for permission to initiate a phase of a clinical trial should also accompany the Investigator's brochure, proposed protocol (Appendix X), case record form, study subject's informed consent document(s) (Appendix V), investigator's undertaking (Appendix VII) and ethics committee clearance, if available, (Appendix VIII).

(e) Reports of clinical studies submitted under items 5 to 8 of Appendix I should be in consonance with the format prescribed in Appendix II of this schedule. The study report shall be certified by a principal investigator or, if no principal investigator is designated, then by each of the investigators participating in the study. The certification should acknowledge the contents of the report, the accurate presentation of the study as undertaken and express agreement with the conclusions. Each page should be numbered.

(v)   Regulatory status in other countries as prescribed in item 9.2 of Appendix I, including information in respect of restrictions imposed, if any, on the use of the drug in other countries, for example, dosage limits, exclusion of certain age groups, warning about adverse drug reactions, etc. (item 9.2 of Appendix I). Likewise, if the drug has been withdrawn in any country by the manufacturer or by regulatory authorities, such information should also be furnished along with the reasons and their relevance, if any, to India. This information must continue to be submitted by the sponsor to the licensing authority during the course of marketing of the drug in India.

(vi) The full prescribing information should be submitted as part of the new drug application for marketing as prescribed in item 10 of Appendix I. The prescribing information (package insert) shall comprise the following sections: generic name; composition; dosage form/s; indications; dose and method of administration; use in special populations (such as pregnant women, lactating women, pediatric patients, geriatric patients, etc.); contraindications; warnings; precautions; drug interactions; undesirable effects; overdose; pharmacodynamic and pharmacokinetic properties; incompatibilities; shelf life; packaging information; and storage and handling instructions. All package inserts, promotional literature, and patient education material subsequently produced are required to be consistent with the contents of the approved full prescribing information. The drafts of label and carton texts should comply with provisions of Rules 96 and 97. After submission and approval by the licensing authority, no changes in the package insert shall be effected without such changes being approved by the licensing authority;

(vii) Complete testing protocol/s for quality control (QC) testing together with a complete impurity profile and release specifications for the product as prescribed in item 11 of Appendix I should be submitted as part of new drug application for marketing. Samples of the pure drug substance and finished product are to be submitted when desired by the regulatory authority.

(2) If the study drug is intended to be imported for the purposes of examination, test or analysis, the application for import of small quantities of drugs for such purpose should also be made on Form 12.

(3) For drugs indicated in life-threatening/serious diseases or diseases of special relevance to the Indian health scenario, the toxicological and clinical data requirements may be abbreviated, deferred or omitted, as deemed appropriate by the licensing authority.

DESIGN AND CONDUCT OF BIOEQUIVALENCE STUDIES FOR ORALLY ADMINISTERED DRUG PRODUCTS

A draft guideline for the conduct of Bioequivalence studies was released by the Directorate General of Health Services in 2003 and after review of the comments from experts, the final "Guidelines for the conduct of Bioavailability and Bioequivalence studies" was released in March 2005 (7). In addition to this guideline, it is mandatory to follow Indian GCP (good clinical practice) (8), pertinent requirements of Schedule Y (9) of the Drugs and Cosmetics Act of India and The

Ethical Guidelines for biomedical research on human subjects issued by Indian Council of Medical Research, India (10).

BE studies are necessary for the following drugs products:

For certain drugs and dosage forms, in vivo documentation of equivalence, through either a BE study, a comparative clinical pharmacodynamic study, or a comparative clinical trial, is regarded as especially important. These include the following:

a. Oral immediate release drug formulations with systemic action when one or more of the following criteria apply:
    i. Indicated for serious conditions requiring assured therapeutic response.
    ii. Narrow therapeutic window/safety margin; steep dose–response curve.
    iii. Pharmacokinetics complicated by variable or incomplete absorption or narrow absorption window, nonlinear pharmacokinetics, presystemic elimination/high first-pass metabolism greater than 70%.
    iv. Unfavorable physicochemical properties, for example, low solubility, instability, metastable modifications, poor permeability, etc..
    v. Documented evidence of BA problems related to the drug or drugs of similar chemical structure or formulations.
    vi. Where a high ratio of excipients to active ingredients exists.
b. Nonoral and nonparenteral drug formulations designed to act by systemic absorption (such as transdermal patches, suppositories, etc.).
c. Sustained or otherwise modified-release drug formulations designed to act by systemic absorption.
d. Fixed dose combination products with systemic action.
e. Nonsolution pharmaceutical products, which are for nonsystemic (oral, nasal, ocular, dermal, rectal, vaginal, etc., applications) and are intended to act without systemic absorption. In these cases, the BE concept is not suitable and comparative clinical or pharmacodynamic studies are required to prove equivalence. There is a need for drug concentration measurements to assess unintended partial absorption. BE documentation is also needed to establish links between:
    i. Early and late clinical trial formulations.
    ii. Formulations used in clinical trials and stability studies, if different.
    iii. Clinical trial formulations and "to be" marketed drug products.
    iv. Other comparisons, as appropriate.

There are no specific guidances for the following products:

- Topical corticosteroid products.
- All other topical products (except transdermals) for external application and not intended to be absorbed, for example, an antifungal cream, topical creams/ointments for use in the vagina).
- Inhalation products.
- Orally administered products not intended to be absorbed into the systemic circulation, for example, sulindac, mesalamine, etc.

Study Design

A BE study should be designed in a manner that any formulation effects can be distinguished from other effects. Typically, if two formulations (one test and the other reference) are to be compared, a two-period, two-sequence cross-over design should be chosen, with two phases of treatment separated by an adequate washout period. Normally the washout period should be equal to or more than five half-lives of the incorporated drug. Other designs such as a parallel design for long half-life drug substances and a replicate design for drug substances with highly variable disposition can be used.

Single-dose studies are generally sufficient although a steady-state study design may be considered under the following circumstances:

a. Drugs that follow dose- or time-dependent pharmacokinetics.
b. Some modified-release products (in addition to a single-dose study).
c. When problems of sensitivity preclude plasma drug concentration measurements after single-dose administration.
d. If intraindividual variability in the plasma drug concentration or disposition is reduced at steady state when compared to variability in a single-dose study.

Studies evaluating food effects are required when there is a possibility that food may affect the BA of the drug. Food effect BA studies focus on effects of food on the release of the drug substance from the drug product as well as on the absorption of the drug substance.

Subjects

Number of Subjects
The number of subjects required for a BE study is determined by considerations such as the error variance from pilot study data, the expected deviation of the test product from the reference product, significance level (usually 0.05) and the power of the study (should be more than 80%). The minimum number of subjects should not be less than 16, unless justified for ethical reasons. Additional subjects can be recruited to allow for possible dropouts or removal from the study. The withdrawn/dropout subjects can be replaced by a substitute (standby) provided that a substitute subject follows all the study requirements.

Sequential or add-on studies are acceptable in specific cases, for example, where a large number of subjects are required or where the results of the study do not convey adequate statistical significance. In all cases, the final statistical analysis must include data from all subjects, and reasons for not including partial data as well as the non–included data must be documented in the final report.

Subject Selection
The selection of subjects and standardization is important to minimize intra- and interindividual variations. The studies should be carried out in healthy male or female volunteers. The choice of gender should be consistent with usage and safety criteria. Women of child-bearing potential should be required to give assurance that they are neither pregnant nor likely to become pregnant until after the study. This is confirmed by pregnancy tests immediately prior to the first and last dose of the study. Women taking contraceptive drugs should not normally be included in the studies. If the drug product is intended to be used in both sexes, attempts should be made to include similar proportions of males and females in the studies. The choice of subject may be narrowed when a drug represents a

potential hazard in one group of users, for example, studies on teratogenic drugs should be conducted only on males.

For a drug product to be used in a geriatric population, subjects of 60 years of age or older should be included.

Volunteers are screened for suitability by means of a comprehensive medical examination including clinical laboratory tests, extensive review of medical history, use of oral contraceptives, alcohol intake, smoking, and use of drugs of abuse. Additional medical investigations may be required before, during, and after the study, based upon the therapeutic class of drug and safety profile.

Use of Patients, *i.e., Instead of Healthy Subjects*
For drugs where the risk of toxicity or side effects is significant, studies are to be carried out in patients whose disease state is stable.

Phenotyping/Genotyping
While designing a study protocol, adequate care should be taken to consider pharmacogenomic issues in the context of the Indian population. Phenotyping and/or genotyping of subjects should be considered for exploratory BA studies and all studies using parallel group design. It may also be considered in crossover studies for safety or pharmacokinetic reasons. If a drug is known to be subject to a major genetic polymorphism, studies could be performed in panels of subjects of known phenotype or genotype for the polymorphism in question.

Standardization of Study Conditions
Standardization of the study environment is necessary to minimize variability. Diet, fluid intake, postdosing postures, exercise, sampling schedules, etc., are to be stated in the protocol and must be complied with during the conduct of the study. The subjects should abstain from smoking, drinking alcohol, coffee, tea, and xanthine-containing foods and beverages and fruit juices during the study and at least for 48 hours before its commencement.

Fasting and Fed Study Considerations
In a single-dose fasting study, the dose should be administered after an overnight fast of at least 10 hours, with continuation for a further successive four hours of fasting after dosing. In a multiple dose fasting study, when the evening dose is to be administered, fasting of two hours before and two hours after dosing is acceptable. When the drug is recommended to be administered along with food or when the drug is a modified-release product, a fed-state study is to be carried out as well as a fasting study.

For fed-state studies, a high-fat breakfast meal should be designed to provide 950 to 1000 kcal. Fifty percent of the calories must be from fat, 15% to 20% from proteins and the rest 30% to 35% from carbohydrates. Because of the vast ethnic and cultural variations in the Indian subcontinent, any single-standard high-fat breakfast is not recommended, but can be specified in the protocol and justified. The high-fat breakfast must be consumed 15minutes before dosing.

Blood/Urine Sample Collection and Times
The duration of blood sample collection in a single-dose study for an immediate release product should be extended for three elimination half-lives. The

sampling should continue until the AUC extrapolated from the time of the last measured drug concentration to infinity time is only a small percentage (normally less than 20%) of the total AUC. Truncated AUC may be used when the presence of enterohepatic recycling does not allow the terminal elimination rate constant to be accurately calculated. There should be at least three sampling points in the absorption phase, three to four around the projected T_{max}, and four points in the elimination phase (intervals between successive sampling points should not be longer than the half-life of the drug, in general). The number of points used to calculate the terminal elimination rate constant should preferably be visually determined by inspection of a semilogarithmic plot of drug concentration versus time. For urinary excretion studies, samples must be collected up to seven or more half-lives.

Parameters To Be Investigated
BA/BE evaluations are based on the measurement of the concentrations of the drug or its metabolite(s). Measurement of the active or inactive metabolites may be necessary in certain situations:

(a) The concentration of the drug may be too low to be accurately measured in the biological matrix.
(b) Unstable drugs.
(c) Drugs with a very short half-life.
(d) The API is a prodrug.

Racemates should be measured by using an achiral method. Measurement of individual enantiomers is only recommended when

(a) the enantiomers exhibit different pharmacokinetic or pharmacodynamic characteristics,
(b) primary efficacy/safety resides with the minor enantiomer, and
(c) at least one of the enantiomers undergoes nonlinear absorption.

Blood/Plasma/Serum Concentration Versus Time Profiles
The drug plasma–time concentration curve is mostly used to assess the rate and extent of absorption of the study drug. These include pharmacokinetic parameters such as C_{max}, T_{max}, AUC_{0-t}, and $AUC_{0-\infty}$. For steady-state studies, C_{max}, C_{min}, and $AUC_{0-\tau}$ and the degree of fluctuation should be calculated.

Urinary Excretion Profiles
When it is not possible to assess the pharmacokinetic parameters in blood, plasma, or serum, urinary excretion profiles may be used to compare the BA of drug products.

Pharmacodynamic Studies
Studies in healthy volunteers or patients using pharmacodynamic parameters may be used for establishing BE between two pharmaceutical products. These studies may become necessary if quantitative analysis of the drug and/or metabolite(s) in plasma or urine cannot be made with sufficient accuracy and sensitivity. Furthermore, pharmacodynamic studies in humans are required if measurements of drug concentrations cannot be used as surrogate end points

for the demonstration of efficacy and safety of the particular pharmaceutical product; for example, for topical products where the drug is not intended to be absorbed into the systemic circulation.

When only pharmacodynamic data are used to demonstrate BE, the applicant should outline what other methods were tried and why they were found unsuitable.

The following requirements should be considered when planning, conducting, and assessing the results from a pharmacodynamic study:

(i) The response measured should be a pharmacological or therapeutic effect, which is relevant to the claims of efficacy and/or safety of the drug.

(ii) The methodology adopted for carrying out the study should be validated for precision, accuracy, reproducibility, and specificity.

(iii) Neither the test nor the reference product should produce a maximal response in the course of the study, since it may be impossible to distinguish differences between formulations given in doses that produce such maximal responses. Investigation of dose–response relationship may become necessary.

(iv) The response should be measured quantitatively under double-blind conditions and be recorded with an instrument-produced or instrument-recorded fashion on a repetitive basis to provide a record of pharmacodynamic events that are a substitute for plasma concentrations.

If such measurements are not possible, recordings on visual-analog scales may be used. In instances, where data are limited to qualitative (categorized) measurements, appropriate special statistical analyses will be required.

(v) Nonresponders should be excluded from the study by prior screening. The criteria by which responders versus nonresponders are identified must be stated in the protocol.

(vi) Where an important placebo effect can occur, comparison between products can only be made by a prior consideration of the placebo effect in the study design. This may be achieved by adding a third period/phase with placebo treatment, in the design of the study.

(vii) A cross-over or parallel study design should be used, as appropriate.

(viii) When pharmacodynamic studies are to be carried out on patients, the underlying pathology and natural history of the condition should be considered in the study design. There should be knowledge of the reproducibility of the baseline conditions.

(ix) In studies where continuous variables can be recorded, the time course of the intensity of the drug action can be described in the same way as in a study where plasma concentrations are measured. From this, parameters can be derived, which describe the area under the effect–time curve, the maximum response and the time when the maximum response occurred.

(x) Statistical considerations for the assessment of the outcomes are in principle, the same as in pharmacokinetic studies.

(xi) A correction for the potential non-linearity of the relationship between dose and area under the effect–time curve should be made on the basis of the outcome of the dose-ranging study.

The conventional acceptance range as applicable to pharmacokinetic studies and BE is not appropriate (too large) in most cases. This range should therefore be defined in the protocol on a case-to-case basis.

Analysis of Samples (Bioanalysis)

Bioanalysis involves the measurement of drug or metabolite in a suitable matrix for making pharmacokinetic assessments. Bioanalytical methodology is divided into two distinct phases, viz., prestudy phase, where method validation is done on the biological matrix (e.g., human plasma) and spiked plasma samples and the study phase, which involves the analysis of subject samples by the use of a validated method where the stability, precision, and accuracy of the assay have been confirmed.

The following are the validation parameters to be established in the prestudy phase, prior to the analysis of subject samples.

(i) *Specificity/selectivity:* The method should be capable of demonstrating that there is no interference with the assay due to the presence of endogenous compounds, degradation products, other drugs or metabolites present in the subject samples.

(ii) *Sensitivity:* Sensitivity of an analytical method is indicated by the lowest limit of quantification (LLOQ) at which the concentration of the analyte is determined with stated precision and accuracy. This is established based on the intra- and interday coefficients of variation (CV), which should not usually be greater than 20%. Concentration values between the LOD and LOQ are identified as BLQ (below quantification limit).

(iii) *Accuracy and Precision:* The accuracy of an analytical method describes the closeness of mean test results obtained by the method to the true value (concentration) of the analyte. Precision is the reproducibility of the individual assays. The precision and accuracy should be established using QC samples prepared in three different concentrations (low, medium, and high). Low QC samples in the vicinity of the lowest concentration to be measured, high QC samples near the C_{max}, and the medium QC samples at an intermediate value to low and high QC samples. The acceptance criteria for intra-assay and interassay precision should not be more than 20% near the LOQ and not more than 15% at other levels. The accuracy of the method should be established in conjunction with precision experiments. The percentage of accuracy at all levels of QC samples should be within ±15% of nominal concentrations.

(iv) *Range and Linearity:* For establishing linearity over a range of concentrations expected in the subject samples, a standard curve should be constructed by using at least five concentrations. More points are necessary in order to establish linearity, when the response function is nonlinear. Extrapolation of concentration beyond calibration curve should not be made.

(v) *Recovery:* The recovery of an analyte in an assay is the detector response obtained from an amount of the analyte added to and extracted from the biological matrix, compared with the detector response obtained for the true concentration of the pure authentic standard. Recovery experiments should be performed by comparing the analytical results for extracted samples at three concentrations (low, medium, and high) with unextracted standards that represent 100% recovery. If the recovery is low, alternate methods should be applied as such methods are prone to be inconsistent. Recovery of an internal standard, when used, should also be established.

(vi) *Stability:* Stability of analytes in the matrix under conditions of the experiments should be established including long-term stability.

At least three freeze–thaw cycles should be assessed. Sorption of drug due to sampling container and stopper should also be investigated. The analytical system stability should be established so that the system remains stable throughout the time course of assay. A suggested design is to run standards at the beginning and end of each run.

Choice of Reference Product

As per the CDSCO guidelines a reference product is a pharmaceutical product, which is identified by the licensing authority (Drugs Controller General of India) as the "designated reference product" and contains the same active ingredient(s) as the new drug (brand/innovator drug product). The designated reference product (http://cdsco.nic.in/listofdrugapprovedmain.html) will normally be the innovator's product. An applicant seeking approval to market a generic equivalent must refer to the designated reference product to which all generic versions must be shown to be bioequivalent. For subsequent new drug product applications in India the licensing authority may, however approve another Indian product (which was studied in comparison with an international innovator product), as designated reference product.

Study Products

The investigational medicinal product (test product) should be manufactured from a production batch. The products should be stored as per storage conditions stated on the label. Randomly selected samples from the shipment by the manufacturer or sponsor should be used for the study. The number of units procured should be twice the number required for the tests to be carried out in vivo as well as for the number of in vitro tests. These samples must be retained (retention samples) under the recommended storage conditions in the original containers for a period of three years after the conduct of study or one year after the period of expiry whichever is earlier. This is to ensure the samples are representative of the ones sent by the sponsor and used for all the tests. The reserve samples should be stored in an area segregated from the area where testing is conducted and with access limited to authorized personnel.

Data Analysis

The main concern in a BE assessment is to limit the consumer's risk, that is, erroneously accepting BE, and also at the same time, minimize the manufacturer's risk, that is, erroneously rejecting BE. This is done by using appropriate statistical methods for data analysis and adequate sample size.

Statistical Analysis

The statistical procedure should be specified in the protocol. In the case of BE studies, the procedures should lead to a decision scheme, which is symmetrical with respect to the two formulations (i.e., leading to the same decision whether the new formulation is compared to the reference product or the reference product to the new formulation). The statistical analysis (e.g., ANOVA) should take into account sources of variation that can be reasonably assumed to have an effect on the response.

The 90% confidence interval (CI) for the ratio of the population means (test/reference) and the two one-sided t test with the null hypothesis of non-BE at the 5% significance level for the parameter under consideration are applied for assessing BE.

To meet the assumption of normality of data underlying the statistical analysis, logarithmic transformation of the data should be carried out for the pharmacokinetic parameters C_{max} and AUC before performing statistical analysis.

The analysis of T_{max} is desirable if it is clinically relevant. The parameter T_{max} should be analyzed using nonparametric methods. In addition to the above, summary statistics such as minimum, maximum, and ratio of pharmacokinetic parameters should be given in the report.

The study protocol should specify the methods for identifying biologically implausible outliers. Post hoc exclusion of outliers is not recommended. A scientific explanation should be provided to justify the exclusion of a volunteer from the analysis.

Acceptance Criteria

Single-Dose Studies
To establish BE, the calculated 90% CI for AUC and C_{max} should fall within the BE range, usually 80% to 125%. This is equivalent to the rejection of the two one-sided t test with the null hypothesis of non-BE at a 5% level of significance. The nonparametric 90% CI for T_{max} should lie within a clinically acceptable range.

Tighter limits for permissible differences in BA may be required for drugs that have

(i)   a narrow therapeutic index,
(ii) a serious, dose-related toxicity,
(iii) a steep dose/effect curve, or
(iv) nonlinear pharmacokinetics within the therapeutic dose range.

A wider acceptance range may be acceptable if it is based on sound clinical justification.

Steady-State Studies Involving Controlled/Modified-Release Dosage Forms
The following special considerations apply to modified-release drug products and for the purpose of these guidelines, modified-release products include the following:

(i)   Delayed release
(ii) Sustained release
(iii) Mixed immediate and sustained release
(iv) Mixed delayed and sustained release
(v)   Mixed immediate and delayed release

Generally, these products should

(i)   act as modified-release formulations and meet the label claim(s),
(ii) preclude the possibility of any dose dumping effect,

(iii) demonstrate a significant difference between the performance of a modified-release product and the conventional release product when used as the reference product,

(iv) provide a therapeutic performance comparable to the reference immediate release formulation administered by the same route in multiple doses (of an equivalent daily amount) or to the reference modified-release formulation,

 (v) produce consistent pharmacokinetic performance between individual dosage units, and

(vi) produce plasma levels that lie within the therapeutic range (where appropriate) for the proposed dosing intervals at steady state.

If all of the above-mentioned conditions are not met but the applicant wishes to submit the formulation for consideration, justification for this should be provided.

Study Parameters

BA data should be obtained for all modified-release drug products although the type of studies required and the pharmacokinetic parameters that should be evaluated may differ depending on the active ingredient involved.

Factors to be considered include whether or not the formulation represents the first market entry of the drug substance, and the extent of accumulation of the drug after repeated dosing.

If the formulation is the first market entry of the drug substance, the product's pharmacokinetic parameters should be determined. If the formulation is a second or subsequent market entry then comparative BA studies using an appropriate reference product, should be performed. For the first generic drug product, the international innovator drug product should be used as the reference product but for subsequent products, it may be the reference product approved earlier, which was equivalent to the international innovator product.

Study Design

The study design should be a single dose or single and multiple dose based on the modified-release products that are likely to accumulate or unlikely to accumulate, both in the fasted and nonfasting state. If the effect of food on the reference product is not known (or it is known that food affects its absorption), two separate two-way cross-over studies, a fasted and a fed study should be carried out.

If it is known with certainty (e.g., from published data) that the reference product is not affected by food, then a three-way cross-over study may be appropriate with

* the reference product in the fasting state,
* the test product in the fasted state, and
* the test product in the fed state.

Requirements for Modified-Release Formulations Unlikely to Accumulate

This section outlines the requirements for modified-release formulations that are used at a dose interval that is not likely to lead to accumulation in the body ($AUC_{0-\tau} / AUC_{0-\infty} \geq 0.8$).

When the modified-release product is the first market entry of that type of dosage form, the reference product should normally be the innovator's immediate release formulation. The comparison should be between a single dose of the modified-release formulation and doses of the immediate-release formulation, which it is intended to replace. The latter must be administered according to the established dosing regimen.

When the modified-release product is the second or subsequent entry on the market, comparison should be with the reference modified-release product for which BE is claimed.

Studies should be performed using single-dose administration in the fasting state as well as following an appropriate meal at a specified time. The following pharmacokinetic parameters should be calculated from plasma (or relevant biological matrix) concentrations of the drug and/or major metabolite(s): $AUC_{0-\tau}$, AUC_{0-t}, $AUC_{0-\infty}$, and C_{max} (where the comparison is with an existing modified-release product), and k_{el}. The 90% CI calculated using log-transformed data for the ratios (test/reference) of the geometric mean AUC (for both $AUC_{0-\tau}$ and AUC_{0-t}) and C_{max} (where the comparison is with an existing modified-release product) should generally be within the range 80% to 125% both in the fasting state and following the administration of an appropriate meal at a specified time before taking the drug.

The pharmacokinetic parameters should support the claimed dose delivery attributes of the modified-release dosage form.

Requirements for Modified-Release Formulations Likely to Accumulate

This section outlines the requirements for modified-release formulations that are used at dose intervals that are likely to lead to accumulation ($AUC_{0-\tau}/AUC_{0-\infty}$ < 0.8).

When a modified-release product is the first market entry of the modified-release type, the reference formulation is normally the innovator's immediate release formulation. Both a single dose and steady-state doses of the modified-release formulation should be compared with doses of the immediate release formulation which it is intended to replace. The immediate-release product should be administered according to the conventional dosing regimen.

Studies should be performed with single-dose administration in the fasting state as well as following an appropriate meal. In addition, studies are required at steady state. The following pharmacokinetics parameters should be calculated from single-dose studies: $AUC_{0-\tau}$, AUC_{0-t}, $AUC_{0-\infty}$, C_{max} (where the comparison is with an existing modified-release product), and k_{el}. The following parameters should be calculated from steady-state studies: $AUC_{0-\tau,(ss)}$, C_{max}, C_{min}, C_{pd}, and degree of fluctuation.

When the modified-release product is the second or subsequent modified-release entry, single-dose and steady-state comparisons should normally be made with the reference modified-release product for which BE is claimed. The 90% CI for the ratio of geometric means (test/reference) of AUC (for both $AUC_{0-\tau}$ and AUC_{0-t}) and C_{max} (where the comparison is with an existing modified-release product) determined using log-transformed data should generally be within the range 80% to 125% when the products are compared after single-dose administration in both the fasting state and the fed state.

The 90% CI for the ratio of geometric means (test/reference drug) for $AUC_{0-\tau(ss)}$, C_{max}, and C_{min} determined using log-transformed data should generally be within the range 80% to 125% when the formulations are compared at steady state.

The pharmacokinetic parameters should support the claimed attributes of the modified-release dosage form. Pharmacodynamic data may reinforce or clarify interpretation of differences in the plasma concentration data. Where these studies do not show BE, comparative efficacy, and safety data may be required for the new product.

Reporting of Results

The BE report should give complete documentation with respect to the protocol, conduct and evaluation of the study. The report should include (as a minimum) the following information:

4.10.1. Table of contents

4.10.2. Title of the study

4.10.3. Names and credentials of responsible investigators

4.10.4. Signatures of the principal and other responsible investigators authenticating their respective sections of the report

4.10.5. Site of the study and facilities used

4.10.6. The period of dates over which the clinical and analytical steps were conducted

4.10.7. Names, batch numbers, and expiry date of the products compared (for the test product, an expiry date may not be available prior to conducting the study)

4.10.8. A signed declaration that this product is identical to that intended for marketing

4.10.9. Results of assays and other pharmaceutical tests (e.g., physical description, dimensions, mean weight, weight uniformity, and comparative dissolution) carried out on the batches of the various products

4.10.10. Full protocol for the study including a copy of the Informed Consent Form and criteria for inclusion/exclusion or withdrawal of subjects

4.10.11. Report of protocol deviations, violations

4.10.12. Documentary evidence that the study was approved by an independent ethics committee and was carried out in accordance with good clinical practice/good laboratory practice.

4.10.13. Demographic data of subjects

4.10.14. Names and addresses of subjects

4.10.15. Details of and justification for protocol deviations

4.10.16. Details of dropout and withdrawals from the study should be fully documented and accounted for

4.10.17 Details of analytical methods used, full validation data, QC data, and criteria for accepting or rejecting assay results

4.10.18 Representative chromatograms (normally 20%) covering the whole concentration range for all, standard and QC samples as well as specimens analyzed

4.10.19. Sampling schedules and deviations of the actual times from the scheduled times

4.10.20. Details of how pharmacokinetic parameters were calculated

4.10.21. Documentation related to statistical analysis

   (i)   Randomization schedule
   (ii)   Plasma concentration and time points of volunteers for test and reference products
   (iii) AUC_{0-t}, $AUC_{0-\infty}$, C_{max}, T_{max}, $K_{el,}$ and $t_{1/2}$ for test and reference products
   (iv)   Logarithmic transformed pharmacokinetic parameters used for BE demonstration
   (v)   ANOVA for AUC_{0-t}, $AUC_{0-\infty}$, and C_{max}
   (vi)   Intersubject, intrasubject, and/or total variability, if possible
   (vii) CIs for AUC_{0-t}, $AUC_{0-\infty}$, C_{max}. CI values should not be rounded off; therefore, to pass a CI range of 80% to 125%, the values should be at least 80.00 and not more than 125.00)
   (viii) Geometric mean, arithmetic mean, ratio of means for AUC_{0-t}, $AUC_{0-\infty}$, C_{max}
   (ix)   Partial AUC, only if it is used
   (x) C_{min}, C_{max}, C_{pd}, $AUC_{0-\tau}$ degree of fluctuation [$(C_{max} - C_{min})/C_{av}$], and swing [$(C_{max} - C_{min})/C_{min}$], if steady-state studies are employed (where C_{pd} is the concentration of predose sample in period 02)

Quality Assurance

Systems and processes established to ensure that the trial is performed and the data are generated in compliance with GCP must be in place. Quality assurance (QA) is validated through in-process QC and in- and postprocess auditing of clinical trial process as well as data.

The sponsor is responsible for the implementation of a system of QA to ensure that the study is performed and the data are generated, recorded, and reported in compliance with the protocol, GCP, and other applicable requirements. Documented standard operating procedures (SOPs) are a prerequisite for QA.

All observations and findings should be verifiable for the credibility of the data and to assure that the conclusions presented are correctly derived from the raw data. Verification processes must therefore be specified and justified.

Statistically controlled sampling may be an acceptable method of data verification in each study. QC must be applied to each stage of data handling to ensure that all data are reliable and have been processed correctly.

Sponsor audits should be conducted by persons independent of those responsible for the study. Investigational sites, facilities, all data and documentation should be available for inspection and audit by the sponsor's auditor as well as by the regulatory authority (ies).

Other considerations for quality documentation are the following:

a.   A meticulous and specified plan for the various steps and procedures for the purpose of controlling and monitoring the study most effectively.
b.   Specifications and instructions for anticipated deviations from the protocol.
c.   Allocation of duties and responsibilities within the research team and their coordination.
d.   Instructions to staff including study description (the way the study is to be conducted and the procedures for drug usage and administration).

e. Addresses and contact numbers, etc., enabling any staff member to contact the research team at any hour.
f. Considerations of confidentiality problems, if any arise.
g. QC of methods and evaluation procedures.

WHEN BIOAVAILABILITY AND BIOEQUIVALENCE STUDIES ARE NOT NECESSARY (WAIVERS)

Under the following circumstances, BE between a new drug product and the reference product may be considered self-evident with no further requirement for documentation:

(a) When new drug product is to be administered parenterally (e.g., intravenous, intramuscular, subcutaneous, intrathecal administration, etc.) as aqueous solutions and that product contains the same active substance(s) in the same concentration and incorporates the same excipients in comparable concentrations as the reference product.
(b) When the new drug product is a solution for oral use, and contains the active substance in the same concentration, and does not contain an excipient that is known or suspected to affect gastrointestinal transit or absorption of the active substance.
(c) When the new drug product is a gas.
(d) When the new drug product is a powder for reconstitution as a solution and the solution meets either criterion (a) or criterion (b) above.
(e) When the new drug product is an otic or ophthalmic or topical product prepared as an aqueous solution and contains the same active substance(s) in the same concentration(s) and essentially the same excipients in comparable concentrations to the reference product.
(f) When the new drug product is an inhalation product or a nasal spray, tested to be administered with or without essentially the same device as the reference product, prepared as aqueous solutions, and contain the same active substance(s) in the same concentration and essentially the same excipients in comparable concentrations as the reference product. Special in vitro testing is required to document device performance comparison between reference inhalation product and the new drug product.

For criteria (e) and (f), the applicant is expected to demonstrate that the excipients in the new drug are essentially the same and in comparable concentrations as those in the reference product. In the event that this information about the reference product cannot be provided by the applicant, in vivo studies need to be performed.

In Vitro Studies

Under the following circumstances BE may be assessed by the use of in vitro dissolution testing.

a. Drugs for which the applicant provides data to substantiate all of the following:
   i. Highest dose strength is soluble in 250 mL of an aqueous media over the pH range of 1 to 7.5 at 37°C.

   ii. At least 90% of the administered oral dose is absorbed on mass balance determination or in comparison to an intravenous reference dose.

   iii. Speed of dissolution as demonstrated by more than 80% dissolution within 15 minutes at 37°C using IP apparatus 1, at 50 rpm or IP apparatus 2, at 100 rpm in a volume of 900 mL or less in each of the following media:

     1. 0.1 N hydrochloric acid or artificial gastric juice (without enzymes).
     2. A pH 4.4 buffer.
     3. A pH 6.8 buffer or artificial intestinal juice (without enzymes).

b. Different strength of the drug manufactured by the same manufacturer, where all of the following criteria are fulfilled:

   i. The qualitative compositions between the strength is essentially the same.

   ii. The ratio of active ingredients and excipients between the strength is essentially the same, or in the case of low strengths, the ratio between the excipients is the same.

   iii. The methods of manufacture is essentially the same.

   iv. An appropriate BE study has been performed on at least one of the strengths of the formulations (usually the highest strengths unless a lower strength is chosen for reasons of safety).

   v. In case of systemic availability—pharmacokinetics have been shown to be linear over the therapeutic dose range.

In vitro dissolution testing may also be suitable to confirm unchanged product quality and performance characteristics with minor formulation or manufacturing changes after approval.

SUMMARY AND CONCLUSIONS

In India, generic medicines are those medicines, which are labeled with their generic names. There is no separate law for registering generic medicines in India. Drugs and drug products are classified as either (*i*) "new drugs" or (*ii*) drugs other than new drugs. If any drug does not fall under the ambit of a new drug, licenses may be issued by the state licensing Authority. India has a two-tier regulatory system, one is CDSCO under the Government of India having certain powers vested in them and each state has its own drug regulatory system having certain powers. CDSCO is authorized to approve new drugs as defined under Rule 122E of the Drugs and Cosmetic Rules following schedule Y of the Drugs and Cosmetics Rules but the state drugs control agency cannot grant new drug licenses. A BA/BE study is needed for manufacturing a new drug (drug product) in India and to import a new drug (drug product) from another country. If the formulation is the first market entry of the drug substance, the product's pharmacokinetic parameters should be determined. If the formulation is a second or subsequent market entry then comparative BA studies using an appropriate reference product, should be performed. For the first generic drug product, the international innovator drug product should be used as the reference product but for subsequent products, it may be the product approved earlier, which has been deemed to be equivalent to the international innovator product.

REFERENCES

1. Pharmacopoeia of India, Vol I, 3rd ed. Ministry of Health and Family Welfare, Government of India, xvi, 1985.
2. The Drugs and Cosmetics Act 1940 & The Drugs and Cosmetics Rules 1945, Ministry of Health and Family Welfare, Government of India, 425–459, 2005.
3. Ghosh A, Hazra A, Mandal SC, New Drugs in India over the past 15 years: Analysis of trends. Natl Med J India 2004; 17(1):10–16.
4. The Patents (Amendment) Act 2005, The Gazette of India: Extraordinary, Part II-Sec. I, 2005:1–18.
5. The Patents (Amendment) Rules, 2005, The Gazette of India: Extraordinary, Part II-Sec. 3 (ii), 2004:60–108.
6. The Drugs and Cosmetics Act 1940 & The Drugs and Cosmetics Rules 1945, Ministry of Health and Family Welfare, Government of India, 97, 2005.
7. Guidelines for Bioavailability & Bioequivalance Studies, Central Drugs Standard Control Organization, Directorate General of Health Services, Ministry of Health & Family Welfare, Government of India, New Delhi, March 9–12, 2005.
8. GCP (Good Clinical Practice) Guidelines issued by CDSCO (Central Drugs Standards Control Organization), Ministry of Health and Family Welfare, India.
9. Schedule Y, and amendment to Drugs & Cosmetic Act & Rules, Ministry of Health and Family Welfare, India, 2005.
10. The Ethical Guidelines for Biomedical research on human subjects issued by Indian Council of Medical Research, India, 2006.

APPENDIX I:DATA TO BE SUBMITTED ALONG WITH THE APPLICATION TO CONDUCT CLINICAL TRIALS/IMPORT/MANUFACTURE OF NEW DRUGS FOR MARKETING IN INDIA

1. Introduction
   A brief description of the drug and the therapeutic class to which it belongs.
2. Chemical and pharmaceutical information
   2.1. Information on active ingredients
        Drug information (generic name, chemical name, or INN)
   2.2. Physicochemical data
        a. Chemical name and structure
           Empirical formula
           Molecular weight
        b. Physical properties
           Description
           Solubility
           Rotation
           Partition coefficient
           Dissociation constant
   2.3. Analytical data
        Elemental analysis
        Mass spectrum
        NMR spectra
        IR spectra
        UV spectra
        Polymorphic identification
   2.4. Complete monograph specification including:
        Identification
        Identity/quantification of impurities

Enantiomeric purity
Assay

2.5. Validation:

Assay method
Impurity estimation method
Residual solvent/other volatile impurities (OVI) estimation method

2.6. Stability studies (for details refer Appendix IX)

Final release specification
Reference standard characterization
Material safety data sheet

2.7. Data on Formulation:

Dosage form
Composition
Master manufacturing formula
Details of the formulation (including inactive ingredients)
In-process quality control check
Finished product specification
Excipient compatibility study
Validation of the analytical method
Comparative evaluation with international brand(s) or approved Indian brands, if applicable
Pack presentation
Dissolution
Assay
Impurities
Content uniformity
pH
Forced degradation studies
Stability evaluation in market intended pack at proposed storage conditions
Packing specifications
Process validation

When the application is for clinical trials only, the INN or generic name, drug category, dosage form, and data supporting stability in the intended container-closure system for the duration of the clinical trial (information covered in items 2.1, 2.3, 2.6, 2.7) are required.

3. Animal Pharmacology (for details refer Appendix IV)

   3.1. Summary
   3.2. Specific pharmacological actions
   3.3. General pharmacological actions
   3.4. Follow-up and Supplemental Safety Pharmacology Studies
   3.5. Pharmacokinetics: absorption, distribution; metabolism; excretion

4. Animal toxicology (for details refer Appendix III)

   4.1. General aspects
   4.2. Systemic toxicity studies
   4.3. Male fertility study
   4.4. Female reproduction and developmental toxicity studies
   4.5. Local toxicity

4.6. Allergenicity/hypersensitivity
4.7. Genotoxicity
4.8. Carcinogenicity
5. Human/Clinical pharmacology (Phase I)
   5.1. Summary
   5.2. Specific pharmacological effects
   5.3. General pharmacological effects
   5.4. Pharmacokinetics, absorption, distribution, metabolism, excretion
   5.5. Pharmacodynamics/early measurement of drug activity
6. Therapeutic exploratory trials (Phase II)
   6.1. Summary
   6.2. Study report(s) as given in Appendix II
7. Therapeutic confirmatory trials (Phase III)
   7.1. Summary
   7.1. Individual study reports with listing of sites and investigators.
8. Special studies
   8.1. Summary
   8.1. Bioavailability/Bio-equivalence.
   8.1. Other studies, for example, geriatrics, pediatrics, pregnant, or nursing women
9. Regulatory status in other countries
   9.1. Countries where the drug is
      a. Marketed
      b. Approved
      c. Approved as IND
      d. Withdrawn, if any, with reasons
   9.2 Restrictions on use, if any, in countries where marketed /approved
   9.2 Free sale certificate or certificate of analysis, as appropriate.
10. Prescribing information
   10.1. Proposed full prescribing information
   10.2. Drafts of labels and cartons
11. Samples and testing protocol/s
   11.1. Samples of pure drug substance and finished product (an equivalent of 50 clinical doses, or more number of clinical doses if prescribed by the licensing authority), with testing protocol/s, full impurity profile, and release specifications.

Notes:
(1) All items may not be applicable to all drugs. For explanation, refer to text of Schedule Y.
(2) For requirements of data to be submitted with application for clinical trials refer to the text of this Schedule.

APPENDIX I-A: DATA REQUIRED TO BE SUBMITTED BY AN APPLICANT FOR PERMISSION TO IMPORT AND/OR MANUFACTURE A NEW DRUG ALREADY APPROVED IN THE COUNTRY

1. Introduction
   A brief description of the drug and the therapeutic class
2. Chemical and pharmaceutical information

2.1 Chemical name, code name or number, if any; nonproprietary or generic name, if any, structure; physicochemical properties

2.2 Dosage form and its composition

2.3 Test specifications
   (a) Active ingredients
   (b) Inactive ingredients

2.4 Tests for identification of the active ingredients and method of its assay

2.5 Outline of the method of manufacture of active ingredients

2.6 Stability data

3. Marketing Information
   3.1 Proposed package insert/promotional literature
   3.2 Draft specimen of the label and carton

4. Special studies conducted with approval of licensing authority
   4.1 Bioavailability/bioequivalence and comparative dissolution studies for oral dosage forms
   4.2 Subacute animal toxicity studies for intravenous infusions and injectables.

APPENDIX II: STRUCTURE, CONTENTS, AND FORMAT FOR CLINICAL STUDY REPORTS

1. Title Page:

   This page should contain information about the title of the study, the protocol code, name of the investigational product tested, development phase, indication studied, a brief description of the trial design, the start and end date of patient accrual, and the names of the sponsor and the participating institutes (investigators).

2. Synopsis (one to two pages): A brief overview of the study from the protocol development to the trial closure should be given here. This section will only summarize the important conclusions derived from the study.

3. Statement of compliance with the "Guidelines for Clinical Trials on Pharmaceutical Products in India—GCP Guidelines" issued by the Central Drugs Standard Control Organization, Ministry of Health, Government of India.

4. List of abbreviations and definitions

5. Table of contents

6. Ethics committee:

   This section should document that the study was conducted in accordance with the ethical principles of Declaration of Helsinki. A detailed description of the ethics committee constitution and date(s) of approvals of trial documents for each of the participating sites should be provided. A declaration should state that EC notifications as per good clinical practice guidelines issued by Central Drugs Standard Control Organization and Ethical Guidelines for Biomedical Research on Human Subjects, issued by Indian Council of Medical Research have been followed.

7. Study team:

   This section will briefly describe the administrative structure of the study (investigators, site staff, sponsor/designates, central laboratory, etc.).

8. Introduction:

A brief description of the product development rationale should be given here.

9. Study objective:

A statement describing the overall purpose of the study and the primary and secondary objectives to be achieved should be mentioned here.

10. Investigational plan:

This section should describe the overall trial design, the subject selection criteria, the treatment procedures, blinding/randomization techniques if any, allowed/disallowed concomitant treatment, the efficacy and safety criteria assessed, the data quality assurance procedures, and the statistical methods planned for the analysis of the data obtained.

11. Trial subjects:

A clear accounting of all trial subjects who entered the study will be given here. Mention should also be made of all cases that were dropouts or protocol deviations. Enumerate the patients screened, randomized, and prematurely discontinued. State reasons for premature discontinuation of therapy in each applicable case.

12. Efficacy evaluation:

The results of evaluation of all the efficacy variables will be described in this section with appropriate tabular and graphical representation. A brief description of the demographic characteristics of the trial patients should also be provided along with a listing of patients and observations excluded from efficacy analysis.

13. Safety evaluation:

This section should include the complete list of all serious adverse events, whether expected or unexpected and adverse events whether serious or not (compiled from data received as per Appendix XI).

The comparison of adverse events across study groups may be presented in a tabular or graphical form. This section should also give a brief narrative of all important events considered related to the investigational product.

14. Discussion and overall conclusion:

Discussion of the important conclusions derived from the trial and scope for further development.

15. List of references
16. Appendices

List of Appendices to the Clinical Trial Report
a. Protocol and amendments
b. Specimen of Case Record Form
c. Investigators' name(s) with contact addresses, phone, email etc.
d. Patient data listings
e. List of trial participants treated with investigational product
f. Discontinued participants
g. Protocol deviations
h. CRFs of cases involving death and life threatening adverse event cases

i. Publications from the trial
j. Important publications referenced in the study
k. Audit certificate, if available
l. Investigator's certificate that he/she has read the report and that the report accurately describes the conduct and the results of the study.

APPENDIX III: ANIMAL TOXICOLOGY (NONCLINICAL TOXICITY STUDIES)

General Principles

Toxicity studies should comply with the norms of good laboratory practice (GLP). Briefly, these studies should be performed by suitably trained and qualified staff employing properly calibrated and standardized equipment of adequate size and capacity. Studies should be done as per written protocols with modifications (if any) verifiable retrospectively. Standard operating procedures (SOPs) should be followed for all managerial and laboratory tasks related to these studies. Test substances and test systems (in vitro or in vivo) should be properly characterized and standardized. All documents belonging to each study, including its approved protocol, raw data, draft report, final report, and histology slides and paraffin tissue blocks should be preserved for a minimum of five years after marketing of the drug.

Toxicokinetic studies (generation of pharmacokinetic data either as an integral component of the conduct of nonclinical toxicity studies or in specially designed studies) should be conducted to assess the systemic exposure achieved in animals and its relationship to dose level and the time course of the toxicity study. Other objectives of toxicokinetic studies include obtaining data to relate the exposure achieved in toxicity studies to toxicological findings and contribute to the assessment of the relevance of these findings to clinical safety, to support the choice of species and treatment regimen in nonclinical toxicity studies, and to provide information, which, in conjunction with the toxicity findings, contributes to the design of subsequent nonclinical toxicity studies.

Systemic Toxicity Studies
Single-Dose Toxicity Studies. These studies (see Appendix I item 4.2) should be carried out in two rodent species (mice and rats) by using the same route as intended for humans. In addition, unless the intended route of administration in humans is only intravenous, at least one more route should be used in one of the species to ensure systemic absorption of the drug. This route should depend on the nature of the drug. A limit of 2 g/kg (or 10 times the normal dose that is intended in humans, whichever is higher) is recommended for oral dosing. Animals should be observed for 14 days after the drug administration, and minimum lethal dose (MLD) and maximum tolerated dose (MTD) should be established. If possible, the target organ of toxicity should also be determined. Mortality should be observed for up to seven days after parenteral administration and up to 14 days after oral administration. Symptoms, signs, and mode of death should be reported, with appropriate macroscopic and microscopic findings where necessary. LD_{10} and LD_{50} should be reported preferably with 95% confidence limits. If LD_{50}s cannot be determined, reasons for the same should be stated.

The dose causing severe toxic manifestations or death should be defined in the case of cytotoxic anticancer agents, and the postdosing observation period should be up to 14 days. Mice should first be used for determination of MTD. Findings should then be confirmed in rat for establishing linear relationship between toxicity and body surface area. In case of nonlinearity, data of the more sensitive species should be used to determine the Phase I starting dose. Where rodents are known to be poor predictors of human toxicity (e.g., antifolates), or where the cytotoxic drug acts by a novel mechanism of action, MTD should be established in non–rodent species.

Repeated-Dose Systemic Toxicity Studies
These studies (see Appendix I, item 4.2) should be carried out in at least two mammalian species, of which one should be a non–rodent species. Dose-ranging studies should precede the 14-, 28-, 90-, or 180-day toxicity studies. Duration of the final systemic toxicity study will depend on the duration, therapeutic indication, and scale of the proposed clinical trial (see item 1.8). If a species is known to metabolize the drug in the same way as humans, it should be preferred for toxicity studies.

In repeated-dose toxicity studies the drug should be administered seven days a week by the route intended for clinical use. The number of animals required for these studies, that is, the minimum number of animals on which data should be available, is shown in item 1.9.

Wherever applicable, a control group of animals given the vehicle alone should be included, and three other groups should be given graded doses of the drug. The highest dose should produce observable toxicity; the lowest dose should not cause observable toxicity, but should be comparable to the intended therapeutic dose in humans or a multiple of it. To make allowance for the sensitivity of the species the intermediate dose should cause some symptoms, but not gross toxicity or death, and should be placed logarithmically between the other two doses.

The parameters to be monitored and recorded in long-term toxicity studies should include behavioral, physiological, biochemical, and microscopic observations. In case of parenteral drug administration, the sites of injection should be subjected to gross and microscopic examination. Initial and final electrocardiogram and fundus examination should be carried out in the non–rodent species.

In the case of cytotoxic anticancer agents, dosing and study design should be in accordance with the proposed clinical schedule in terms of days of exposure and number of cycles. Two rodent species may be tested for initiating Phase I trials. A non–rodent species should be added if the drug has a novel mechanism of action, or if permission for Phases II, III, or marketing is being sought.

For most compounds, it is expected that single-dose tissue distribution studies with sufficient sensitivity and specificity will provide an adequate assessment of tissue distribution and the potential for accumulation. Thus, repeated-dose tissue distribution studies should not be required uniformly for all compounds and should only be conducted when appropriate data cannot be derived from other sources. Repeated-dose studies may be appropriate under certain circumstances based on the data from single-dose tissue distribution studies, toxicity studies, and toxicokinetic studies. The studies may be most appropriate for compounds that have an apparently long half-life, incomplete elimination, or unanticipated organ toxicity.

Notes:

(i) Single-dose toxicity study: Each group should contain at least five animals of either sex. At least four graded doses should be given. Animals should be exposed to the test substance in a single bolus or by continuous infusion or several doses within 24 hours. Animals should be observed for 14 days. Signs of intoxication, effect on body weight, gross pathological changes should be reported. It is desirable to include histopathology of grossly affected organs, if any.

(ii) Dose-ranging study: Objectives of this study include the identification of target organ of toxicity and establishment of MTD for subsequent studies.

(a) Rodents: Study should be performed in one rodent species (preferably rat) by the proposed clinical route of administration. At least four graded doses including control should be given, and each dose group as well as the vehicle control should consist of a minimum of five animals of each sex. Animals should be exposed to the test substance daily for 10 consecutive days. Highest dose should be the MTD of single-dose study. Animals should be observed daily for signs of intoxication (general appearance, activity, and behavior, etc.), and periodically for the body weight and laboratory parameters. Gross examination of viscera and microscopic examination of affected organs should be done. (b) Non–rodent species: One male and one female are to be taken for ascending Phase MTD study. Dosing should start after initial recording of cage-side and laboratory parameters. Starting dose may be three to five times the extrapolated effective dose or MTD (whichever is less), and dose escalation in suitable steps should be done every third day after drawing the samples for laboratory parameters. Dose should be lowered appropriately when clinical or laboratory evidence of toxicity is observed. Administration of test substance should then continue for 10 days at the well-tolerated dose level following which, samples for laboratory parameters should be taken. Sacrifice, autopsy, and microscopic examination of affected tissues should be performed as in the case of rodents.

(iii) 14- to 28-day repeated-dose toxicity studies: One rodent (6–10/sex/group) and one non–rodent (2–3/sex/group) species are needed. Daily dosing by proposed clinical route at three dose levels should be done with highest dose having observable toxicity, mid-dose between high and low dose, and low dose. The doses should preferably be multiples of the effective doses and free from toxicity. Observation parameters should include cage-side observations, body weight changes, food/water intake, blood biochemistry, hematology, and gross and microscopic studies of all viscera and tissues.

(iv) 90-day repeated-dose toxicity studies: One rodent (15–30/sex/group) and one non–rodent (4–6/sex/group) species are needed. Daily dosing by proposed clinical route at three graded dose levels should be done. In addition to the control a "high-dose-reversal" group and its control group should be also included. Parameters should include signs of intoxication (general appearance, activity, behavior, etc.), body weight, food intake, blood biochemical parameters, hematological values, urine analysis, organ weights, gross and microscopic study of viscera and tissues. Half the animals in "reversal" groups (treated and control) should be sacrificed after 14 days of stopping the treatment. The remaining animals should be sacrificed after 28 days of stopping the treatment or after the recovery of signs and/or clinical pathological changes—whichever comes later, and evaluated for the parameters used for the main study.

(v) 180-day repeated-dose toxicity studies: One rodent (15–30/sex/group) and one non–rodent (4–6/sex/group) species are needed. At least four groups, including control, should be taken. Daily dosing by proposed clinical route at three graded dose levels should be done. Parameters should include signs of intoxication, body weight, food intake, blood biochemistry, hematology, urine analysis, organ weights, gross and microscopic examination of organs and tissues.

Male Fertility Study

One rodent species (preferably rat) should be used. Dose selection should be done from the results of the previous 14- or 28-day toxicity study in rat. Three dose groups, the highest one showing minimal toxicity in systemic studies, and a control group should be taken. Each group should consist of six adult male animals. Animals should be treated with the test substance by the intended route of clinical use for minimum 28 days and maximum 70 days before they are paired with female animals of proven fertility in a ratio of 1:2 for mating.

Drug treatment of the male animals should continue during pairing. Pairing should be continued till the detection of vaginal plug or 10 days, whichever is earlier. Females thus getting pregnant should be examined for their fertility index after day 13 of gestation. All the male animals should be sacrificed at the end of the study. Weights of each testis and epididymis should be separately recorded. Sperms from one epididymis should be examined for their motility and morphology. The other epididymis and both testes should be examined for their histology.

Female Reproduction and Developmental Toxicity Studies

These studies (see Appendix I, item 4.4) need to be carried out for all drugs proposed to be studied or used in women of child-bearing age. Segment I, II, and III studies (see later) are to be performed in albino mice or rats, and Segment II study should include albino rabbits also as a second test species.

On the occasion, when the test article is not compatible with the rabbit (e.g., antibiotics that are effective against gram-positive anaerobic organisms and protozoas) the Segment II data in the mouse may be substituted.

Female Fertility Study (Segment I)

The study should be done in one rodent species (rat preferred). The drug should be administered to both males and females, beginning a sufficient number of days (28 days in males and 14 days in females) before mating. Drug treatment should continue during mating and, subsequently, during the gestation period. Three graded doses should be used; the highest dose (usually the MTD obtained from previous systemic toxicity studies) should not affect general health of the parent animals. At least 15 males and 15 females should be used per dose group. Control and the treated groups should be of similar size. The route of administration should be the same as intended for therapeutic use.

Dames should be allowed to litter and their medication should be continued till the weaning of pups. Observations on body weight, food intake, clinical signs of intoxication, mating behavior, progress of gestation/parturition periods, length of gestation, parturition, postpartum health, and gross pathology (and histopathology of affected organs) of dames should be recorded. The pups from both treated and control groups should be observed for general signs of intoxication, sex-wise distribution in different treatment groups, body weight, growth parameters, survival, gross examination, and autopsy. Histopathology of affected organs should be done.

Teratogenicity Study (Segment II)

One rodent (preferably rat) and one non–rodent (rabbit) species are to be used. The drug should be administered throughout the period of organogenesis, using three dose levels as described for Segment I. The highest dose should cause minimum maternal toxicity and the lowest one should be proportional to the proposed dose for clinical use in humans or a multiple of it. The route of administration should be the same as intended for human therapeutic use.

The control and the treated groups should consist of at least 20 pregnant rats (or mice) and 12 rabbits, on each dose level. All fetuses should to be subjected to gross examination, one of the fetuses should be examined for skeletal abnormalities and the other half for visceral abnormalities. Observation parameters should include (dames) signs of intoxication, effect on body weight, effect on food intake, examination of uterus, ovaries and uterine contents, number of corpora lutea, implantation sites, resorptions (if any); and for the fetuses, the total number, gender, body length, weight and gross/visceral/skeletal abnormalities, if any.

Perinatal Study (Segment III)

This study is specially recommended if the drug is to be given to pregnant or nursing mothers for long periods or where there are indications of possible adverse effects on fetal development. One rodent species (preferably rat) is needed. Dosing at levels comparable to multiples of human dose should be done by the intended clinical route. At least four groups (including control), each consisting of 15 dames should be used. The drug should be administered throughout the last trimester of pregnancy (from day 15 of gestation) and then the dose that causes low fetal loss should be continued throughout lactation and weaning. Dames should then be sacrificed and examined as described later.

One male and one female from each litter of F_1 generation (total 15 males and 15 females in each group) should be selected at weaning and treated with vehicle or test substance (at the dose levels already described) throughout their periods of growth to sexual maturity, pairing, gestation, parturition, and lactation. Mating performance and fertility of F_1 generation should thus be evaluated to obtain the F_2 generation whose growth parameters should be monitored till weaning. The criteria of evaluation should be the same as described earlier (3.4.1).

Animals should be sacrificed at the end of the study and the observation parameters should include (dames) body weight, food intake, general signs of intoxication, progress of gestation/parturition periods, and gross pathology (if any); and for pups, the clinical signs, sex-wise distribution in dose groups, body weight, growth parameters, gross examination, survival, and autopsy (if needed) and where necessary, histopathology.

Local Toxicity

These studies (see Appendix I, item 4.5) are required when the new drug is proposed to be used by some special route (other than oral) in humans. The drug should be applied to an appropriate site (e.g., skin or vaginal mucous membrane) to determine local effects in a suitable species. Typical study designs for these studies should include three dose levels and untreated and/ or vehicle control, preferably use of two species, and increasing group size with increase in duration of treatment. Where dosing is restricted due to anatomical or humane reasons, or the drug concentration cannot be increased beyond a certain level due to the problems of solubility, pH or tonicity, a clear statement to this effect should be given. If the drug is absorbed from the site of application, appropriate systemic toxicity studies will also be required.

Notes:

(i) Dermal toxicity study: The study should be done in rabbit and rat. Daily topical (dermal) application of test substance in its clinical dosage form should be done. Test material should be applied on shaved skin covering not less than 10% of the total body surface area. Porous gauze dressing should be used to hold liquid material in place. Formulations with different concentrations (at least three) of test substance, several fold higher than the clinical dosage form should be used. Period of application may vary from 7 to 90 days depending on the clinical duration of use. Where skin irritation is grossly visible in the initial studies, a recovery group should be included in the subsequent repeated-dose study. Local signs (erythema, edema, and eschar formation) as well as histological examination of sites of application should be used for evaluation of results.

(ii) Photoallergy or dermal phototoxicity: It should be tested by Armstrong/Harber Test in guinea pig. This test should be done if the drug or a metabolite is related to an agent causing photosensitivity or the nature of action suggests such a potential (e.g., drugs to be used in treatment of leucoderma). Pretest in eight animals should screen four concentrations (patch application for 2 hours ± 15 minutes) with and without UV exposure (10 J/cm^2). Observations recorded at 24 and 48 hours should be used to ascertain highest nonirritant dose. Main test should be performed with 10 test animals and 5 controls. Induction with the dose selected from pretest should use 0.3 mL/patch for 2 hour ± 15 minutes followed by 10 J/cm^2 of UV exposure. This should be repeated on day 0, 2, 4, 7, 9, and 11 of the test. Animals should be challenged with the same concentration of test substance between day 20 and 24 of the test with a similar 2-hour application followed by exposure to 10 J/cm^2 of UV light. Examination and grading of erythema and edema formation at the challenge sites should be done 24 and 48 hours after the challenge. A positive control like musk ambrett or psoralin should be used.

(iii) Vaginal toxicity test: Study is to be done in rabbit or dog. Test substance should be applied topically (vaginal mucosa) in the form of pessary, cream, or ointment. Six to ten animals per dose group should be taken. Higher concentrations or several daily applications of test substance should be done to achieve multiples of daily human dose. The minimum duration of drug treatment is seven days (more according to clinical use), subject to a maximum of 30 days. Observation parameters should include swelling, closure of introitus, and histopathology of vaginal wall.

(iv) Rectal tolerance test: For all preparations meant for rectal administration this test may be performed in rabbits or dogs. Six to ten animals per dose group should be taken. Formulation in volume comparable to human dose (or the maximum possible volume) should be applied once or several times daily, per rectally, to achieve administration of multiples of daily human dose. The minimum duration of application is seven days (more according to clinical use), Subject to a maximum of 30 days. Size of suppositories may be smaller, but the drug content should be several folds higher than the proposed human dose. Observation parameters should include clinical signs (sliding on backside), signs of pain, blood and/or mucus in feces, condition of anal region/sphincter, gross, and (if required) histological examination of rectal mucosa.

(v) Parenteral drugs: For products meant for intravenous or intramuscular or subcutaneous or intradermal injection the sites of injection in systemic toxicity studies should be specially examined grossly and microscopically. If needed, reversibility of adverse effects may be determined on a case-to-case basis.

(vi) Ocular toxicity studies (for products meant for ocular instillation): These studies should be carried out in two species, one of which should be the albino rabbit, which has a sufficiently large conjunctival sac. Direct delivery of drug onto the cornea in case of animals having small conjunctival sacs should be ensured. Liquids, ointments, gels, or soft contact lenses (saturated with drug) should be used. Initial single-dose application should be done to decide the exposure concentrations for repeated-dose studies and the need to include a recovery group. Duration of the final study will depend on the proposed length of human exposure subject to a maximum of 90 days. At least two different concentrations exceeding the human dose should be used for demonstrating the margin of safety. In acute studies, one eye should be used for drug administration and the other kept as control. A separate control group should be included in repeated-dose studies.

Slit-lamp examination should be done to detect the changes in cornea, iris, and aqueous humor. Fluorescent dyes (sodium fluorescein, 0.25–1.0%) should be used for detecting the defects in surface epithelium of cornea and conjunctiva. Changes in intraocular tension should be monitored by a tonometer. Histological examination of eyes should be done at the end of the study after fixation in Davidson's or Zenker's fluid.

(vii) Inhalation toxicity studies: The studies are to be undertaken in one rodent and one non–rodent species using the formulation that is to be eventually proposed to be marketed. Acute, subacute, and chronic toxicity studies should be performed according to the intended duration of human exposure. Standard systemic toxicity study designs (described earlier) should be used. Gases and vapors should be given in whole body exposure chambers; aerosols are to be given by nose-only method. Exposure time and concentrations of test substance (limit dose of 5 mg/L) should be adjusted to ensure exposure at levels comparable to multiples of intended human exposure. Three dose groups and a control (plus vehicle control, if needed) are required. Duration of exposure may vary subject to a maximum of 6 h/d and five days a week. Food and water should be withdrawn during the period of exposure to test substance.

Temperature, humidity, and flow rate of exposure chamber should be recorded and reported. Evidence of exposure with test substance of particle size of 4 μm (especially for aerosols) with not less that 25% being 1 μm should be provided. Effects on respiratory rate, findings of bronchial lavage fluid examination, and histological examination of respiratory passages and lung tissue should be included along with the regular parameters of systemic toxicity studies or assessment of margin of safety.

Allergenicity/Hypersensitivity

Standard tests include guinea pig maximization test (GPMT) and local lymph node assay (LLNA) in mouse. Any one of the two may be done.

Notes:

(i) Guinea pig maximization test: The test is to be performed in two steps: first, determination of maximum nonirritant and minimum irritant doses, and second, the main test. The initial study will also have two components. To determine the intradermal induction dose, four dose levels should be tested by the same route in a batch of four males and four females (two of each sex should be given Freund's adjuvant). The minimum irritant dose should be used for induction. Similarly, a topical minimum irritant dose should be determined for challenge. This should be established in two males and two females. A minimum of six males and six females per group should be used in the main study. One test and one control group should be used. It is preferable to have one more positive control group. Intradermal induction (day 1) coupled with topical challenge (day 21) should be done. If there is no response, re-challenge should be done 7 to 30 days after the primary challenge. Erythema and edema (individual animal scores as well as maximization grading) should be used as evaluation criteria.

(ii) Local lymph node assay: Mice used in this test should be of the same sex, either only males or only females. Drug treatment is to be given on ear skin. Three graded doses, the highest being maximum nonirritant dose plus vehicle control should be used. A minimum of six mice per group should be used. Test material should be applied on ear skin on three consecutive days and on day 5, the draining auricular lymph nodes should be dissected out five hours after IV ^{3}H-thymidine or bromo-deoxy-uridine (BrdU). Increase in ^{3}H-thymidine or BrdU incorporation should be used as the criterion for evaluation of results.

Genotoxicity

Genotoxic compounds, in the absence of other data, shall be presumed to be transspecies carcinogens, implying a hazard to humans. Such compounds need not be subjected to long-term carcinogenicity studies. However, if such a drug is intended to be administered for chronic illnesses or otherwise over a long period of time—a chronic toxicity study (up to one year) may be necessary to detect early tumorigenic effects.

Genotoxicity tests are in vitro and in vivo tests conducted to detect compounds, which induce genetic damage directly or indirectly. These tests should enable hazard identification with respect to damage to DNA and its fixation.

The following standard test battery is generally expected to be conducted:

(i) A test for gene mutation in bacteria.
(ii) An in vitro test with cytogenetic evaluation of chromosomal damage with mammalian cells or an in vitro mouse lymphoma tic assay.
(iii) An in vivo test for chromosomal damage using rodent hematopoietic cells.

Other genotoxicity tests, for example, tests for measurement of DNA adducts, DNA strand breaks, DNA repair or recombination serve as options in addition to the standard battery for further investigation of genotoxicity test results obtained in the standard battery. Only under extreme conditions in which one or more tests comprising the standard battery cannot be employed for technical reasons, alternative validated tests can serve as substitutes provided sufficient scientific justification should be provided to support the argument that a given standard battery test is not appropriate.

Both in vitro and in vivo studies should be done. In vitro studies should include Ames' Salmonella assay and chromosomal aberrations (CA) in cultured cells. In vivo studies should include micronucleus assay (MNA) or CA in rodent bone marrow. Data analysis of CA should include analysis of "gaps.'

Cytotoxic anticancer agents: Genotoxicity data are not required before Phase I and II trials. But these studies should be completed before applying for Phase III trials.

Notes:

Ames' Test (Reverse mutation assay in Salmonella): *S. typhimurium* tester strains such as TA98, TA100, TA102, TA1535, TA97, or *Escherichia coli* WP2 *uvrA* or *E. coli* WP2 *uvrA* (pKM101) should be used.

(i) In vitro exposure (with and without metabolic activation, S9 mix) should be done at a minimum of five log dose levels. "Solvent" and "positive" control should be used. Positive control may include 9-amino-acridine, 2-nitrofluorine, sodium azide, and mitomycin C, respectively, in the tester strains mentioned above. Each set should consist of at least three replicates. A 2.5-fold (or more) increase in number of revertants in comparison to spontaneous revertants would be considered positive.

(ii) In vitro cytogenetic assay: The desired level of toxicity for in vitro cytogenetic tests using cell lines should be greater than 50% reduction in cell number or culture confluency. For lymphocyte cultures, an inhibition of mitotic index by greater than 50% is considered sufficient. It should be performed in CHO cells or on human lymphocyte in culture. In vitro exposure (with and without metabolic activation, S9 mix) should be done using a minimum of three log doses. "Solvent" and "positive" control should be included. A positive control such as cyclophosphamide with metabolic activation and mitomycin C for without metabolic activation should be used to give a reproducible and detectable increase clastogenic effect over the background, which demonstrates the sensitivity of the test system. Each set should consist of at least three replicates. Increased number of aberrations in metaphase chromosomes should be used as the criteria for evaluation.

(iii) In vivo micronucleus assay: One rodent species (preferably mouse) is needed. Route of administration of test substance should be the same as intended for humans. Five animals per sex per dose groups should be used. At least three dose levels, plus "solvent" and "positive" control should be tested. A positive control like mitomycin C or cyclophosphamide should be used. Dosing should be done on day 1 and 2 of study followed by sacrifice of animals six hours after the last injection. Bone marrow from both the femora should be taken out, flushed with fetal bovine serum (20 minutes), pelleted and smeared on glass slides. Giemsa–MayGruenwald staining should be done and increased number of micronuclei in polychromatic erythrocytes (minimum 1000) should be used as the evaluation criteria.

(iv) In vivo cytogenetic assay: One rodent species (preferably rat) is to be used. Route of administration of test substance should be the same as intended for humans. Five animals/sex/dose groups should be used. At least three dose levels, plus "solvent" and "positive" control should be tested. Positive control may include cyclophosphamide. Dosing should be done on day 1 followed by intraperitoneal colchicine administration at 22 hours. Animals should be sacrificed two hours after colchicine administration. Bone marrow from both the femora should be taken out, flushed with hypotonic saline (20 minutes), pelleted and resuspended in Carnoy's fluid. Once again the cells should be pelleted and dropped on clean glass slides with a Pasteur pipette. Giemsa staining should be done and increased number of aberrations in metaphase chromosomes (minimum 100) should be used as the evaluation criteria.

Carcinogenicity (see Appendix I, Item 4.8)

Carcinogenicity studies should be performed for all drugs that are expected to be clinically used for more than six months as well as for drugs used frequently in an intermittent manner in the treatment of chronic or recurrent conditions. Carcinogenicity studies are also to be performed for drugs if there is concern about their carcinogenic potential emanating from previous demonstration of carcinogenic potential in the product class that is considered relevant to humans or where structure–activity relationship suggests carcinogenic risk or when there is evidence of preneoplastic lesions in repeated-dose toxicity studies or when long-term tissue retention of parent compound or metabolite(s) results in local tissue reactions or other pathophysiological responses. For pharmaceuticals developed to treat certain serious diseases, licensing authority may allow carcinogenicity testing to be conducted after marketing permission has been granted.

In instances where the life expectancy in the indicated population is short (i.e., less than two to three years), no long-term carcinogenicity studies may be required. In cases where the therapeutic agent for cancer is generally successful and life is significantly prolonged there may be later concerns regarding secondary cancers. When such drugs are intended for adjuvant therapy in tumor-free patients or for prolonged use in non–cancer indications, carcinogenicity studies may be/are needed. Completed rodent carcinogenicity studies are not needed in advance of the conduct of large-scale clinical trials, unless there is special concern for the patient population.

Carcinogenicity studies should be done in a rodent species (preferably rat). Mouse may be employed only with proper scientific justification. The selected strain of animals should not have a very high or very low incidence of sponta-neous tumors.

At least three dose levels should be used. The highest dose should be sub-lethal, and it should not reduce the lifespan of animals by more than 10% of expected normal. The lowest dose should be comparable to the intended human therapeutic dose or a multiple of it, for example, 2.5 times; to make allowance for the sensitivity of the species. The intermediate dose to be placed logarithmically between the other two doses. An untreated control and (if indicated) a vehicle control group should be included. The drug should be administered seven days a week for a fraction of the lifespan comparable to the fraction of human lifespan over which the drug is likely to be used therapeutically. Generally, the period of dosing should be 24 months for rats and 18 months for mice.

Observations should include macroscopic changes observed at autopsy and detailed histopathology of organs and tissues. Additional tests for carcino-genicity (short-term bioassays, neonatal mouse assay, or tests employing trans-genic animals) may also be done depending on their applicability on a case to case basis.

Note:
Each dose group and concurrent control group not intended to be sacrificed early should contain at least 50 animals of each sex. A high-dose satellite group for evaluation of pathology other than neoplasia should contain 20 animals of each sex while the satellite control group should contain 10 animals of each sex. Observation parameters should include signs of intoxication, effect on body weight, food intake, clinical chemistry parameters, hematology parameters, urine analysis, organ weights, gross pathology, and detailed histopathology. Comprehensive descriptions of benign and malignant tumor development, time of their detection, site, dimensions, histological typing, etc. should be given.

Animal Toxicity Requirements for Clinical Trials and Marketing of a New Drug
Nonclinical toxicity testing and safety evaluation data of an IND needed for the conduct of different phases of clinical trials

Note:
Refer Appendix III (points 1.1 through 1.7 and Tables 1.8 and 1.9) for essential features of study designs of the nonclinical toxicity studies listed below.

For Phase I Clinical Trials
Systemic toxicity studies

  i. Single-dose toxicity studies
 ii. Dose-ranging studies
iii. Repeat-dose systemic toxicity studies of appropriate duration to support the duration of proposed human exposure

Male fertility study
In vitro genotoxicity tests
Relevant local toxicity studies with proposed route of clinical application (dura-tion depending on proposed length of clinical exposure)
Allergenicity/hypersensitivity tests (when there is a cause for concern or for par-enteral drugs, including dermal application)

Photoallergy or dermal phototoxicity test (if the drug or a metabolite is related
to an agent causing photosensitivity or the nature of action suggests such a
potential)

For Phase II Clinical Trials

Provide a summary of all the nonclinical safety data (listed earlier) already submitted while obtaining the permissions for Phase I trial, with appropriate
references

In case of an application for directly starting a Phase II trial—complete details of
the nonclinical safety data needed for obtaining the permission for Phase I
trial, as per the list provided above must be submitted.

Repeat-dose systemic toxicity studies of appropriate duration to support the
duration of proposed human exposure

In vivo genotoxicity tests

Segment II reproductive/developmental toxicity study (if female patients of
child-bearing age are going to be involved)

For Phase III Clinical Trials

Provide a summary of all the nonclinical safety data (listed earlier) already submitted while obtaining the permissions for Phase I and II trials, with appropriate
references.

In case of an application for directly initiating a Phase III trial, complete
details of the nonclinical safety data needed for obtaining the permissions for
Phase I and II trials, as per the list already provided must be provided.

i.   Repeat-dose systemic toxicity studies of appropriate duration to support the
     duration of proposed human exposure
ii. Reproductive/developmental toxicity studies
iii. Segment I (if female patients of child-bearing age are going to be involved)
iv. Segment III (for drugs to be given to pregnant or nursing mothers or where
     there are indications of possible adverse effects on fetal development)
     Carcinogenicity studies (when there is a cause for concern or when the
     drug is to be used for more than six months)

For Phase IV Clinical Trials

Provide a summary of all the nonclinical safety data (listed earlier) already submitted while obtaining the permissions for Phase I, II, and III trials, with appropriate references.

In case an application is made for initiating the Phase IV trial, complete
details of the nonclinical safety data needed for obtaining the permissions for
Phase I, II, and III trials, as per the list already provided must be submitted.

Application of Good Laboratory Practices (GLP)

The animal studies are conducted in an accredited laboratory. Where the safety
pharmacology studies are part of toxicology studies, these studies should also be
conducted in an accredited laboratory.

APPENDIX IV: ANIMAL PHARMACOLOGY

General Principles

Specific and general pharmacological studies should be conducted to support use of therapeutics in humans. In the early stages of drug development enough information may not be available to rationally select study design for safety assessment. In such a situation, a general approach to safety pharmacology studies can be applied. Safety pharmacology studies are studies that investigate potential undesirable pharmacodynamic effects of a substance on physiological functions in relation to exposure within the therapeutic range or above.

Specific Pharmacological Actions

Specific pharmacological actions are those which demonstrate the therapeutic potential for humans.

The specific studies that should be conducted and their design will be different based on the individual properties and intended uses of investigational drug. Scientifically validated methods should be used. The use of new technologies and methodologies in accordance with sound scientific principles should be preferred.

General Pharmacology

Essential Safety Pharmacology

Safety pharmacology studies need to be conducted to investigate the potential undesirable pharmacodynamic effects of a substance on physiological functions in relation to exposure within the therapeutic range and above. These studies should be designed to identify undesirable pharmacodynamic properties of a substance that may have relevance to its human safety; to evaluate adverse pharmacodynamic and/or pathophysiological effects observed in toxicology and/or clinical studies; and to investigate the mechanism of the adverse pharmacodynamic effects observed and/or suspected.

The aim of the essential safety pharmacology is to study the effects of the test drug on vital functions. Vital organ systems such as cardiovascular, respiratory, and central nervous systems should be studied. Essential safety pharmacology studies may be excluded or supplemented based on scientific rationale. Also, the exclusion of certain test(s) or exploration(s) of certain organs, systems or functions should be scientifically justified.

Cardiovascular System

Effects of the investigational drug should be studied on blood pressure, heart rate, and the electrocardiogram. If possible in vitro, in vivo, and/or ex vivo methods including electrophysiology should also be considered.

Central Nervous System

Effects of the investigational drug should be studied on motor activity, behavioral changes, coordination, sensory and motor reflex responses, and body temperature.

Respiratory System
Effects of the investigational drug on respiratory rate and other functions such as tidal volume and hemoglobin oxygen saturation should be studied.

Follow-Up and Supplemental Safety Pharmacology Studies
In addition to the essential safety pharmacological studies, additional supplemental and follow-up safety pharmacology studies may need to be conducted as appropriate. These depend on the pharmacological properties or chemical class of the test substance, and the data generated from safety pharmacology studies, clinical trials, pharmacovigilance, experimental *in vitro* or *in vivo* studies, or from literature reports.

Follow-Up Studies for Essential Safety Pharmacology
Follow-up studies provide additional information or a better understanding than that provided by the essential safety pharmacology.

Cardiovascular System
These include ventricular contractility, vascular resistance, and the effects of chemical mediators, their agonists and antagonists on the cardiovascular system.

Central Nervous System
These include behavioral studies, learning and memory, electrophysiology studies, neurochemistry, and ligand-binding studies.

Respiratory System
These include airway resistance, compliance, pulmonary arterial pressure, blood gases, and blood pH.

Supplemental Safety Pharmacology Studies
These studies are required to investigate the possible adverse pharmacological effects that are not assessed in the essential safety pharmacological studies and are a cause for concern.

Urinary System
These include urine volume, specific gravity, osmolality, pH, proteins, cytology and blood urea nitrogen, creatinine, and plasma proteins estimation.

Autonomic Nervous System
These include binding to receptors relevant for the autonomic nervous system, and functional response to agonist or antagonist responses in vivo or in vitro, and effects of direct stimulation of autonomic nerves and their effects on cardiovascular responses.

Gastrointestinal System
These include studies on gastric secretion, gastric pH measurement, gastric mucosal examination, bile secretion, gastric emptying time in vivo, and ileocecal contraction in vitro.

Other Organ Systems

Effects of the investigational drug on organ systems not investigated elsewhere should be assessed when there is a cause for concern. For example dependency potential, skeletal muscle, immune function, and endocrine functions may be investigated.

Conditions Under Which Safety Pharmacology Studies Are Not Necessary

Safety pharmacology studies are usually not required for locally applied agents, for example, dermal or ocular, in cases when the pharmacology of the investigational drug is well known, and/or when systemic absorption from the site of application is low. Safety pharmacology testing is also not necessary, in the case of a new derivative having similar pharmacokinetics and pharmacodynamics.

Timing of Safety Pharmacology Studies in Relation to Clinical Development

Prior to First Administration in Humans

The effects of an investigational drug on the vital functions listed in the essential safety pharmacology should be studied prior to first administration in humans. Any follow-up or supplemental studies identified should be conducted if necessary, based on a cause for concern.

During Clinical Development

Additional investigations may be warranted to clarify observed or suspected adverse effects in animals and humans during clinical development.

Before Applying for Marketing Approval

Follow-up and supplemental safety pharmacology studies should be assessed prior to approval unless not required, in which case this should be justified. Available information from toxicology studies addressing safety pharmacology endpoints or information from clinical studies can replace such studies.

Application of Good Laboratory Practices

The animal studies are to be conducted in an accredited laboratory. Where the safety pharmacology studies are part of toxicology studies, these studies should also be conducted in an accredited laboratory.

APPENDIX V: INFORMED CONSENT

Checklist for Study Subject's Informed Consent Documents

Essential Elements

1. Statement that the study involves research and explanation of the purpose of the research
2. Expected duration of the subject's participation

3. Description of the procedures to be followed, including all invasive procedures
4. Description of any reasonably foreseeable risks or discomforts to the subject
5. Description of any benefits to the subject or others reasonably expected from research. If no benefit is expected subject should be made aware of this.
6. Disclosure of specific appropriate alternative procedures or therapies available to the subject.
7. Statement describing the extent to which confidentiality of records identifying the subject will be maintained and who will have access to subject's medical records
8. Trial treatment schedule(s) and the probability for random assignment to each treatment (for randomized trials)
9. Compensation and/or treatment(s) available to the subject in the event of a trial-related injury
10. An explanation about whom to contact for trial related queries, rights of subjects and in the event of any injury
11. The anticipated prorated payment, if any, to the subject for participating in the trial
12. Subject's responsibilities on participation in the trial
13. Statement that participation is voluntary, that the subject can withdraw from the study at any time and that refusal to participate will not involve any penalty or loss of benefits to which the subject is otherwise entitled
14. Any other pertinent information

Additional Elements That May Be Required
a. Statement of foreseeable circumstances under which the subject's participation may be terminated by the investigator without the subject's consent.
b. Additional costs to the subject that may result from participation in the study.
c. The consequences of a subject's decision to withdraw from the research and procedures for orderly termination of participation by subject.
d. Statement that the subject or subject's representative will be notified in a timely manner if significant new findings develop during the course of the research which may affect the subject's willingness to continue participation will be provided.
e. A statement that the particular treatment or procedure may involve risks to the subject (or to the embryo or fetus, if the subject is or may become pregnant), which are currently unforeseeable.
f. Approximate number of subjects enrolled in the study.

Informed Consent form to participate in a clinical trial

Study Title:
Study Number:
Subject's Initials: Subject's Name
Date of Birth/Age:

Please initial box (Subject)

I confirm that I have read and understood the information sheet dated   []
_____ for the above study and have had the opportunity to ask questions.

I understand that my participation in the study is voluntary and that I   []
am free to withdraw at any time, without giving any reason, without my
medical care or legal rights being affected.

I understand that the Sponsor of the clinical trial, others working on the   []
Sponsor's behalf, the Ethics Committee and the regulatory authorities
will not need my permission to look at my health records both in respect
of the current study and any further research that may be conducted in
relation to it, even if I withdraw from the trial. I agree to this access. However, I understand that my identity will not be revealed in any information released to third parties or published.

I agree not to restrict the use of any data or results that arise from this   []
study provided such a use is only for scientific purpose(s)

I agree to take part in the above study.   []

Signature (or Thumb impression) of the Subject/Legally Acceptable
Representative:
Date
Signatory's Name:
Signature of the Investigator: Date:
Study Investigator's Name:
Signature of the Witness: Date:
Name of the Witness:

APPENDIX VII: UNDERTAKING BY THE INVESTIGATOR

1. Full name, address and title of the principal investigator (or investigator(s)
   when there is no principal investigator)
2. Name and address of the medical college, hospital, or other facility where
   the clinical trial will be conducted: Education, training and experience
   that qualify the investigator for the clinical trial [Attach details including Medical Council registration number, and/or any other statement(s) of
   qualification(s)]
3. Name and address of all clinical laboratory facilities to be used in the study.
4. Name and address of the ethics committee that is responsible for approval
   and continuing review of the study.
5. Names of the other members of the research team (Co- or subinvestigators)
   who will be assisting the investigator in the conduct of the investigation (s).
6. Protocol title and study number (if any) of the clinical trial to be conducted
   by the investigator.
7. Commitments:

(i)   I have reviewed the clinical protocol and agree that it contains all the necessary information to conduct the study. I will not begin the study until all necessary ethics committee and regulatory approvals have been obtained.

(ii) I agree to conduct the study in accordance with the current protocol. I will not implement any deviation from or changes of the protocol without agreement by the Sponsor and prior review and documented approval/favorable opinion from the ethics committee of the amendment, except where necessary to eliminate an immediate hazard(s) to the trial subjects or when the change(s) involved are only logistical or administrative in nature.

(iii) I agree to personally conduct and/or supervise the clinical trial at my site.

(iv) I agree to inform all subjects that the drugs are being used for investigational purposes and I will ensure that the requirements relating to obtaining informed consent and ethics committee review and approval specified in the GCP guidelines are met.

(v)   I agree to report to the Sponsor all adverse experiences that occur in the course of the investigation(s) in accordance with the regulatory and GCP guidelines.

(vi) I have read and understood the information in the investigator's brochure, including the potential risks and side effects of the drug.

(vii) I agree to ensure that all associates, colleagues and employees assisting in the conduct of the study are suitably qualified and experienced and they have been informed about their obligations in meeting their commitments in the trial.

(viii) I agree to maintain adequate and accurate records and to make those records available for audit/inspection by the sponsor, ethics committee, licensing authority, or their authorized representatives, in accordance with regulatory and GCP provisions. I will fully cooperate with any study related audit conducted by regulatory officials or authorized representatives of the sponsor.

(ix) I agree to promptly report to the ethics committee all changes in the clinical trial activities and all unanticipated problems involving risks to human subjects or others.

(x)   I agree to inform all unexpected serious adverse events to the sponsor as well as the ethics committee within seven days of their occurrence.

(xi) I will maintain confidentiality of the identification of all participating study patients and assure security and confidentiality of study data.

(xii) I agree to comply with all other requirements, guidelines, and statutory obligations as applicable to clinical investigators participating in clinical trials

1. Signature of investigator with date

APPENDIX VIII: ETHICS COMMITTEE

The number of persons in an Ethics Committee should have at least seven members. Ethics committee should appoint, from among its members, a chairperson (who is from outside the institution) and a member secretary. Other members should be a mix of medical/nonmedical, scientific and nonscientific persons, including lay public, to reflect the different viewpoints.

For review of each protocol the quorum of ethics committee should be at least five members with the following representations:

(a) Basic medical scientists (preferably one pharmacologist).
(b) Clinicians
(c) Legal expert
(d) Social scientist/representative of nongovernmental voluntary agency/philosopher/ethicist/theologian or a similar person
(e) Lay person from the community.

In any case, the ethics committee must include at least one member whose primary area of interest/specialization is nonscientific and at least one member who is independent of the institution/trial site. Besides, there should be appropriate gender representation on the ethics committee. If required, subject experts may be invited to offer their views. Further, based on the requirement of research area, for example, HIV AIDS, genetic disorders, etc., specific patient groups may also be represented in the ethics committee as far as possible.

Only those ethics committee members who are independent of the clinical trial and the sponsor of the trial should vote/provide opinion in matters related to the study.

Example of Format for Approval of Ethics Committee

To:
The Institutional Ethics Committee/Independent Ethics Committee (state name of the committee, as appropriate) reviewed and discussed your application to conduct the clinical trial entitled " . . . . . . " on . . . . (date).
The following documents were reviewed:

a. Trial Protocol (including protocol amendments), dated____Version no:
b. Patient Information Sheet and Informed Consent Form (including updates if any) in English and/or vernacular language.
c. Investigator's Brochure, dated____, Version no.____
d. Proposed methods for patient accrual including advertisement (s) etc. proposed to be used for the purpose.
e. Principal Investigator's current CV.
f. Insurance Policy/Compensation for participation and for serious adverse events occurring during the study participation.
g. Investigator's Agreement with the Sponsor.
h. Investigator's Undertaking (Appendix VII).

The following members of the ethics committee were present at the meeting held on (date, time, place).
Chairman of the Ethics Committee:
Member secretary of the Ethics Committee:
Name of each member with designation:
APPROVAL GRANTED: YES/NO

The Institutional Ethics Committee/ Independent Ethics Committee expects to be informed about the progress of the study, any SAE occurring in the course of the study, any changes in the protocol and patient information/informed consent and asks to be provided a copy of the final report.
Signed: _____ Member Secretary, Ethics Committee. Date:

APPENDIX IX: STABILITY TESTING OF NEW DRUGS

Stability testing is to be performed to provide evidence on how the quality of a drug substance or formulation varies with time under the influence of various environmental factors such as temperature, humidity and light, and to establish shelf life for the formulation and recommended storage conditions.

Stability studies should include testing of those attributes of the drug substance that are susceptible to change during storage and are likely to influence quality, safety, and/or efficacy. In case of formulations the testing should cover, as appropriate, the physical, chemical, biological, and microbiological attributes, preservative content (e.g., antioxidant, antimicrobial preservative), and functionality tests (e.g., for a dose delivery system).

Validated stability-indicating analytical procedures should be applied. For long-term studies, frequency of testing should be sufficient to establish the stability profile of the drug substance.

In general, a drug substance should be evaluated under storage conditions that test its thermal stability and, if applicable, its sensitivity to moisture. The storage conditions and the length of studies chosen should be sufficient to cover storage, shipment, and subsequent use.

Stress testing of the drug substance should be conducted to identify the likely degradation products, which in turn establish the degradation pathways, evaluate the intrinsic stability of the molecule, and validate the stability, indicating power of the analytical procedures used. The nature of the stress testing will depend on the individual drug substance and the type of formulation involved.

Stress testing may generally be carried out on a single batch of the drug substance. It should include the effect of temperatures, humidity where appropriate, oxidation, and photolysis on the drug substance.

Data should be provided for (a) photostability on at least one primary batch of the drug substance as well as the formulation, as the case may be and (b) the susceptibility of the drug substance to hydrolysis across a wide range of pH values when in solution or suspension.

Long-term testing should cover a minimum of 12 months' duration on at least three primary batches of the drug substance or the formulation at the time of submission and should be continued for a period of time sufficient to cover the proposed shelf life. Accelerated testing should cover a minimum of six months duration at the time of submission.

In case of drug substances, the batches should be manufactured to a minimum of pilot scale by the same synthetic route and using a method of manufacture that simulates the final process to be used for production batches. In case of formulations, two of the three batches should be at least pilot scale and the third one may be smaller. The manufacturing process(es) used for primary batches should simulate that to be applied to production batches and should provide products of the same quality and meeting the same specifications as that intended for marketing.

The stability studies for drug substances should be conducted either in the same container—closure system as proposed for storage and distribution or in a container—closure system that simulates the proposed final packaging. In case of formulations, the stability studies should be conducted in the final container—closure system proposed for marketing.

Stability Testing of New Drug Substances and Formulations

(i) General study storage conditions for testing drug substances and formulations

Study	Study conditions	Duration of study
Long term	30°C ± 2°C/65% RH ± 5% RH	12 months
Accelerated	40°C ± 2°C/75% RH ± 5% RH	6 months

If at any time during six months testing under the accelerated storage condition, such changes that occur cause the product to fail in complying with the prescribed standards, additional testing under an intermediate storage condition should be conducted and evaluated against significant change criteria.

(ii) Refrigerated study conditions for testing drug substances and formulations

Study	Study conditions	Duration of study
Long term	5°C ± 3°C	12 months
Accelerated	25°C ± 2°C/60% RH ± 5% RH	6 months

(iii) Deep freeze study conditions for testing drug substances and formulations

Study	Study conditions	Duration of study
Long term	−20°C ± 5°C	12 months

(iv) Drug substances intended for storage below −20°C shall be treated on a case-by-case basis.

(v) Stability testing of the formulation after constitution or dilution, if applicable, should be conducted to provide information for the labeling on the preparation, storage condition, and in-use period of the constituted or diluted product. This testing should be performed on the constituted or diluted product through the proposed in-use period.

APPENDIX X: CONTENTS OF THE PROPOSED PROTOCOL FOR CONDUCTING CLINICAL TRIALS

1. Title Page
   a. Full title of the clinical study
   b. Protocol/study number, and protocol version number with date
   c. The IND name/number of the investigational drug
   d. Complete name and address of the sponsor and contract research organization if any
   e. List of the investigators who are conducting the study, their respective institutional affiliations, and site locations
   f. Name(s) of clinical laboratories and other departments and/or facilities participating in the study.

2. Table of Contents
   A complete table of contents including a list of all appendices.
   i) Background and introduction
      a. Preclinical experience
      b. Clinical experience
         Previous clinical work with the new drug should be reviewed here and a description of how the current protocol extends existing data should be provided. If this is an entirely new indication, how this drug was considered for this should be discussed. Relevant information regarding pharmacological, toxicological and other biological properties of the drug/biological/medical device, and previous efficacy and safety experience should be described.
   ii) Study Rationale
      This section should describe a brief summary of the background information relevant to the study design and protocol methodology. The reasons for performing this study in the particular population included by the protocol should be provided.
   iii) Study objective(s) (primary as well as secondary) and their logical relation to the study design.
3. Study Design
   a. Overview of the study design: Including a description of the type of study (i.e., double-blind, multicenter, placebo-controlled, etc.), a detail of the specific treatment groups and number of study subjects in each group and investigative site, subject number assignment, and the type, sequence and duration of study periods.
   b. Flow chart of the study
   c. A brief description of the methods and procedures to be used during the study.
   d. Discussion of study design: This discussion details the rationale for the design chosen for this study.
4. Study Population
   The number of subjects required to be enrolled in the study at the investigative site and by all sites along with a brief description of the nature of the subject population required is also mentioned.
5. Subject eligibility
   a. Inclusion criteria
   b. Exclusion criteria
6. Study assessments—plan, procedures and methods to be described in detail
7. Study conduct stating the types of study activities that would be included in this section would be: medical history, type of physical examination, blood or urine testing, electrocardiogram (ECG), diagnostic testing such as pulmonary function tests, symptom measurement, dispensation and retrieval of medication, subject cohort assignment, adverse event review, etc.
   Each visit should be described separately as Visit 1, Visit 2, etc.
   Discontinued subjects: Describes the circumstances for subject withdrawal, dropouts, or other reasons for discontinuation of subjects. State how dropouts would be managed and if they would be replaced

Describe the method of handling of protocol waivers, if any. The person(s) who approves all such waivers should be identified and the criteria used for specific waivers should be provided.

Describes how protocol violations will be treated, including conditions where the study will be terminated for noncompliance with the protocol.

8. Study treatment
   a. Dosing schedule (dose, frequency, and duration of the experimental treatment) Describe the administration of placebos and/or dummy medications if they are part of the treatment plan. If applicable, concomitant drug(s), their doses, frequency, and duration of concomitant treatment should be stated.
   b. Study drug supplies and administration: A statement about who is going to provide the study medication and that the investigational drug formulation has been manufactured following all regulations. Details of the product stability, storage requirements and dispensing requirements should be provided.
   c. Dose modification for study drug toxicity: Rules for changing the dose or stopping the study drug should be provided.
   d. Possible drug interactions
   e. Concomitant therapy: The drugs that are permitted during the study and the conditions under which they may be used are detailed here. Describe the drugs that a Subject is not allowed to use during parts of or the entire study. If any washout periods for prohibited medications are needed prior to enrollment, these should be described here.
   f. Blinding procedures: A detailed description of the blinding procedure if the study employs a blind on the Investigator and/or the Subject
   g. Unblinding procedures: If the study is blinded, the circumstances in which unblinding may be done and the mechanism to be used for unblinding should be given
9. Adverse events (see Appendix XI): Description of expected adverse events should be given.
   Procedures used to evaluate an adverse event should be described.
10. Ethical Considerations: Give the summary of the following:
    a. Risk/benefit assessment:
    b. Ethics committee review and communications
    c. Informed consent process
    d. Statement of subject confidentiality including ownership of data and coding procedures
11. Study monitoring and supervision: A description of study monitoring policies and procedures should be provided along with the proposed frequency of site monitoring visits, and who is expected to perform monitoring.
    Case record form (CRF) completion requirements, including who gets which copies of the forms and any specifics required in filling out the forms CRF correction requirements, including who is authorized to make corrections on the CRF and how queries about study data are handled and how errors, if any, are to be corrected should be stated.
    Investigator study files, including what needs to be stored following study completion should be described.

12. Investigational product management
   a. Give investigational product description and packaging (stating all ingredients and the formulation of the investigational drug and any placebos used in the study)
   b. The precise dosing required during the study
   c. Method of packaging, labeling, and blinding of study substances
   d. Method of assigning treatments to subjects and the subject identification code numbering system
   e. Storage conditions for study substances
   f. Investigational product accountability: describe instructions for the receipt, storage, dispensation, and return of the investigational products to ensure a complete accounting of all investigational products received, dispensed, and returned/destroyed.
   g. Describe policy and procedure for handling unused investigational products.

13. Data Analysis:

   Provide details of the statistical approach to be followed including sample size, how the sample size was determined, including assumptions made in making this determination, efficacy endpoints (primary as well as secondary) and safety endpoints.

   Statistical analysis: Give complete details of how the results will be analyzed and reported along with the description of statistical tests to be used to analyze the primary and secondary endpoints defined earlier. Describe the level of significance, statistical tests to be used, and the methods used for missing data; method of evaluation of the data for treatment failures, noncompliance, and subject withdrawals; rationale and conditions for any interim analysis if planned.

   Describe statistical considerations for pharmacokinetic (PK) analysis, if applicable.

14. Undertaking by the Investigator (see Appendix VII).

15. Appendices: Provide a study synopsis, copies of the informed consent documents (patient information sheet, informed consent form, etc.); CRF and other data collection forms; a summary of relevant preclinical safety information and any other documents referenced in the clinical protocol.

APPENDIX XI: EXAMPLE OF FORMAT AND CONTENT OF CASE REPORT FORM FOR SERIOUS ADVERSE EVENTS OCCURRING IN A CLINICAL TRIAL

1. Patient details: (Initials and other relevant identifier (hospital/OPD record number, etc.)*

   Gender
   Age and/or date of birth
   Weight
   Height

2. Suspected drug(s)

   Generic name of the drug*
   Indication(s) for which suspect drug was prescribed or tested
   Dosage form and strength

Daily dose and regimen (specify units—e.g., mg, mL, mg/kg)
Route of administration
Starting date and time of day
Stopping date and time, or duration of treatment

3. Other treatment(s)

Provide the same information for concomitant drugs (including nonprescription/OTC drugs) and non–drug therapies, as for the suspected drug(s).

4. Details of suspected adverse drug reaction(s)

Full description of reaction(s) including body site and severity, as well as the criterion (or criteria) for regarding the report as serious. In addition to a description of the reported signs and symptoms, whenever possible, describe a specific diagnosis for the reaction. *
Start date (and time) of onset of reaction
Stop date (and time) or duration of reaction
De-challenge and re-challenge information
Setting (e.g., hospital, out-patient clinic, home, nursing home)

5. Outcome

Information on recovery and any sequelae; results of specific tests and/or treatment that may have been conducted
For a fatal outcome, cause of death and a comment on its possible relationship to the suspected reaction; Postmortem findings to be reported where relevant.
Other information: anything relevant to facilitate assessment of the case, such as medical history including allergy, drug or alcohol abuse; family history; findings from special investigations, etc.

6. Details about the Investigator*

Name
Address
Telephone number
Profession (specialty)
Date of reporting the event to licensing authority
Date of reporting the event to ethics committee overseeing the site
Signature of the investigator
Note: Information marked with an asterisk (*) must be provided.

7 Japan

Juichi Riku

Meiji Pharmaceutical University, Tokyo, Japan

INTRODUCTION

Generic drug products make up a relatively small proportion of the total pharmaceutical market in Japan. Generic drug products accounted for as little as 17.2% of the market by volume in 2007 and around 6.2% by value. These figures pale in comparison with countries such as the United States, Germany, and the United Kingdom where generic product penetration levels are more than half of the total pharmaceutical products volume.

Over the past few years, the Japanese government has launched a series of initiatives designed to boost the use of generic drug products. In 2007, the Japanese government officially announced a specific program to raise the generic product volume-based share over the next five years from 17% to more than 30% by the year 2012. They implemented a Japanese version of a system for generic substitution in April 2008. Under the reforms, pharmacists are allowed to substitute the original branded product with a generic product with the patient's consent if the prescribing doctor does not mark the "substitution not allowed" box on the prescription form accompanied by signature. The introduction of the new prescription form is a significant milestone in encouraging the dispensing of generic products by pharmacists.

However, not all doctors and pharmacists have been in favor of the generic substitution initiatives because they remain dubious of generic drug products, commonly questioning their quality and bioequivalence (BE) to the brand products. It seems that most of their lingering suspicions about generic products stem from misconceptions and a lack of understanding about rigorous multistep approval reviews of generic products in Japan.

Approval Process and Review of Generic Product Applications

The Minister of Health, Labor, and Welfare (MHLW) grants approvals to manufacture and market generic drug products as well as the originator products in accordance with the Pharmaceutical Affairs Law in Japan.

Applications are reviewed according to the submission of the product's

(a) name
(b) ingredients and their quantities
(c) specifications and test methods
(d) manufacturing method and process
(e) storage conditions and expiration date
(f) dosage and administration
(g) indications, etc.

The quality, BE, and therapeutic equivalence of the generic drug product are examined and reviewed for approval of the product as specified by the MHLW ordinances subject to the "equivalence reviews." The following data must be included with the application for marketing approval.

Specifications and test methods
Stability
Bioequivalence

All documents and data are subject to the "compliance reviews" (1) through written documentation and on-site inspections in accordance with "Standard for Collections and Preparation of Approval Review Data" (2). The MHLW may have the equivalence review of the generic drug product and compliance review performed by an independent administrative organization, the Pharmaceuticals and Medical Devices Agency (PMDA).

The generic drug product must undergo both paper and on-site good manufacturing practices (GMP) reviews or inspections for each product before approval, in precisely the same way as the original drug product was reviewed.

The MHLW will not grant an approval for a generic product unless that product is the same quality as the originator drug, and both formulations are pharmaceutically equivalent and bioequivalent to be considered therapeutically equivalent. Therefore, the PMDA will rigorously conduct an Equivalence and Compliance Review of the generic drug product in comparison with the originator drug product. The common technical document (CTD), comprising a set of specifications for inclusion in the application dossier for the registration of medicines and designed to be used across Europe, Japan, and the United States does not necessarily apply for use with generic products. Generally, it takes about a year on average for the generic drug product to be approved after the time of application.

Patent and Exclusivity

The patent term is 20 years from the time of application as a rule in Japan, but it can be extended for a maximum of 5 years. There is no limit to the number of patent extensions, unlike the United States and the European Union (EU). Applications for approval of generic drug products are acceptable prior to patent expiration of the brand product based on the "Bolar provision" (3) but approval will not be granted until the patent term has expired. Although the "Bolar provision" has not yet been written into law in Japan, the Japanese Supreme Court ruled in 1999 that research and development on generic drug products during the life of a patent is not considered to be an infringement. A system called "re-examination" is in place in Japan where an application for a generic drug product will, however, not be permitted until the re-examination period of the innovator drug product, 4 to 10 years, normally 8 years from its approval date for new chemical entities, has expired. The purpose of the Japanese re-examination is to verify the efficacy and safety of innovator drug products through postmarketing surveillance.

BIOEQUIVALENCE STUDIES

Historical Background

In 1971, the MHLW required submission of BE data in support of applications for generic drug products for the first time, and the Guideline for Bioequivalence

Studies for Generic Drugs (1971) was released, in which large animals, such as dogs and rabbits, could be used in BE studies, but humans were not required. In 1981, the 1971 Guideline was revised to require the use of humans in BE studies on the grounds that animal experiments were considered unreliable to extrapolate the results for BE to humans. However, this 1981 Guideline did not resolve all issues such as those related to validation, methods for statistical analysis, partial acceptance of animal tests, etc.

In 1997, a thoroughly revised Guideline "The Guideline for Bioequivalence Studies of Generic Products 1997" (1997 Guideline) (4) was introduced by the MHLW. Since then, BE studies for generic drug products must be conducted in accordance with the 1997 Guideline and follow-on guidelines. The Japanese National Institute of Health Sciences (NIHS) has responsibility for the preparation of guidelines and has issued several statutory guidelines for BE studies of generic drug products through the MHLW as follows:

Guidelines
(a) Guideline for Bioequivalence Studies of Generic Products, 1997 (4)

Oral immediate-release (conventional) dosage forms
Enteric-coated products
Oral controlled-release dosage forms
Other dosage forms
Dosage forms exempt from BE studies

(b) Guideline for Bioequivalence Studies for Different Strengths of Oral Solid Dosage Forms, 2000 (5).

Different strengths of oral solid dosage forms with the same dosage and administration as previously approved products, for example, 5 mg tablet as well as a 10 mg tablet

(c) Guideline for Bioequivalence Studies for Additional Dosage Forms of Oral Solid Dosage Forms, 2001 (6).

Additional dosage forms of oral solid dosage forms with the same dosage and administration as previously approved products, for example, addition of 5 mg capsule to 5 mg tablet.

(d) Guideline for Bioequivalence Studies of Generic Products for Topical Dermal Application, 2003 (7).

Generic products for topical dermal application without systemic action, such as ointments, skin patches, etc., including steroids, and others.

(e) Guideline for Bioequivalence Studies for Formulation Changes of Oral Solid Dosage Forms, 2000 (8).

Postapproval changes in the components and composition of oral solid dosage forms other than the active ingredients.

(f) Draft Guideline for Bioequivalence Studies for Changes in Manufacturing of Oral Solid Dosage Forms: Immediate-release (Conventional) and Enteric-Coated Products, 2002 (9).

Postapproval changes in manufacturing of oral solid dosage forms.
This draft guideline has not been officially implemented as of November, 2009.

(g) Revision of Guideline for Bioequivalence Studies of Generic Products, 2006 (10).

Partial revision of Guidelines (a), (b), (d), and (e).

In establishing the Japanese guideline for BE studies for generic products in 1997, the MHLW intended to harmonize it with international guidelines as much as possible, such as the WHO guideline (11) published in 1996 in the WHO Technical Repot Series as a fundamental concept. However, some significant differences in some test methods, such as dissolution testing, scale-up and postapproval changes (SUPAC), the use of the biopharmaceutical classification system (BCS) and biowaivers, etc., exist between the Japanese and other major countries' guidelines, although there are similarities in many respects.

The 1997 Guideline describes the principles and procedures relating to BE studies for generic drug products. The main objective of a BE study is to assure the therapeutic equivalence of a generic drug product (test) to the innovator product (reference) by comparing the BAs between an innovator product and generic drug product. If this is not feasible, pharmacological effects supporting efficacy ("pharrmacodynamic studies") or therapeutic effectiveness associated with the major indications ("clinical studies") should be compared.

For oral drug products, in Japan, dissolution testing of these dosage forms is required together with BE studies and plays an important role in selecting appropriate subjects for the in vivo study and supporting the in vivo equivalence when the dissolution of test and reference products is similar. Such use of dissolution testing is specific to the Japanese guidelines. Japanese experts have expressed their opinion in this regard that any difference in bioavailability (BA) between test and reference products are probably due to a difference in dissolution in the gastrointestinal (GI) tract, which is more than likely related to a difference in formulation characteristics, such as excipients and manufacturing method, amongst others. In most other countries, a difference in BA between a test and reference product is generally based only on data from human studies since apart from biowaivers granted under certain specified conditions, data from dissolution tests are not employed to assess BA/BE. The argument for the use of dissolution data to complement and aid in the assessment of BA/BE in Japan extends to the contention that dissolution tests are generally more sensitive to a difference in formulation than data from human subjects. Dissolution tests, for example, can be performed under various simulated GI conditions where pH or agitation intensity can be changed and the influence thereof investigated. If a test product shows slower dissolution than the reference product under certain conditions, the test product may result in lower BA in some subjects, such as, for example, in achlorhydric subjects. In such cases, BE should carefully be assessed using appropriate subjects, such as achlorhydric subjects, or in an appropriate target population. Use of dissolution testing for BA/BE in Japan clearly marks a distinct difference between Japanese BA/BE guidelines and those of most other countries.

Oral Immediate-Release Dosage Forms and Enteric-Coated Products

An appropriate BE study protocol including the required number of subjects and sampling intervals, etc., should be determined according to preliminary studies.

Figures 1–3 show decision trees for BE studies in human subjects.

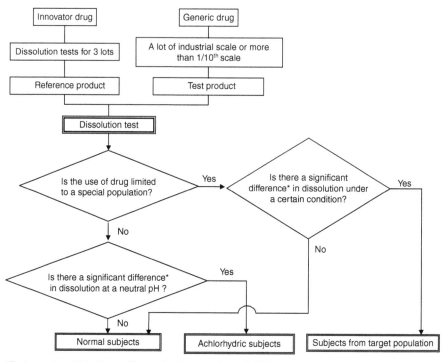

*Each meaning of "significant difference" is described in the Guideline

FIGURE 1 Bioequivalence test for oral immediate-release dosage forms and enteric-coated products.

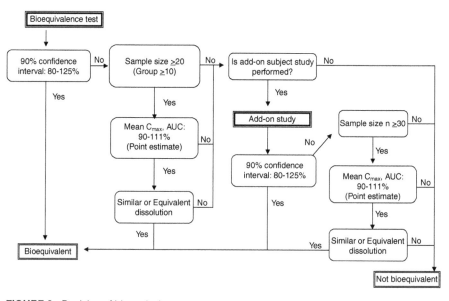

FIGURE 2 Decision of bioequivalence.

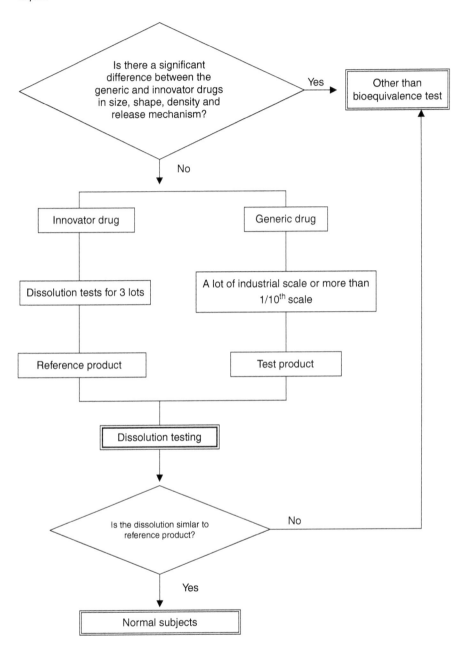

FIGURE 3 Bioequivalence test for oral controlled-release dosage forms.

Test Methods

Choice of Reference and Test Products

Dissolution tests, the details of which are hereinafter described in "Dissolution Tests" section (vide infra) should be performed on three lots of the innovator product available on the domestic Japanese market to select an appropriate reference product using the Japanese Pharmacopoeial (12) paddle method at 50 rpm. Among the three lots, the one which shows intermediate dissolution rate patterns amongst the lots tested should be selected as the reference product. The generic drug product (test) should be taken from the "Biobatch" lot, produced on an industrial scale. This batch should be at least 1/10th or larger of the intended production batch. A difference in drug content or potency between the test and reference products should be less than 5%.

Study Design

In principle, cross-over studies should be employed with random assignment of individual subjects to each group. Parallel designs can be employed for a drug with a long half-life, for example, approximately 72 hours and longer.

Number of Subjects

A sufficient number of subjects for assessing BE should be included but a minimum number has not been specified in the guidelines. If BE cannot be demonstrated because of an insufficient number, an add-on study can be performed using not less than half the number of subjects in the initial study. Provision for an add-on must be clearly stated a priori in the study protocol and if undertaken, no further add-ons are subsequently permissible.

Selection of Subjects

Healthy adult human subjects (volunteers) should be employed, usually male in practice, although not specified in the guidelines. When the test and reference products show a significant difference in dissolution around pH 6.8, subjects with low gastric acidity (achlorhydric subjects) should be employed unless the application of the drug is limited to a special population. Since achlorhydria is quite common in Japan, and significant differences in BE may not be shown in subjects with normal gastric acidity, it is important to confirm BE in subjects with lesser gastric acidity, such as around pH 6.8. This rule does not apply to enteric-coated products. If the use of the drug is limited to a special population and the test and reference products show a significant difference in dissolution even under one of the conditions of the dissolution test, the in vivo test should be performed using subjects from the target population.

When it is unfavorable to use healthy subjects because of potent pharmacological action or adverse effects, patients should be employed. If the clearance of the drug differs to a large extent amongst subjects due to genetic polymorphism, subjects with higher clearance should be employed.

Single- Versus Multiple-Dose Studies

Single-Dose Studies. One dose unit or the usual clinical dose should generally be administered. A higher dose which does not exceed the maximum dose of the dosage regimen may be employed if analytical difficulties relating to detection sensitivity exist. Generally, BE studies should be performed using single-dose studies. If necessary, multiple-dose studies may be employed only if such products are intended to be chronically administered to patients.

Immediate-release dosage forms are usually administered to subjects with 100 to 200 mL water (normally 150 mL) after fasting for at least 10 hours. Fasting should continue for at least 4 hours postdose. If a postprandial dose is specified in the dosage regimen and if BA in the fasting state is very low or a high incidence of severe adverse events is anticipated, a fed study could be used. For a fed study of immediate-release or enteric-coated dosage forms, a low-fat diet of 700 kcal or less containing not more than 20% by energy of the lipid should be employed. Although fasting studies are generally used for most oral dosage forms, for oral controlled-release products both fasted and fed studies are required and a "high-fat" diet is recommended for such products (see "Test Methods" section later in the chapter). The meal should be eaten within 20 minutes, and the drug products administered according to the dosing regimen or 30 minutes after the meal, if the dosing time is not indicated in the regimen.

Multiple-Dose Studies. Drugs should, in principle, be administered under fasting conditions as in single-dose studies and when repeatedly given, administration should be between meals (more than 2 hours after a meal) at regular intervals. C_{max} and AUC_τ are used as BE metrics in multiple-dose studies.

Measurement of Drug Substances

(a) Biological fluids to be sampled: Generally, blood samples (plasma/serum) should be employed although urine may be used if there is a rationale to justify its use.

(b) Sampling schedule: Blood samples should be taken at a frequency sufficient to adequately assess C_{max}, AUC, and any other relevant parameters. There should be at least seven sampling points including zero time, 1 point before C_{max}, 2 points around C_{max}, and 3 points during the elimination phase. Sampling should be continued until the area under the blood concentration–time curve at time t (AUC_t) is >80% of AUC_∞, normally more than three times the elimination half-life after t_{max}. However, when the elimination half-life of the parent drug or active metabolite(s) is extremely long (>72 hours), blood samples should be collected for at least 72 hours.

(c) Drug substances to be measured: Generally, the parent compound should be measured although major active metabolite(s) may be measured instead of the parent under certain circumstances. For example, if the parent compound cannot be measured, in the case of chiral drug compounds, a chiral compound that contributes to major pharmacological effects should be measured. However, stereoselective analysis is not required if stereoselective differences in pharmacokinetics is not a concern.

(d) Assay: Analytical methods should be fully validated with respect to specificity, accuracy, precision, linearity, limit of quantitation (LOQ) limit, and stability of the analyte. Several references have been cited in the MHLW's Q&A (13), but neither specified as formal guidelines nor included in Japanese BE guidelines.

Washout Periods
Washout periods in cross-over studies between administration of test and reference products should usually be more than five times the elimination half-life of the parent compound or active metabolites.

Bioequivalence Acceptance Range
The acceptance range for BE is 0.80 to 1.25 as the geometric mean ratios of the average values of AUC and C_{max} of the test to reference product, using 90% confidence intervals. These standards must be met on log-transformed parameters calculated from the measured data.

Parameters to be Assessed
When blood samples are used, AUC_t and C_{max} should be subjected to the BE assessment used in single-dose studies.

Logarithmic Transformation
The pharmacokinetic parameters, C_{max} and AUC_t for immediate-release dosage forms and C_{max} and AUC_τ for oral controlled-release products should be statistically analyzed after logarithmic transformation.

Statistical Analysis
The 90% confidence interval or two one-sided t tests with a significance level of 5% should be used. When an add-on subject study is performed and there are no fundamental differences between the two studies in formulation, design, assay and subjects, data from the initial and add-on subject studies can be pooled and statistically analyzed.

   It is interesting to note that in Japan, even though the 90% confidence interval may be outside the 0.8 to 1.25 acceptance range, test products may be accepted as bioequivalent, **provided** the following **three** conditions have been satisfied:

(1) The total sample size of the initial BE study is not less than 20 ($n = 10$/group) or pooled sample size of the initial and add-on subject studies is not less than 30.
(2) The differences in average logarithmic values of AUC and C_{max} between two products are between log (0.90) to log (1.11).
(3) Dissolution behavior of test and reference products is evaluated to be similar under all dissolution testing conditions.

   Additional metrics such as AUC_∞, t_{max}, MRT, k_{el}, etc., should also be subjected to statistical assessment, and if a significant difference is detected in these reference or secondary parameters between reference and test products, an explanation must be given to justify that such a difference(s) is/are not considered to affect the clinical outcomes.

Dissolution Tests
Dissolution tests should be performed, using suitably validated dissolution test methods and analytical assay.

Testing Time

Dissolution testing should generally be performed for two hours in pH 1.2 medium and six hours in other test fluids. However, testing can be terminated whenever the average dissolution rate of the reference product reaches 85%.

Testing Conditions

Apparatus: Paddle apparatus specified in the JP (12).
Test solutions

(a) Products containing acidic drugs
   (i) Test at 50 rpm in the following three types of dissolution media: pH 1.2, 5.5 to 6.5, 6.8 to 7.5, and also in water as well, that is, a single medium between 5.5 and 6.5, and one between 6.8 and 7.5.
   (ii) Test at 100 rpm in either one of the following three types of dissolution media: pH 1.2, 5.5 to 6.5, 6.8 to 7.5
(b) Products containing neutral or basic drugs, and coated products
   (i) Test at 50 rpm in the following three types of dissolution media: pH 1.2, 3.0 to 5.0, 6.8, and also in water as well in each of these media.
   (ii) Test at 100 rpm in either one of the following three types of dissolution media: pH 1.2, 3.0 to 5.0, 6.8
(c) Products containing low-solubility drugs
   (i) Test at 50 rpm in the following dissolution media: (1) pH 1.2, (2) pH 4.0, (3) pH 6.8, (4) Water, (5) pH 1.2 + polysorbate 80, (6) pH 4.0 + polysorbate 80, (7) pH 6.8 + polysorbate 80 and at 100 rpm in either one of (5) or (6) or (7) above.
(d) Enteric-coated products
   (i) Test at 50 rpm in the following dissolution media: (1) pH 1.2, (2) pH 6.0, (3) pH 6.8, and at 100 rpm in (2) above.

Similarity of Dissolution Profiles

Average dissolution rates of test products should be compared with those of reference products. If the f_2 function is used for evaluation, the judgment is based on the annexure appended in the guideline. It should be emphasized that the judgment of similarity in dissolution rate does not mean BE but rather based on the thought that there might be less possibility that test products are not bioequivalent if the dissolution rates of test products are similar to the reference products.

   If the results meet any one of the following acceptance criteria under all testing conditions, the products are judged to be similar.

Acceptance Criteria

The average dissolution from the reference product reaches 85% or greater within 15 minutes: the average dissolution from the test product also reaches 85% or greater within 15 minutes or does not deviate by more than 15% from that of the reference product at 15 minute.

   The average dissolution from the reference product reaches 85% or greater between 15 and 30 minutes: the average dissolution from the test product does not deviate by more than 15% from that of the reference product at two time points when the average dissolution from the reference product is around 60% and 85%.

The average dissolution from the reference product does not reach 85% or greater within 15 minutes: the criteria are specified in details in the guideline (10).

PHARMACODYNAMIC STUDIES

These studies are performed to establish the equivalence of products using pharmacological activity in humans as an assessment index and may be applied to drug products that do not produce measurable concentrations of the parent drug or active metabolite in blood or urine or whose BA does not reflect therapeutic effectiveness.

CLINICAL STUDIES

These studies are performed to establish the therapeutic equivalence of drugs using clinical endpoints as assessment index. If BE studies and pharmacodynamic studies are neither possible nor inappropriate, this type of study can be used.

ORAL CONTROLLED-RELEASE DOSAGE FORMS

Test methods, testing conditions, and assessment of BE studies for oral controlled-release dosage forms are similar to those of immediate-release oral dosage forms and enteric-coated products. Figure 3 shows a decision tree of BE for oral controlled-release dosage forms.

Reference and Test Products

A generic controlled-release dosage forms should not significantly differ from the reference product in size, shape, specific gravity, and release mechanism. The dissolution characteristics of the test product must be similar to that of the reference, as described later.

Test Methods

Bioequivalence studies should be performed using single-dose studies in both the fasting and fed states. In fed studies, a high-fat diet of 900 kcal or more containing 35% or more lipid content should be used. The meal should be eaten within 20 minutes, and drugs administered within 10 minutes thereafter.

When the possibility of severe adverse events may occur after dosing in the fasting state, the fasting dose studies can be replaced with postprandial dose studies with the low-fat meal employed in the study for immediate-release oral dosage forms and enteric-coated products (see "Single- Versus Multiple-Dose Studies" section)

Dissolution Tests

Testing Time

Dissolution testing should generally be performed over a 24-hour period but may be terminated after two hour in pH 1.2 medium and whenever the average dissolution rate of the reference product reaches 85%.

Testing Conditions
The test should be carried out under the following conditions using the paddle apparatus and either the rotating basket or disintegration testing apparatus.

 Paddle apparatus:
 Test at 50 rpm using the following dissolution media: pH1.2, (2) pH 3.0–5.0, (3) pH 6.8–7.5, (4) water, (5) 1% polysorbate 80 added to pH 6.8–7.5 medium and at 100 rpm and 200 rpm in dissolution medium of pH 6.8–7.5.

 Acceptance criteria for similarity and equivalence of dissolution behavior
 If the results meet any one of the following criteria under all testing conditions, the test products are judged to be similar or equivalent to the reference product when the average dissolution from the reference product reaches 80% or greater at the time specified under at least one testing condition (see "Similarity of Dissolution Profiles" section). If the f_2 function is used for evaluation, the judgment is based on the Attachment in the Guideline.

Acceptance Criteria

(1) When the average dissolution from the reference product reaches 80% or greater at the time specified and the average dissolution from the test product does not deviate by more than 15% (in this case, judged as similar) or 10% (in this case, judged as equivalent) from those of the reference product at three time points when the average dissolution from the reference product is around 30%, 50%, and 80%.

(2) When the average dissolution from the reference product reaches 50% or greater but does not reach 80% at the time specified and the average dissolution from the test product does not deviate by more than 12% (in this case, judged as similar) or 8% (in this case, judged as equivalent) from those of the reference product at the appropriate time point when the average dissolution from the reference product is half the average dissolution and at the time specified.

(3) When the average dissolution from the reference product does not reach 50% at the time specified and the dissolution from the test product does not deviate by more than 9% (in this case, judged as similar) or 6% (in this case, judged as equivalent) from those of the reference product at the appropriate time point when the average dissolution from the reference product is half the dissolution and at the time specified.

OTHER DOSAGE FORMS

Bioequivalence studies for products for topical application to the skin should follow the *Guidance for Bioequivalence Studies of Generic Products for Topical Dermal Use 2003* (7) and "Revision of Guideline for Bioequivalence Studies of Generic Products, 2006" (10).

Products for Topical Dermal Use

The following are amongst the tests to be considered for the assessment of the BE of topical dermal drug products:

Dermatopharmacokinetic test
Pharmacological test
Test for measuring unabsorbed amount

Pharmacokinetic test
Clinical trial
In vitro efficacy test
Animal test

The most suitable test from those already mentioned should be selected by considering the characteristics of the topical product. If other alternative appropriate tests are available, they may be employed.

The testing procedure, analytical method, stability of drugs during storage and analysis, etc., should be validated. Detailed standard operating procedures (SOPs) should be prepared for each test, involving the application and removal of products, recovery of the sample, measurements of pharmacological response, skin stripping procedures, and analytical method, because the testing procedures for topical products are generally complicated.

Dermatopharmacokinetics Test

This test is used to assess BE by comparing the amount of active ingredient from the test and reference products penetrating into the *stratum corneum*. Topical drugs are generally distributed into the *stratum corneum* and reach the epidermal cells. Thus, BA in the skin can be estimated by measuring the amount of the drug in the *stratum corneum* by means of skin stripping using adhesive tapes. This method is applicable to topical drugs whose site of action is the *stratum corneum* itself or deeper.

Parameters for BE assessment include drug recovery amount or the mean drug concentration at a steady state in the *stratum corneum*. The parameters should be logarithmically transformed and the 90% confidence interval of the difference in mean values of parameters between reference and test products should be calculated by a parametric method.

Pharmacological Test

This test is used to assess BE by using the pharmacological response of the topical product, which correlates with clinical efficacy or BA. Topically applied corticosteroids produce a vasoconstrictor effect depending on the drug uptake into the skin that results in skin blanching. This pharmacological response correlates with clinical efficacy of topical corticosteroids and can be used as a measure to assess the BE of topical corticosteroid products. The AUEC (area under the effect curve) values are used to assess BE of such products. When an instrumental method is used such as a, chromameter, to measure the degree of skin blanching, the AUEC values should not be log-transformed. When a visual assay is employed to measure skin blanching, the 90% confidence interval of the difference in mean AUEC values between test and reference products is calculated by a nonparametric method or parametric method after the logarithmic transformation of the AUEC values (14).

Test for Measuring Unabsorbed Drug

This test is used to estimate the amount of drug absorbed into the skin from the amount remaining in the product following application. However, the use of this test is generally limited because the drug uptake from topical products in general is usually very low, making it difficult to estimate the uptake precisely, although this test may be useful if precise measurements can be made. The amount distributed into the skin following application of the product is used as a measure of BE. In principle, the data should be logarithmically transformed, and the 90% confidence interval of the difference of the mean parameter values between test and reference products is calculated by a parametric method.

Pharmacokinetic Test
This test is used to assess BE by using pharmacokinetic parameters from blood concentration–time curves after product application. Pivotal tests should be performed according to the 1997 Guideline (4). AUC or C_{ss} (steady-state drug concentration) is used as the parameter to assess BE. In principle, the data should be logarithmically transformed, and the 90% confidence interval confidence of the difference in mean parameter values between test and reference products should be calculated by a parametric method.

Clinical Endpoint Test
This test is used to assess BE by using a suitable clinical endpoint response, which should be selected by considering the clinical property of the drug. This test should be performed using a statistically sufficient number of patients. The appropriate acceptable range for BE should be established for each drug to judge the BE of clinical efficacy of reference and test products.

In Vitro Efficacy Test
This test is used to assess BE by using in vitro activity as the index. The test may be applicable for topical drugs that are not intended to penetrate the *stratum corneum*, such as bactericides, disinfectants, and antiseptics whose active site is on the surface of the skin or which are applied to superficial affected sites. The in vitro efficacy tests do not include drug release tests for topical drugs where only physicochemical parameters are measured.

Animal Tests
This test is used to assess BE by using a pharmacological response produced on the skin. The test may be applicable for topical drugs such as bactericides, disinfectants, antiseptics, hemostatics, and wound repair agents whose active site is the surface of the skin and which are not intended to penetrate the *stratum corneum*. The appropriate acceptable range for BE should be established for each drug to assess BE.

ADDITIONAL DOSAGE FORMS, DIFFERENT STRENGTHS, AND PRODUCT CHANGES (ORAL DOSAGE FORMS)

Additional Dosage Forms
Bioequivalence studies for additional dosage forms of oral solid dosage forms for example, such as the addition of a 5-mg capsule to an approved product series of tablets, should follow the "Guideline for Bioequivalence Studies for Additional Dosage Forms of Oral Solid Dosage Forms 2001" (6). This guideline does not apply, as a rule, to oral controlled-release dosage forms. In principle, BE studies should follow the 1997 Guideline except for the following:

(a) For enteric-coated formulations, postprandial dose administration is required in addition to a fasting study.

(b) When a difference in the time of onset of action might affect the clinical usefulness of a drug, acceptance criterion for BE should be determined for t_{max} in addition to the usual criteria of C_{max} and AUC. According the MHLW Q&A (15), products may be considered to be equivalent if the difference in mean t_{max} between reference and test products is less than 20 minutes except for anti-inflammatory agents where C_{max} and AUC_t where an even narrower difference in mean t_{max} between reference and test products may be required in addition to meeting the 90% confidence interval criteria for C_{max} and AUC.

Different Strengths

Bioequivalence studies for different strengths of solid oral dosage forms, such as additional strength(s) of the same dosage form, for example adding a 5-mg tablet to a series of approved tablets of different strengths, should follow the "Guideline for Bioequivalence Studies for Different Strengths of Oral Solid Dosage Forms 2000" (5) and the 2006 revision (10), which applies to products that contain the same active ingredient, in the same dosage form and used in the same dosage regimen as the product already approved but differing in strength.

Reference Product

The reference product is the original innovator product that has previously been approved in Japan on the basis of therapeutic efficacy and safety data established by clinical trials, and is selected from three batches of the innovator product on the domestic market. Dissolution tests are generally performed on three lots of the innovator available on the domestic market to select the appropriate batch for use as the reference product in the biostudy. Among the three lots, the one which shows intermediate dissolution patterns between the lots should be selected as the reference product. It should be noted that the reference product need not always be the innovator product but can sometimes also be an approved generic product.

For example, where a range of innovator products, such as say, a 10-mg and 20-mg tablet are available on the market in Japan, and only a 10-mg generic product has previously been approved on the basis of a BE study. Should the same applicant wish to market a 20-mg generic tablet, either the 20-mg innovator product or the 10 mg previously approved generic can be used as a reference product.

Types of Formulation Changes

The type of formulation change is classified according to five different levels, A, B, C, D, and E, and is decided by the degree of the change in formulation from the original formulation, as shown in Table 1 in the case of immediate-release uncoated products. The tests required for BE assessment differ depending on the type of change in formulation from the approved product, as shown in Table 2. This classification is based on the US FDA SUPAC-IR guideline (16).

Level A: When a small change in the amount of excipients (e.g., starch, lactose) is unlikely to alter the quality and/or BA of the product and the ratios of compositions are identical between test and reference products, in vivo studies are not required for any products if the test and reference

TABLE 1 Level of Change in Excipients in Immediate-Release Uncoated Products

Level	Total Excipient Changes/Original Formulation (w/w%)
A	Trace change (color/flavor, etc.)
B	5.0
C	10
D	15

products are shown to be equivalent with respect to dissolution at a specified pH. If the test and reference products are not equivalent with respect to their dissolution, BE tests should be performed according to the 1997 Guideline.

Levels B, C, and D: When a formulation change in which the ratios of compositions are not identical, the changes should be determined according to levels of changes in individual excipients and categorized excipients. No in vivo studies are required for any products if the test and reference products are equivalent with respect to dissolution at multiple pHs at Level B. Products at Levels C and D are classified on the basis of a therapeutic window range and dissolution rate of the products, where the required tests differ (Table 2). If the test and reference products are not equivalent with respect to dissolution or do not meet the criteria described in the guideline, BE tests should be performed.

Level E: Bioequivalence tests should be performed according to the 1997 Guideline (4).

TABLE 2 Level of Formulation Change and Tests

Level	Dosage form	Therapeutic range	Solubility	Dissolution	Test
A	–	–	–	–	A single-dissolution test
B	–	–	–	–	Multiple-dissolution tests
C	IR, DR	Not narrow	Not low	–	Multiple-dissolution tests
	IR, DR	Not narrow	Low	–	In vivo test
	IR, DR	Narrow	Not low	≥85%, 30 min	Multiple-dissolution tests
				<85%, 30 min	In vivo test
	IR, DR	Narrow	Low	–	In vivo test
	CR	Not narrow	–	–	Multiple-dissolution tests
	CR	Narrow	–	–	In vivo test
D	IR	Not narrow	Not low	≥85%, 30 min	Multiple-dissolution tests
	Other IR DR, CR	–	–	–	In vivo test
E	–	–	–	–	In vivo test

IR: immediate release (conventional), DR: delayed release (enteric coated), CR: controlled release.

INTRAVENOUS AQUEOUS SOLUTIONS

Intravenous aqueous solutions are exempted from BE studies. However, BE studies for injectables including intravenous suspension/emulsion solutions other than intravenous aqueous solutions should follow the usual BE studies required for extravascular dosage forms.

WAIVERS OF IN VIVO BIOEQUIVALENCE STUDIES

Human BE studies may be waived in certain circumstances such as assessment of additional dosage forms (see "Additional Dosage Forms" section)/different strengths (see "Different Strengths" section) of oral solid dosage forms and intravenous aqueous solutions (see "Intravenous Aqueous Solutions" section). However, there exists a major difference between the Japanese guidelines and the SUPAC-IR in the use of the biopharmaceutical classification system (BCS). Japanese guidelines do not use the BCS to grant waivers on the basis of solubility and permeability requirements. Conditions for waivers of in vivo studies vary with the dissolution characteristics upon which greater importance is placed in Japan (17).

ACCEPTANCE OF FOREIGN BE DATA

According to a notice in 1998 by the MHLW (18), data on human BE studies conducted overseas and submitted for the approval of a generic product in Japan, requires that sufficient appropriate data should be submitted to assess the possibility of extrapolating foreign acquired data to Japanese ("bridging studies") requirements. However, the MHLW suggests that all BE studies should be conducted preferably in Japan as a basic policy. Furthermore, the reference product used in BE studies intended for the Japanese market must be the domestic innovator product that has been approved for use in Japan.

SUMMARY AND CONCLUSIONS

Current BE methods in Japan and other countries are designed to provide assurance of therapeutic equivalence of generic drug products with the innovator drug product; thereby justifying generic substitution. Generic drug products must be pharmaceutically equivalent and bioequivalent to be considered therapeutically equivalent for approval. In Japan, the requirement that applications for generic drug products must contain information showing that a generic drug product is bioequivalent to the innovator drug product was initiated by the MHLW in 1971. The MHLW announced the fundamental guideline for BE studies for generic products in 1997, and then followed this by several additional derived guidelines.

Although the Japanese guidelines are harmonized to international guidelines in many ways, there exist some significant differences in some parts, such as placing more importance on dissolution testing, and not using BCS to permit biowaivers, etc. (19).

It is however clear that regulatory quality and BE evaluation of generic products in Japan is quite rigorous through the Equivalence Reviews and the Compliance Reviews.

It is hoped that the use of generic products will increase as a number of obstacles are being overcome step by step in Japan, with a better understanding of the Japanese assessment of BE and the associated requirements of quality, safety, and efficacy.

REFERENCES

1. Ministry of Health, Labor and welfare. Enforcement of the partial amendments to Japanese Pharmaceutical Affairs Law. Yakuhatsu Notification No.421, March 27, 2007.
2. Japanese Pharmaceutical Affairs Law. Article 14, Paragraph 3, June, 1996.
3. Annette Cunningham. Bolar Provision: A global history and the future for Europe. http://www.genericsweb.com/index.php?object_id = 238.
4. Ministry of Health, Labor and Welfare. Guideline for Bioequivalence Studies of Generic Products, 1997. http://www.nihs.go.jp/drug/be-guide(e)/Generic/be97E.html.
5. Ministry of Health, Labor and Welfare. Guideline for Bioequivalence Studies for Different Strengths of Oral Solid Dosage Forms, 2000. http://www.nihs.go.jp/drug/be-guide(e)/strength/strength.html.
6. Ministry of Health, Labor and Welfare. Guideline for Bioequivalence Studies for Additional Dosage Forms of Oral Solid Dosage Forms, 2001.
7. Ministry of Health, Labor and Welfare. Guideline for Bioequivalence Studies of Generic Products for Topical Dermal Application, 2003. http://www.nihs.go.jp/drug/be-guide(e)/Topical_BE-E.pdf.
8. Ministry of Health, Labor and Welfare. Guideline for Formulation Changes of Oral Solid Dosage Forms, 2000. http://www.nihs.go.jp/drug/be-guide(e)/form/form-change.PDF.
9. Ministry of Health, Labor and Welfare. Draft Guideline for Bioequivalence Studies for Changes in Manufacturing of Oral Solid Dosage Forms: Immediate-release (Conventional) and Enteric Coated Products, 2002.
10. Ministry of Health, Labor and Welfare. Revision of Guideline for Bioequivalence Studies of Generic Products, 2006.
11. WHO Expert Committee. WHO Technical Report Series No. 863: Multisource (generic) pharmaceutical products: WHO guidelines on registration requirements to establish interchangeability, 1996. http://apps.who.int/medicinedocs/es/d/Js5516e/19.html.
12. Ministry of Health, Labor and Welfare. Japanese Pharmacopoeia 15th 2006. English ed. 116. http://jpdb.nihs.go.jp/jp15e/JP15.pdf.
13. Ministry of Health, Labor and Welfare. Q & A on Revision of Guideline for Bioequivalence Studies of Generic Products, Shinsakanrika, Administrative Communication, November 24, 2006.
14. Hauschke D, Steinijans VW, Diletti E. A distribution—Free procedure for the statistical analysis of bioequivalence studies. J Clin Pharmacol Ther Toxicol 1990; 28(2):72–78.
15. Ministry of Health, Labour and Welfare. Q & A on Guideline for Bioequivalence Studies for Additional Dosage Forms of Oral Solid Dosage Forms, 2001. Shinsakanrika, Administrative Communication, May 31, 2001.
16. Center for Drug Evaluation and Research. Guidance for Industry, Immediate Release Solid Oral Dosage Forms—Scale-Up and Postapproval Changes: Chemistry, Manufacturing, and Controls, In Vitro Dissolution Testing, and In Vivo Bioequivalence Documentation. Center for Drug Evaluation and Research (CDER), November 1995, CMC5. Federal Register, vol. 60, no. 230. http://www.fda.gov/downloads/Drugs/GuidanceComplianceRegulatoryInformation/Guidances/ucm070636.pdf
17. Nobuo Aoyagi. Japanese guidance on bioavailability and bioequivalence. Eur J Drug Metab Pharmacokinet 2000; 27(1):28–31.
18. Ministry of Health, Labor and Welfare. Q & A on Guideline for Bioequivalence Studies of Generic Products, 1997, Shinsakanrika, Administrative Communication, October 30, 1998.
19. Kiyoto Nakai, Masahiko Fujita, and Hiroyuki Ogata. International harmonization of bioequivalence studies and issues shared in common. Yakugaku Zasshi 2000; 120(11):1193–1200.

8 South Africa

Isadore Kanfer and Roderick B. Walker

Faculty of Pharmacy, Rhodes University, Grahamstown, South Africa

Michael F. Skinner

Biopharmaceutics Research Institute, Rhodes University, Grahamstown, South Africa

INTRODUCTION

The Medicines Control Council (MCC) in South Africa is a statutory body that was established in terms of the Medicines and Related Substances Control Act (MRSCA), 101 of 1965, to oversee the regulation of medicines in South Africa. Applicants are required to submit evidence of quality, safety, and efficacy for new drugs and medicinal products as well as for the registration of generic (multisource) medicinal products. In the latter instance, bioequivalence (BE) data can be used as a surrogate measure of safety and efficacy. To facilitate the registration process for generic medicines, guidelines have been prepared to serve as a recommendation to applicants wishing to submit data in support of the registration of such medicines (1–4).

The mandate of the South African MCC is to safeguard and protect the public through ensuring that all medicines that are sold and used in South Africa are safe, therapeutically effective, and consistently meet acceptable standards of quality. In this respect, all submissions must provide the necessary data for **quality, safety, and efficacy** to register an interchangeable multisource (generic) pharmaceutical product (medicine) and thereby infer that it is therapeutically equivalent.

Several important definitions and specific terms have been described in the relevant Act (5) as well as in the associated guidelines. For the most part, these considerations parallel the requirements for multisource interchangeability as defined within the United States and the European Union (EU). However, a notable exception is that the reference (comparator) product used in the BE study need not be the domestic innovator product. This provision permits BE data generated between the test (generic) product and a foreign reference product to be submitted as proof of safety and efficacy. The implication here is that a BE study undertaken with the primary objective of gaining market access in a foreign country is simply "piggy-backed" in the dossier for South African registration as a secondary consideration. This has resulted in the development of a market for the sale of dossiers where the same BE data are used by several generic drug companies to gain access to the South African market without the need to undertake the necessary studies to establish interchangeability between that generic and the innovator products being sold in South Africa.

Historical Background

Proof of safety and efficacy of generic medicines have in the past been based upon requirements described in "official" notices or circulars issued by the MCC. In fact, registration requirements for generic medicines, particularly with respect to safety and efficacy, have not been well defined until fairly recently. For example, in some instances BE testing was mentioned as a requirement whereas in others this requirement has been optional depending on the interpretation of the MCCs issued notices and/or circulars. In many instances, only in vitro dissolution testing was required on the basis of Circular 14/95 that was first issued in the early 1990s and subsequently updated in 1995 (6).

Part 2.2.1 of Circular14/95 (6) (see Appendix 1) provided for the use of comparative dissolution studies of the test and reference products as a proof of efficacy. The dissolution requirements were contingent on there being a United States Pharmacopeia (USP) monograph for the active ingredient, which included dissolution requirements and provided that the active was not included in a compiled list, which contained 96 drugs and drug combinations listed in alphabetic order (Appendix 1). No reasons were given why those specific compounds were on the list. The assay results of both the reference and test products were required as well as content uniformity test results and dissolution testing in three media where required as follows:

i) The first medium must be that specified in the USP monograph.
ii) The other two media shall generally be from the following, spanning a wide pH range and not the same as the medium specified in the USP:
   a) An acidic medium e.g., Gastric fluid USP
   b) Water
   c) An alkaline medium e.g., intestinal fluid USP

Dissolution testing was required to be carried out according to the USP dissolution requirements specified in the monograph using at least six units of the product, the apparatus, medium volume, and rotation speed. For the other two media, a further six units of both the test and reference products were required to be tested using the specified USP dissolution apparatus, medium volume, and rotation speed specified in the monograph.

All dissolution testing was required to be multipoint studies and the results presented in a tabulated and graphical form. However, no indication was given regarding acceptance criteria for the declaration of BE. In the case where the active is insoluble in the other two media, a motivation could be submitted to the MCC for the omission of further testing in the other two media.

Proof of efficacy of vitamins or vitamins and mineral combinations and also for phenolphthalein and sucralfate products was accepted on the basis of disintegration testing where the disintegration test needed to be carried out according to the USP disintegration test for nutritional supplements for the vitamins or the general disintegration test for the other substances.

The requirements for proof of efficacy for the following types of products are listed here:

1) For all products with an antacid or acid neutralizing claim, the acid neutralizing capacity test included in the USP was required.

2) For all products with a bacteriostatic/bactericidal/antiseptic claim, microbial growth inhibition zones could be used.
3) For all creams, ointments and gels, proof of release by membrane diffusion as proof of efficacy was recommended.
4) For inhalations, the Anderson sampler or equivalent apparatus could be used for particle size distribution.
5) For cortisone containing creams and ointments, the "Blanching test" was specified for use with these products (it is interesting to note that as bioequivalence testing was not mandated per se in that circular that the Blanching test was the only comparative test that was required to be undertaken in human subjects).

This guideline (i.e., Circular 14/95) made provision for the use of any other method, which an applicant wished to submit, provided the rationale for the particular method was included.

It is interesting to note that there are still products on the South African market, which were registered according to the requirements described in Circular 14/95 (6). A matter of concern here is that in many cases the innovator product used for the comparison may no longer be on the market and therefore a generic product that is the market leader and as such required to be used as the reference product in BE testing may well be a product that has never been assessed for safety and efficacy in a BE study.

It was only in the early 2000s that the requirements for the registration and market approval of generic medicines were published as official guidelines. During 2002, legislation that required that multisource (generic) medicines registered in South Africa must be offered to all patients when a physician prescribes an innovator product was introduced.

In terms of the Medicines and Related Substances Control Act, 1965 (Act No. 101 of 1965) as amended by Act No. 90 of 1997 (5) and Act 59 of 2002 (7), provision is made for generic substitution in Section 22F.

"Subject to subsections (2), (3), and (4) of the Act, a pharmacist or a person licensed in terms of Section 22C shall:

a) *inform all members of the public who visit the pharmacy or any other place where dispensing takes place, as the case may be, with a prescription for dispensing, of the benefits of the substitution for a branded medicine by an interchangeable multi-source medicine; and*
b) *dispense an interchangeable multi-source medicine prescribed by a medical practitioner, dentist, practitioner, nurse or other person registered under the Health Professions Act, 1974, unless expressly forbidden by the patient to do so."*

Furthermore, subsection (4) states that "A pharmacist shall not sell an interchangeable multi-source medicine

* *if the person prescribing the medicine has written in his or her own hand on the prescription the words 'no substitution' next to the item prescribed;*
* *if the retail price of the interchangeable multi-source medicine is higher than that of the prescribed medicine; or*
* *where the product has been declared not substitutable by the council."* (4)

The above guidelines therefore preferentially mandate the use of generic medicines wherever possible. In light of the above guidelines, it is clear that the intention of this legislation was to make medicines more affordable and available to the wider South African public.

A notable anomaly is the reference made to a negative list (4) (see earlier, i.e., *where the product has been declared not substitutable by the council"*) wherein various products are listed as "nonsubstitutable." The MCC provides regulations and guidelines on how to conduct a BE assessment of generic medicines and included are acceptance criteria for such products that are used to declare their eligibility for registration and marketing on the basis of an acceptable safety and efficacy profile. Hence, when a generic product is proven to be bioequivalent with the reference product, it is clearly expected that there are no mitigating or other factors that can assume otherwise. It is therefore inconsistent that once a generic product has been shown to be bioequivalent, further measures taken a posteriori, to include the product on a nonsubstitutable list, are used to disallow substitution of that medicine for the innovator product when a BE study has shown it to be bioequivalent/therapeutically equivalent. Such an action by the MCC to renege on requirements for substitution, which have been met, questions the validity of the regulations, guidances and policies. It would be more appropriate that sponsors be prospectively informed that a product would/could not be considered substitutable since that would obviate the efforts and expense of undertaking a BE trial on such products.

DESIGN AND CONDUCT OF BIOEQUIVALENCE STUDIES FOR ORALLY ADMINISTERED PHARMACEUTICAL PRODUCTS

Study Design
This is described under Section 3.1 of the Biostudies guidelines (2), which is introduced with the statement "The study should be designed in such a way that the formulation effect can be distinguished from other effects. If the number of formulations to be compared is two, a balanced two-period, two-sequence crossover design is considered to be the design of choice." Other designs such as well-established parallel designs for very long half-life substances may also be considered provided the study design and the statistical analyses are scientifically sound.

Generally, single-dose studies are recommended but provision is made to also allow steady-state studies provided such a study can be justified.

Fed or Fasting Conditions
The guidelines state that BE studies for immediate-release dosage forms should be performed under fasting conditions, unless food effects influence/affect bioavailability (BA). If the reference product dosage directions specifically state that the medicine must be administered with food, a food-effect study is required. If the dosage directions for the reference product state either "with or without food" or make no statement with respect to food, then a fasting study only will suffice.

In the case of modified-release dosage forms, both a fasted study and a fed study are required to demonstrate any possible influence of food and to exclude any possibility of dose dumping. Since meals that are high in total calories and fat content are more likely to affect the gastrointestinal (GI) physiology and thereby result in a larger effect on BA, the use of high-calorie and high-fat meals during food-effect studies is recommended.

Number of Subjects

The number of subjects should be justified on the basis of providing at least 80% power of meeting the acceptance criteria. The minimum number of subjects should not be less than 12. If 12 subjects do not provide 80% power, more subjects should be included.

A minimum of 20 subjects is required for modified-release oral dosage forms.

Add-On Subjects

If the BE study was performed with the appropriate size but BE cannot be demonstrated because of a result of a larger than expected random variation or a relative difference, an add-on subject study can be performed using not more than the number of subjects in the initial study. Combining is acceptable only in the case when the same protocol was used and preparations from the same batches were used.

Provision for an add-on study must be included a priori in the protocol and carried out strictly according to the study protocol and standard operating procedures (SOPs), and must be given appropriate statistical treatment, including consideration of consumer risk.

Dropouts and Withdrawals

A sufficient number of subjects should be initially entered into the study to allow for possible dropouts or withdrawals. Although dropouts generally should not be replaced since replacement of subjects could complicate the statistical model and analysis, replacements may be acceptable provided that this intention had been indicated in the protocol, the reasons for withdrawal (e.g., adverse drug reaction, personal reasons) clearly stated and inclusion of replacements motivated and accordingly justified. In addition, it should be stated in the protocol whether samples from extra subjects will be assayed if not required for statistical analysis.

Subject Selection

BE studies should normally be performed with healthy volunteers. The inclusion/exclusion criteria should be clearly stated in the protocol.

In general, the following subject characteristics are required:

i)   **Sex**—Both male and female subjects can be included but the risk to women of childbearing potential should be considered on an individual basis.
ii) **Age**—Subjects should be between 18 and 55 years.
iii) **Body mass**—Subjects should have a body mass within the normal range according to accepted normal values for the body mass index (BMI = mass in kg divided by height in meters squared, i.e., kg/m^2), or within 15% of the ideal body mass, or any other recognized reference.
iv) **Informed consent**—All subjects participating in the study should be capable of giving informed consent.

v) **Medical screening**—Subjects should be screened for suitability by means of clinical laboratory tests, an extensive review of medical history, and a comprehensive medical examination. Depending on the active pharmaceutical ingredients (APIs) therapeutic class and safety profile, special medical investigations may have to be carried out before, during and after the completion of the study.

vi) **Smoking/drug and alcohol abuse**—Subjects should preferably be non-smokers and without a history of alcohol or drug abuse. If moderate smokers are included they should be identified as such and the possible influences of their inclusion on the study results should be discussed in the protocol.

Inclusion of Patients

If the API under investigation is known to produce adverse effects and the pharmacological effects or risks are considered unacceptable for healthy volunteers, it may be necessary to use patients instead, under suitable precautions and supervision. In this case the applicant should justify the use of patients instead of healthy volunteers.

Genetic Phenotyping

Phenotyping and/or genotyping of subjects may be considered in crossover studies (e.g., BE, dose proportionality, food interaction studies) for safety or pharmacokinetic reasons. If an API is known to be subject to major genetic polymorphism, studies could be performed in cohorts of subjects of known phenotype or genotype for the polymorphism in question.

STANDARDIZATION OF STUDY CONDITIONS

Standardization of the diet, fluid intake, and exercise is recommended.

Dosing

The time of day for ingestion of doses should be specified.

Fluid Intake at Dosing

As fluid intake may profoundly influence the gastric transit of orally administered dosage forms, the volume of fluid administered at the time of dosing should be constant (e.g., 200 mL).

Food and Fluid Intake

In fasted studies the period of fasting prior to dosing should be standardized and supervised. All meals and fluids taken after dosing should also be standardized in regard to composition and time of administration and in accordance with any specific requirements for each study.

Concomitant Medication

Subjects should not take other medicines for a suitable period prior to, and during, the study and should abstain from food and drinks, which may interact with circulatory, GI, liver, or renal function (e.g., alcoholic or xanthine-containing beverages or certain fruit juices).

Posture and Physical Activity

The BA of an active moiety from a dosage form can be dependent upon GI transit times and regional blood flows, hence posture and physical activity may need to be standardized.

SAMPLE COLLECTION AND SAMPLING TIMES

In most cases the drug/API may be measured in serum or plasma. However, in some cases, whole blood or urine may be more appropriate for analysis.

Sampling Frequency

The sampling schedule should be planned to provide an adequate estimation of C_{max} and to cover the plasma drug concentration time curve long enough to provide a reliable estimate of the extent of absorption. This is generally achieved if the AUC derived from measurements is at least 80% of the AUC extrapolated to infinity.

   If a reliable estimate of terminal half-life is necessary, it should be obtained by collecting at least three to four samples above the limit of quantitation (LOQ) during the terminal log-linear phase.

   For long half-life drugs/APIs (>24 hours), the study should cover a minimum of 72 hours, unless 80% is recovered before 72 hours. The guidance states that "for moieties demonstrating high inter-subject variability in distribution and clearance the use of AUC truncation warrants caution. In these circumstances sampling periods beyond 72 hours may be required."

   To allow accurate estimation of relevant parameters, sampling points should be chosen such that the plasma concentration versus time profiles can be adequately defined.

Blood Sampling

a) The blood sampling frequency and duration should be sufficient to account for at least 80% of the known AUC to infinity (AUC_∞), usually approximately three terminal half-lives of the drug/API.

b) For most drugs/APIs 12 to 18 samples including a predose sample should be collected per subject per dose.

c) Sample collection should be spaced such that the maximum concentrations of drug/API in blood (C_{max}) and the terminal elimination rate constant (K_{el}) can be estimated.

d) At least three to four samples above the LOQ should be obtained during the terminal log-linear phase to estimate K_{el} by linear regression analysis.

e) The actual clock time when samples are collected, as well as the elapsed time relative to drug/API administration should be recorded.

   If drug/API concentrations in blood are too low to be detected and a substantial amount (>40%) of the drug/API is eliminated unchanged in the urine, then urine may serve as the biological fluid to be sampled.

Urine Sampling

a) Volumes of each sample should be measured immediately after collection and included in the report.

b) Urine should be collected over an extended period and generally no less than seven times the terminal elimination half-life, so that the amount excreted to infinity (Ae_∞) can be estimated.
c) Sufficient samples should be obtained to permit an estimate of the rate and extent of renal excretion. For a 24-hour study, sampling times of 0 to 2, 2 to 4, 4 to 8, 8 to 12, and 12 to 24 hours postdose are usually appropriate.
d) The actual clock time when samples are collected, as well as the elapsed time relative to API administration should be recorded.

CHARACTERISTICS TO BE INVESTIGATED

Moieties To Be Measured

Parent or Metabolite(s)
The evaluation of BA and BE should, in general, be based on measured concentrations of the parent compound (i.e., the active). The determination of moieties should be measured in biological fluids to take into account both concentration and activity.

The guideline recommends that for a BA study, both the parent active and its major active metabolites should be measured, if analytically feasible. This is necessary to assess both the concentration and the relative contribution of both the active parent and its major active metabolite(s) in the biological fluid to the clinical safety and/or efficacy of the active component(s).

However, for BE determinations, measurement of only the active parent released from the dosage form, rather than any metabolite(s), is generally recommended. The rationale for this recommendation is that the concentration–time profile of the active parent is more sensitive to changes in formulation performance than a metabolite, which is more reflective of metabolite formation, distribution, and elimination.

Notwithstanding, it is important to state a priori in the study protocol which chemical entities (pro-drug, API, metabolite) will be analyzed in the samples.

In some situations, however, measurements of an active or inactive metabolite may be necessary instead of the parent compound. Instances where this may be necessary are as follows:

a) If the concentration of the API is too low to be accurately measured in the biological matrix.
b) If there is a major difficulty with the analytical method.
c) If the parent compound is unstable in the biological matrix.
d) If the half-life of the parent compound is too short, thus, giving rise to significant variability.

Justification for not measuring the parent compound should be submitted by the applicant and BE determinations on the basis of metabolites should be justified in each case.

The following examples are given in the guideline:

• The measurement of concentrations of therapeutically active metabolite is acceptable if the substance studied is a prodrug.

- If an active metabolite is formed as a result of gut wall or other presystemic metabolic process(es) and the metabolite contributes meaningfully to safety and/or efficacy, either the metabolite or the parent concentrations must be measured and assessed in accordance with the protocol.

The guideline emphasizes that it is important to note that measurement of one analyte, either the active pharmaceutical ingredient or metabolite, allows the risk of making a Type-I error (the consumer risk) to remain at the 5% level. If more than one of several analytes is selected retrospectively as the BE determinant, then the consumer and producer risks change (8). Furthermore, when measuring active metabolites, the washout period and sampling times may need to be adjusted to adequately characterize the pharmacokinetic profile of the metabolite.

Enantiomers Versus Racemates

Although, generally the measurement of the racemate only using an achiral assay is recommended, measurement of individual enantiomers in BE studies is recommended only when all of the following conditions are met:

a) The enantiomers exhibit different pharmacodynamic characteristics.
b) The enantiomers exhibit different pharmacokinetic characteristics.
c) Primary efficacy and safety activity resides with the minor enantiomer.
d) Nonlinear absorption is present (as expressed by a change in the enantiomer concentration ratio with change in the input rate of the drug/API) for at least one of the enantiomers.

Pharmaceutical Products with Complex Mixtures of APIs

Reference is made to certain pharmaceutical products that may contain complex active substances (i.e., active moieties or APIs) that are mixtures of multiple synthetic and/or natural source components). It is stated that some or all of the components of these complex active mixtures cannot be characterized with regard to chemical structure and/or biological activity. The guideline then indicates that

> quantification of all active or potentially active components in pharmacokinetic studies to document BA and BE is neither encouraged nor desirable. BA and BE studies should rather be based on a small number of markers of rate and extent of absorption.

PHARMACOKINETIC PARAMETERS

Blood/Plasma/Serum Concentration Versus Time Profiles

The following BA parameters are required to be estimated:

a) AUC_t, AUC_∞, C_{max}, t_{max} for plasma concentration versus time profiles.
b) AUC_∞, C_{max}, C_{min}, fluctuation (% PTF), and swing (% swing) for studies conducted at steady state.
c) Any other justifiable characteristics as referred to in Appendix I.
d) The method of estimating AUC values should be specified.

Urinary Excretion Profiles

Justification should be given when these data are to be used to estimate the rate of absorption. Sampling points should be chosen so that the cumulative urinary excretion profiles can be adequately defined to allow accurate estimation of relevant parameters.

The following BA parameters are required to be estimated:

a) Ae_t, Ae_∞ as appropriate for urinary excretion studies.
b) Any other justifiable characteristics as referred to in Appendix I.
c) The method of estimating AUC values should be specified.

PHARMACODYNAMIC STUDIES

If pharmacodynamic parameters/effects are used as BE criteria, the applicant must submit justification for their use. In addition

a) a dose–response relationship should be demonstrated.
b) sufficient measurements should be taken to provide an appropriate pharmacodynamic response profile.
c) the complete dose–effect curve should remain below the maximum physiological response.
d) all pharmacodynamic measurements/methods should be validated with respect to specificity, accuracy and reproducibility.

BIOANALYSIS

Bioanalysis of all analytes must be conducted according to good laboratory practice (GLP) and cGMP. All analytical methods used should be fully validated and documented. The following characteristics of the assay need to be addressed:

a) Stability of stock solutions.
b) Stability of the analyte(s) in the biological matrix under processing conditions and during the entire period of storage.
c) Specificity.
d) Accuracy.
e) Precision.
f) Limits of detection (LOD) and quantification (LOQ).
g) Response function.
h) Robustness and ruggedness.

Separately prepared quality control (QC) samples should be analyzed with processed test samples at intervals based on the total number of samples.

All procedures must be performed according to preestablished SOPs and all relevant procedures, and formulae used to validate the bioanalytical method should be submitted and discussed.

Any modification of the bioanalytical method, before and during analysis of study samples may require adequate re-validation, and all modifications should be reported and the scope of re-validation justified.

STUDY PRODUCTS

Reference Products

Reference to another guideline entitled Pharmaceutical and Analytical (P&A) Guideline (1) is made under this section. The requirements relating to pharmaceutical and analytical information are provided in this guideline, including elements of pharmaceutical and biological availability in Part 2A—Basis for Registration and Overview of Application. Although there is a Part 2C section, Quality Overall Summary (QOS), PART 2B is conspicuously absent. Perusal of the P&A guideline reveals the following:

Under Section 2.1.3 headed as Study Products and subsection (b) headed Reference Products (comparators) (*see also Biostudies and Dissolution Guidelines*), a statement is made that

> *products containing chemical entities/active moieties that are not registered in South Africa cannot be used as reference products in efficacy and safety studies submitted in support of an application.*

It is also stated that the reference product may contain a different chemical form from that of the proposed generic product (pharmaceutical alternative) provided it is confirmed that the safety/efficacy profile is not altered. Furthermore, if well known (e.g., hydrochloride, maleate, nitrate, stearate), reference to a pharmacopoeia accepted by the council (i.e., MCC) may be acceptable. This is in direct conflict with the provisions of the Medicines and Related Substances Act as amended (5), wherein it is unambiguously indicated that such comparisons do not qualify for assessment as interchangeable medicines, since they are not pharmaceutically equivalent and thus cannot be declared bioequivalent In order to make provision for a pharmaceutical alternative, the Act would need to be amended accordingly to incorporate the new definition, since the guideline is enabled by the Act and not vice versa.

Product strengths not available in South Africa may be applied for and/or used in biostudies provided that the dose range is approved/registered in South Africa.

The reference product should be an innovator product registered by the MCC and should *preferably* be procured in South Africa. An exception is an *"old medicine"* that may be used as a reference product when no other such product has been registered provided that it is available on the South African market. If more than one such product is available the market leader should be used as the reference (e.g., IMS database) and the applicant has to submit evidence to substantiate the market leadership claim.

Quite extraordinarily, several options for the selection of the reference product are given and listed in order of preference:

(i) The innovator product registered and procured in South Africa; or
(ii) The innovator product for which a marketing authorization has been granted by the health authority of a country with which the council aligns itself (see General Information guideline 3.1.4 (9), and which is to be purchased from that market, or

(iii) A product from the latest edition of the WHO International comparator products for equivalent assessment of interchangeable multisource (generic) products QAS/05.143.[a] or;

(iv) In the case that no innovator product can be identified—within the context of (i) to (iii) above, the choice of the reference must be made carefully and must be comprehensively justified by the applicant.

Although it is noted that "A product that has been approved based on comparison with a non-domestic reference product may or may not be interchangeable with currently marketed domestic products," the significance and implication of the foregoing statement is unclear? The guidelines state that in all cases, the choice of reference product should be justified by the applicant and the country of origin of the reference product should be reported together with lot number and expiry date. Recently, an amended guideline[b] has included the following requirements for applicants using a foreign reference product.

(i) *The name and address of the manufacturing site where the foreign reference product is manufactured.*

(ii) *The qualitative formulation of the foreign reference product.*

(iii) *Copies of the immediate container label as well as the carton or outer container label of the foreign reference product.*

(iv) *For modified release, evidence of the mechanism of modified release of the foreign reference product.*

(v) *The method of manufacture of the foreign reference product if claimed by the applicant to be the same.*

(vi) *Procurement information of the foreign reference product*
 • *Copy of licensing agreement/s*
 • *Distribution arrangements/agreement/s*
 • *Copy of purchase invoice (to reflect date and place of purchase)*

It is interesting to note that no other specific conditions for the use of a foreign (non-domestic) reference product are mentioned until Section 5.1.2 of the Biostudies guideline (2). It is therefore highly likely that different generic products may well be or have been approved following comparison to different non-domestic reference products without comparisons having been made between different nondomestic reference products (the implications for generic substitution and therapy, in general are obvious). Section 4.2 of the Dissolution guidelines and "Foreign Reference Products" section of this chapter describe the provisions made for the use of a foreign reference product and the comparative dissolution testing required to determine whether such a foreign reference product complies with the specified requirements for use in BE studies.

The choice of reference products for combination products makes reference to the Biostudies (2) and Dissolution guidelines (3) and states that such products

---

[a] http://www.who.int/medicines/services/expertcommittees/pharmprep/QAS05_143_ Comparator_Rev%201.pdf. Accessed June 10, 2009.

[b] Pharmaceutical and Analytical: Medicines Control Council, Department of Health, 2.02 P&A Apr09 v3.doc. http://www.mccza.com/documents/2.02_P&A_Apr09_v3.doc. Accessed June 11, 2009.

should, in general, be assessed with respect to BA and BE of individual active substances:

- either single entity products administered concurrently (in the case of new clinically justifiable combinations), or
- using an existing combination as the reference provided that the combination was registered on clinical and not bioequivalence data.

In the former instance, immediate-release oral dosage forms containing a single API may be used as the reference and that these reference products may include Old medicines.

The irregular permission of the use of a foreign (non-domestic) reference product goes entirely against the mandate and intention of the Medicines and Related Substances Control Act, 1965 (Act No. 101 of 1965) as amended by Act No. 90 of 1997 (5) and Act 59 of 2002 (7), where provision is made for generic substitution in Section 22F for the use of interchangeable generic medicines wherever possible. Since it is generally unknown whether the formulation of a foreign reference product is identical to that innovator's product being sold in the South Africa market, in the absence of comparative BA data between the same generic product and the domestic reference product, that generic product cannot be considered to be therapeutically equivalent to the domestic reference product. Consequently, such generic products, in spite of the recommendations in the South African guidelines, cannot be considered interchangeable. Furthermore, when a generic product intended for South African registration has been compared with a foreign reference product, compliance with the conditions stipulated in the South African guidelines for the declaration of BE is therefore questionable.

Retention Samples
A sufficient number of retention samples of both test and reference products used in the BE study should be kept for one year in excess of the accepted shelf life, or two years after completion of the trial or until approval, whichever is longer, to allow re-testing if required by the MCC.

DATA ANALYSIS

Statistical Analysis
The guideline states that "The statistical method for testing relative bioavailability (i.e. average bioequivalence) is based upon the 90% confidence interval for the ratio of the population means (Test/Reference) for the parameters under consideration. Pharmacokinetic parameters derived from measures of concentration, e.g. AUC_t, AUC_∞ and C_{max} should be analysed using ANOVA. Data for these parameters should be transformed prior to analysis using a logarithmic transformation."

If appropriate to the evaluation, the analysis technique for t_{max} should be nonparametric and should be applied to untransformed data.

In addition to the appropriate 90% confidence intervals, summary statistics such as geometric and arithmetic means, SD and% RSD, as well as ranges for pharmacokinetic parameters (minimum and maximum), should be provided. A disk with raw data formatted appropriately for evaluation where the formatting is described in "Pharmacokinetic and Statistical Report" section (vide infra) or Section 3.9.3 (a) of the Biostudies guideline (2).

Acceptance Range for Pharmacokinetic Parameters

All pharmacokinetic parameters to be tested, the procedure for testing and the acceptance ranges should be stated a priori in the protocol.

a) Single-dose studies—The acceptance criteria are as follows:
   i) AUC_t ratio—The 90% confidence interval for the test/reference ratio should lie within the acceptance interval of 0.80 to 1.25 (80–125%).

   Provision is made for the use of alternative methods, such as for example, scaled average BE (ABE) for highly variable drugs but must be justified and based on sound scientific principles. The use of alternative methods must be clearly stated a priori in the protocol and cannot be added retrospectively.

   ii) C_{max} ratio—The 90% confidence interval for the test/reference ratio should lie within an acceptance interval of 75% to 133%, calculated using log-transformed data, except for narrow therapeutic range APIs when an acceptance interval of 80% to 125% will apply. It is interesting to note that a wider acceptance range has been recommended for all products with the exception only in the case of narrow therapeutic range APIs. No consideration has been given to other classes of drug products where the risk of accepting such products as bioequivalent using the wider acceptance range may have serious consequences. Furthermore, the implications of using this wider acceptance interval to declare BE with a foreign reference product conjure up food for thought.

   As previously indicated, in the case of highly variable APIs, a wider interval or other appropriate measure may be acceptable, but should be stated a priori and justified in the protocol.

b) Steady-state studies
   i) Immediate-release dosage forms—The guidance states that the acceptance criteria are the same as for single-dose studies but using AUC_∞ instead of AUC_t. No explanation is given for such an unusual choice for AUC where, clearly, AUC during a dosage interval at steady state (AUC_τ or AUC_{ss}) has generally been considered as the parameter of choice for steady-state studies (10). It is also unclear whether only AUC is required or whether the determination and assessment of $C_{max\ (ss)}$ is to be included?

   ii) Controlled-/modified-release dosage forms—The acceptance criteria are as follows:
   - AUC_∞ ratio—Once again, no explanation has been given for the use of this parameter. The 90% confidence interval for the test/reference ratio should lie within the acceptance interval of 0.80 to 1.25 (80–125%).
   - $C_{max\ (ss)}$ and $C_{min\ (ss)}$—The 90% confidence interval for the test/reference ratio should lie within the acceptance interval of 0.75 to 1.33 (75–133%) calculated using log-transformed data.
   - % Swing and % PTF—The 90% confidence interval for the test/reference ratio should lie within the acceptance interval of 0.80 to 1.25 (80–125%) calculated using log-transformed data.

STUDY REPORT
Complete documentation is required to be submitted including the protocol, conduct, and evaluation and evidence of compliance with GCP, GLP, and cGMP.

Clinical Report
The following information must be included in the clinical section of the BE study report:

a) A statement indicating the independence of the ethics committee.
b) Documented proof of ethical approval of the study.
c) A complete list of the members of the ethics committee, their qualifications, and affiliations.
d) Names and affiliations of the all investigator(s), the site of the study and the period of its execution.
e) The names and batch numbers of the products being tested.
f) The name and addresses of the applicants of both the reference and the test products.
g) Expiry date of the reference product and the date of manufacture of the test product used in the study.
h) Assay and dissolution profiles for test and reference products.
i) Certificate of analysis of the API used in the test product biobatch.
j) A summary of adverse events, which should be accompanied by a discussion on the influence of these events on the outcome of the study.
k) A summary of protocol deviations (sampling and nonsampling), which should be accompanied by a discussion on the influence of these adverse events on the outcome of the study.
l) Subjects who dropout or are withdrawn from the study should be identified and their withdrawal fully documented and accounted for.

Analytical Report
The following must be included in the analytical section of the BE report:

a) Validation report.
b) All individual subject concentration data.
c) Calibration data, that is, raw data and back-calculated concentrations for standards, as well as calibration curve parameters, for the entire study.
d) Quality control samples for the entire study.
e) Chromatograms from analytical runs for 20% of all subjects (or for a minimum of four subjects, whichever is the greater) including chromatograms for the associated standards and QC samples.
f) A summary of protocol deviations, which should be accompanied by a discussion on the influence of these deviations on the outcome of the study. Protocol deviations should be justified.

Pharmacokinetic and Statistical Report
The following information must be included:

a) All individual plasma concentration versus time profiles presented on a linear/linear as well as log/linear scale (or, if appropriate, cumulative urinary excretion data presented on a linear/linear scale).

These data should be submitted in hard copy and also formatted electronically in a format compatible for processing by SAS software. Individual subject data should be in rows and arranged in columns, which reflect the subject number, phase number, sequence, formulation, and sample concentration versus time data.

b) The method(s) and programmes used to derive the pharmacokinetic parameters from the raw data.
c) A detailed ANOVA and/or nonparametric analysis, the point estimates and corresponding confidence intervals for each parameter of interest.
d) Tabulated summaries of pharmacokinetic and statistical data.
e) The statistical report should contain sufficient detail to enable the statistical analysis to be repeated, for example, individual demographic data, randomization scheme, individual subject concentration versus time data, values of pharmacokinetic parameters for each subject, descriptive statistics of pharmacokinetic parameters for each formulation and period.

Quality Assurance

The study report should contain a signed quality assurance (QA) statement confirming release of the document. The applicant should indicate whether the site(s) (clinical and analytical) where the study was performed was subjected to a prestudy audit to ascertain its/their status of GCP and GLP and/or cGMP conditions. All audit certificates should clearly indicate the date of audit and the name(s), address(es) and qualifications of the auditor(s) and, in addition, an independent monitor's statement must be included.

Mention is made in the guideline that the applicant must demonstrate that the excipients in the pharmaceutically equivalent product are essentially the same and in comparable concentrations as those in the reference product. Interestingly, no mention is made of the use of a pharmaceutical alternative as provided in the guideline. In the event that this information about the reference product cannot be provided by the applicant, it is incumbent upon the applicant to perform in vivo or in vitro studies to demonstrate that the differences in excipients do not affect product performance.

BIOAVAILABILTY AND BIOEQUIVALENCE REQUIREMENTS

Orally Administered Pharmaceutical Products Intended for Systemic Action

Solutions
Biowaivers apply to products falling under this category.

Pharmaceutically equivalent solutions for oral use (including syrups, elixirs, tinctures, or other soluble forms but not suspensions) containing the active pharmaceutical ingredient in the same molar concentration as the comparator product, and containing only excipient(s) known to have no effect on GI transit, GI permeability, and hence absorption or stability of the active pharmaceutical ingredient in the GI tract are considered to be equivalent without the need for further documentation.

Pharmaceutically equivalent powders for reconstitution as solution, meeting the solution criteria mentioned earlier, are considered to be equivalent without the need for further documentation.

Suspensions
BE for a suspension should be treated in the same way as for immediate-release solid oral dosage forms.

Immediate-Release Products—Tablets and Capsules
BE studies are generally required for these dosage forms. Although the guideline states that in vivo BE studies should be accompanied by in vitro dissolution profiles on all strengths of each product, no indication is given relating to the dissolution conditions. However, a statement is made regarding waivers for in vivo BA and BE studies for immediate-release solid oral dosage forms and reference to the conditions of the dissolution tests required is given. The reference relates to Section 5 of the Biostudies guideline (2) and is described in Waivers of In Vivo Bioequivalence Studies for Oral Solid Dosage Forms section (vide infra).

Modified-Release Products
BE studies (single dose) are required for these dosage forms, which include delayed-release products and extended- (controlled-)release products (as defined in the P&A guideline). Fasted as well as fed studies are necessary and multiple-dose studies are generally not recommended.

Fixed-Dose Combination Products (Including Copackaged Products)
Combination products should in general be assessed with respect to BA and BE of APIs either separately (in the case of a new combination) or as an existing combination. In the case of a new combination the study should be designed in such a way that the possibility of a pharmacokinetic and/or pharmacodynamic active–active interaction could be detected.

The guideline states that the approval of an FDC in general, will be considered in accordance with the WHO Technical report series 929[c] "Guidelines for registration of fixed-dose combination medicinal products 2005" or the latest revision.

Fixed-dose combinations for antiretroviral compounds will be considered in accordance with the FDA "Guidance for Industry: Fixed Dose Combinations, Co-Packaged Drug Products, and Single-Entity Versions of Previously Approved Antiretrovirals for the Treatment of HIV" October 2006[d] or the latest revision.

Miscellaneous Oral Dosage Forms
Rapidly dissolving pharmaceutical products, such as buccal and sublingual dosage forms, should be tested for in vitro dissolution and in vivo BA and/or BE. Chewable tablets should also be evaluated for in vivo BA and/or BE. Chewable

---

[c] http://www.who.int/medicines/publications/pharmprep/en/index.html. Accessed June 10, 2009.

[d] http://www.fda.gov/ForConsumers/ByAudience/ForPatientAdvocates/HIVandAIDS Activities/ucm124426.htm. Accessed June 10, 2009.

tablets (as a whole) should be subject to in vitro dissolution because a patient, without proper chewing, might swallow them. In general, in vitro dissolution test conditions for chewable tablets should be the same as for non-chewable tablets of the same API/moiety.

Medicines Intended for Local Action

This section covers nonsolution pharmaceutical products, which are for non-systemic use (oral, nasal, ocular, dermal, rectal, vaginal, etc., application) and are intended to act without systemic absorption. In these cases, BE is established through comparative clinical or pharmacodynamic, dermatopharmacokinetic studies and/or in vitro studies. In certain cases, active concentration measurement may still be required for safety reasons in order to assess unintended systemic absorption.

Parenteral Solutions

The applicant must demonstrate that the excipients in the pharmaceutically equivalent (no mention made of pharmaceutical alternative dosage forms?) product are essentially the same and in comparable concentrations as those in the reference product. In the event that this information about the reference product cannot be provided by the applicant, it is incumbent upon the applicant to perform in vivo or in vitro studies to demonstrate that the differences in excipients do not affect product performance. The nature of such studies are however, not disclosed.

Aqueous Solutions

Aqueous solutions to be administered by parenteral routes (intravenous, intramuscular, subcutaneous) containing the same active pharmaceutical ingredient(s) in the same molar concentration and the same or similar excipients in comparable concentrations as the comparator product are considered to be equivalent without the need for further documentation.

Certain excipients (e.g., buffer, preservative, antioxidant) may be different provided the change in these excipients is not expected to affect the safety and/or efficacy of the medicine product.

Powders for Reconstitution

Pharmaceutically equivalent products that are powders for reconstitution as solution meeting the criterion for Aqueous Solutions above are considered to be equivalent without the need for further documentation. No mention is made of the requirements, if any, for pharmaceutical alternatives.

Other

BE studies are required for all other parenterals and for intramuscular dosage forms, monitoring is required until at least 80% of the AUC_∞ has been covered.

TOPICAL PRODUCTS

The guideline states that "Pharmaceutically equivalent topical products prepared as aqueous solutions containing the same active pharmaceutical ingredient(s) in the same molar concentration and essentially the same excipients in comparable concentrations are considered to be equivalent without the need for further documentation."

Reference is again made to the need to demonstrate that the excipients in the pharmaceutically equivalent product are essentially the same and in comparable concentrations as those in the reference product.

Local Action

For topical preparations containing corticosteroids intended for application to the skin and scalp, the human vasoconstrictor test (blanching test) is recommended for BE assessment of such products. Either visual or chromameter data are acceptable but in each case all data must be validated.

For simple topical solutions with bacteriostatic, bactericidal, antiseptic, and/or antifungal claims, a biowaiver based on appropriate validated in vitro test methods, for example, microbial growth inhibition zones, is acceptable.

For all other topical formulations, clinical data (comparative clinical efficacy) are required. Proof of release by membrane diffusion is not acceptable as proof of efficacy, unless data are presented that show a correlation between release through a membrane and clinical efficacy.

The guideline also states that whenever systemic exposure resulting from locally applied/locally acting medicinal products entails a risk of systemic adverse reactions, systemic exposure should be measured.

Systemic Action

A BE study is always required for other locally applied products with systemic action, for example, transdermal products.

PRODUCTS INTENDED FOR OTHER ROUTES OF ADMINISTRATION

Applicants must demonstrate that the excipients in the pharmaceutically equivalent product are essentially the same and used in comparable concentrations as those in the reference product.

Otic and Ophthalmic Products

Pharmaceutically equivalent otic or ophthalmic products prepared as aqueous solutions and containing the same active pharmaceutical ingredient(s) in the same molar concentration and essentially the same excipients in comparable concentrations are considered to be equivalent without the need for further documentation.

Certain excipients (e.g., preservative, buffer, substance to adjust tonicity or thickening agent) may be different provided use of these excipients is not expected to effect safety and/or efficacy of the product.

Aerosols, Nebulizers, and Nasal Sprays

BE assessment is not required for pharmaceutically equivalent solutions for aerosol or nebulizer inhalation or nasal sprays, tested to be administered with or without essentially the same device, prepared as aqueous solutions, containing the same active pharmaceutical ingredient(s) in the same concentration and essentially the same excipients in comparable concentrations.

The pharmaceutical product may include different excipients provided its use is not expected to affect safety and/or efficacy of the product.

Particle size distribution may be used in support of proof of efficacy for inhalations. The Anderson sampler or equivalent apparatus should be used. In addition, appropriate information should be submitted to provide evidence of clinical safety and efficacy.

Gases
Pharmaceutically equivalent gases are considered to be equivalent without the need for further documentation.

WAIVERS OF IN VIVO BIOEQUIVALENCE STUDIES FOR ORAL SOLID DOSAGE FORMS
Biowaivers are considered based on the following:

In Vitro Studies—Dissolution Profile Comparison
Under this section, reference is made to another guideline, the Dissolution guideline (3).

Comparative dissolution profiles, in three media should be carried out on the test and the reference product and tested for similarity. The f_2 similarity factor should be used to compare dissolution profiles from different products and/or strengths of a product. An f_2 value ≥ 50 indicates a sufficiently similar dissolution profile such that further in vivo studies are not necessary. For an f_2 value <50, it may be necessary to conduct an in vivo study. However, when both test and reference products dissolve 85% or more of the label amount of the API in ≤ 15 minutes similarity is accepted without the need to calculate f_2 values.

Proportionally Similar Formulations
Section 2.11 of the Biostudies guideline (2) states:

> *"Pharmaceutical products are considered proportionally similar in the following cases:*
>
> *2.11.1 When all APIs and inactive pharmaceutical ingredients (IPIs) are in exactly the same proportion between different strengths (e.g., a 100 mg strength tablet has all API and IPIs exactly half of a 200 mg strength tablet and twice that of a 50 mg strength tablet).*
>
> *2.11.2 When the APIs and IPIs are not in exactly the same proportion but the ratios of IPIs to the total mass of the dosage form are within the limits defined by the Post-registration Amendment guideline.*
>
> *2.11.3 When the pharmaceutical products contain a low concentration of the APIs (e.g., less than 5%) and these products are of different strengths but are of similar mass.*
>
> *The difference in API content between strengths may be compensated for by mass changes in one or more of the IPIs provided that the total mass of the pharmaceutical product remains within 10% of the mass of the pharmaceutical product on which the bioequivalence study was performed. In addition, the same IPIs should be used for all strengths, provided that the changes remain within the limits defined by the Post-registration Amendment guideline."*

Hence, a prerequisite for qualification for a biowaiver based on dose-proportionality of formulations is that

- the multisource product at one strength has been shown to be bioequivalent to the corresponding strength of the reference product and

- the further strengths of the multisource product are proportionally similar in formulation to that of the studied strength.

When both of these criteria are met and the dissolution profiles of the further dosage strengths are shown to be similar to the one of the studied strength on a percentage released vs. time basis, a biowaiver procedure can be considered for the further strengths.

Furthermore, according to the Dissolution guidelines (3), when a biowaiver is requested for different strengths of test/multisource products which are

- proportionally formulated (see Biostudies guidelines 2.11 and 5.1.1),
- manufactured by the same manufacturer at the same manufacturing site, and
- an appropriate BE study has been performed on at least one of the strengths of the formulation (usually the highest strength unless a lower strength is chosen for reasons of safety), either (a) or (b) below applies:

a) *If the reference product used in the biostudy is an innovator product procured in South Africa, then dissolution profiles generated for the test and other strength multisource products being applied for (i.e., lower and higher strengths) should be compared as described in Section 3 of this guideline for each of the specified media. When sink conditions do not exist in one or more media, the profiles of the higher and lower strengths may not be similar in those media due to saturation, in which case supporting data may be generated with the local innovator of the same strength.*

b) *If the reference product used in the biostudy is a foreign innovator product, (not procured in SA) and where the local innovator product and the foreign innovator product show similar dissolution profiles in the three specified media then, for the proportionally similar strength(s), dissolution profiles of the test and local innovator reference products respectively (i.e., strength vs. same strength) should be compared as described in Section 3 of this guideline for each of the specified media.*

Immediate-Release Tablets

When the pharmaceutical product is the same dosage form but of a different strength and is proportionally similar in its API and inactive pharmaceutical ingredients (IPIs), a biowaiver may be acceptable.

Modified-Release Products

Beaded Capsules—Lower Strength

For extended-release beaded capsules where the strength differs only in the number of beads containing the API, a single-dose, fasting BE study should be carried out on the highest strength. A biowaiver for the lower strength based on dissolution studies can be requested.

Dissolution profiles in support of a biowaiver should be generated for each strength by using the recommended dissolution test methods described in Section 3 of the Dissolution guideline (3).

Dissolution of test and reference products should be conducted in each of the following three media:

- Acidic media such as 0.1 N HCl
- pH 4.5 buffer
- pH 6.8 buffer

If both the test and reference products show more than 85% dissolution within 15 minutes, the profiles are considered similar (no calculations required). If not, the f_2 value must be calculated. If $f_2 \geq 50$, the profiles are normally regarded similar such that further in vivo studies are not necessary. Note that only one measurement should be considered after 85% dissolution of both products has occurred and excluding point zero.

The similarity factor (f_2) is a logarithmic reciprocal square-root transformation of the sum of squared errors, and is a measurement of the similarity in the percentage (%) dissolution between the two curves, viz.:

$$f_2 = 50 \log\{[1 + (1/n) \sum\nolimits_{t=1^n} (Rt - Tt)^2]^{-0.5}.100\}$$

where n is the number of time points, R_t is the dissolution value of the reference batch at time t, and T_t is the dissolution value of the test batch at time t.

The dissolution profile of two products, that is, of the test and reference products (using 12 units each) should be determined and for f_2 calculations, a minimum of three time points (excluding point zero) must be used, and only one measurement included after 85% dissolution of both products has occurred.

Generally, f_2 values greater than 50 (50–100) ensure sameness or equivalence of the two curves and, thus, of the performance of the test and reference products. For curves to be considered similar, f_2 values should be close to 100.

This model-independent method is considered to be most suitable for dissolution profile comparisons when three to four or more dissolution time points are available. The following recommendations should also be considered:

i) The dissolution measurements of the test and reference batches should be made under exactly the same conditions. The dissolution time points for both profiles should be the same (e.g., 10, 15, 20, 30, 45, 60 minutes, etc.).
ii) Only one measurement should be considered after 85% dissolution of both products have occurred.
iii) To allow use of mean data, the percent coefficient of variation (CV) at the earlier time points (e.g., 15 minutes) should not be more than 20%, and at other time points should not be more than 10%.

Tablets—Lower Strength
For extended-release tablets when the pharmaceutical product is

a) in the same dosage form but in a different strength,
b) is proportionally similar in its APIs and IPIs, and
c) has the same drug/API release mechanism.

An in vivo BE determination of one or more lower strengths may be waived on the basis of dissolution testing as previously described. Dissolution profiles should be generated on all the strengths of the test and the reference products.

When the highest strength (generally, as usually the highest strength is used unless a lower strength is chosen for reasons of safety) of the multisource product is bioequivalent to the highest strength or dose[e] of the reference product,

---

[e] Dose included in the dosage range of the MCC-approved package insert of the innovator product registered in South Africa.

and other strengths are proportionally similar in formulations and the dissolution profiles are similar between the dosage strengths, biowaivers can be considered for the lower/other strengths.

Foreign Reference Products

BE studies submitted where a foreign reference product has been used requires demonstration of equivalence between the foreign product and the innovator product marketed in South Africa. If the reference product is not the current innovator product available in the South African market, then the reference product may be procured from another country provided that it complies with the requirements specified in the P&A guideline as described in Section 2.1.3 (b) of that guideline and also in "Reference Products" section. Dissolution profiles of the test and reference products should be compared for similarity as described in Section 3 of the Dissolution guideline (3) for each of the three specified media irrespective of the solubility profile, viz.:

- Acidic media such as 0.1 N HCl
- pH 4.5 buffer
- pH 6.8 buffer

Postregistration/Approval Amendments

The Biostudies guideline comments primarily on registration requirements for multisource pharmaceutical products, however, it also states that in vitro dissolution testing may also be suitable to confirm similarity of product quality and performance characteristics with minor formulation or manufacturing changes after approval.

Section 4.3 of the Dissolution guideline describes the types of dissolution testing required when amendments are made to pharmaceutical products, manufacturing procedures, and other associated processes including change of site. Three different cases are described, viz.:

a) Case A
Dissolution testing should be conducted as a release test according to the original submission, or in accordance with compendial requirements, for that product.
b) Case B
Dissolution testing should be conducted as a multipoint test in the application/compendial medium at intervals such as 10, 15, 20, 30, 45, 60, and 120 minutes, or until an asymptote is reached for the proposed and currently registered formulation.
c) Case C
Dissolution testing should be conducted as a multipoint test in the three dissolution media as specified above for the proposed, and currently registered formulations, at intervals such as 10, 15, 20, 30, 45, 60, and 120 minutes, or until either 85% dissolution is obtained, or an asymptote is reached.

The various types of amendments are also described as follows:

a) Type A
In the event that the Type A change made is such that it is unlikely to have an effect on the quality and performance of a dosage form, Case A dissolution testing is appropriate.

b) Type B
In the event that the changes, which were made, could have a significant impact on the quality and performance of a dosage form, Case B or C dissolution testing is appropriate. Profiles of the currently used product and the proposed product should be proven to be similar, according to the requirements as describe in this Guideline.

c) Type C
In the case of changes that are likely to have a significant impact on formulation quality and performance, in vivo bioequivalence testing should be conducted unless otherwise justified. Case B or Case C dissolution testing may also be required. Biowaivers may also be considered if a proven in vitro/in vivo correlation (IVIVC) has been established.

BIOPHARMACEUTICS CLASSIFICATION SYSTEM
Biowaivers may be granted on the basis of the biopharmaceutics classification system (BCS) under the following conditions:

Dosage forms containing APIs, which are highly soluble and highly permeable (i.e., BCS class 1), and are rapidly dissolving are eligible for a biowaiver based on the BCS provided

- the dosage form is *rapidly dissolving* (as defined in the Dissolution guideline, i.e., no less than 85% of the labeled amount of the API dissolves in 30 minutes) and
- the dissolution profile of the multisource product is similar to that of the reference product at pH 1.2, pH 4.5, and pH 6.8 buffer using the paddle method at 75 rpm or the basket method at 100 rpm (as described in the Dissolution guideline) and meets the criteria of dissolution profile similarity, $f_2 \geq 50$ (or equivalent statistical criterion).

If both the reference and the multisource dosage forms are *very rapidly dissolving*, that is, 85% or more dissolution at 15 minutes or less in all three media under the above test conditions, the two products are deemed equivalent and a profile comparison is not necessary.

CONCLUSIONS
A unique and potentially problematic situation currently exists in South Africa. Generic substitution has been mandated as a means to ensure that accessibility to affordable medicines is achieved. Prescribers and dispensers of medicines must therefore inform patients that prescribed medicines may be available at a lower price than the innovator/Brand product and recommend accordingly. In other words, where an approved generic medicine exists, the generic medicine will be substituted for the innovator/brand product unless the dispenser is prohibited from doing so by the patient. Clearly, the substitution must be based on the premise that approved generic medicines have been deemed to be interchangeable following assessment by the MCC. Of particular concern is the fact that most generic medicines approved and marketed in South Africa do not comply with internationally accepted requirements for interchangeability as many/most have not been assessed by comparison with the innovator/brand product available on the South African market but rather against a "foreign" reference product as permitted in the national guidelines. A foreign reference product, although being supplied by the same innovator/brand company, may not be the same (identical formulation, manufacture etc.) as the innovator/brand

product being sold on the South African market as it well-recognized that in some cases, innovator/brand products may and are formulated differently for different markets. For example, Tegretol XR® tablets, a prolonged action carbamazepine product, is marketed in the United States as a nondisintegrating dosage form OROS®). The same innovator is listed as the manufacturer of prolonged action carbamazepine dosage forms in various other countries where those dosage forms are also tablets but which disintegrate in aqueous fluid and in South Africa is marketed as Tegretol CR®. The release mechanisms and formulation of the United States reference listed product and the product marketed by the same innovator in South Africa are clearly different. Hence, in the absence of specific confirmatory data, a nondomestic innovator/brand product used as the reference product in a BE study involving a generic medicine intended for a particular domestic market cannot be assumed to be bioequivalent to the domestic innovator/brand product. The only data required by the MCC to show that the foreign reference product is the "same" as the reference product marketed in South Africa are dissolution profiles comparing f_2 values between the foreign and domestic reference products conducted in three different dissolution media at pH 1.5, 4.5, and 6.8. Furthermore, applicants are able to conduct such comparisons for all classes of APIs irrespective of properties such as solubility, permeability, potency, therapeutic index (e.g., narrow) amongst others and no risk assessment is apparently required. Therefore a generic product that has been shown to be bioequivalent to a nondomestic reference may well be usable or "prescribable" but in the absence of the necessary comparative BA data showing BE between that same generic product and the domestic reference product, that generic product cannot be considered to be interchangeable.

It appears that the MCC, by making provision for a foreign or non-domestic reference product to be used in a BE study has inadvertently created a two-tier system for the approval of generic medicines in South Africa. The top tier can therefore be considered to consist of generic products approved on the basis of comparison with the domestic innovator/brand product as the reference, whereas another (second or lower) tier includes those generic products approved on the basis of a comparison of the generic with a nondomestic innovator/brand as the reference in a BE study. In addition, the latter tier probably also includes all other generic medicines which have been approved on the basis of in vitro testing only, apart from those products which incorporate an API classified as Class 1 according to the BCS (11,12).

A similar situation of ill-conceived interchangeability exists for approved generic controlled-/modified-release dosage forms as well as non-oral dosage forms such as topical products for local use, inhalation products and various other such generic products which are not intended to be absorbed into the systemic circulation.

In summary, it should be emphasized that only when a generic medicine has been shown to be bioequivalent with the domestic innovator/brand product used as reference, will substitution be acceptable and appropriate. Clearly, on the basis of "similar" BA, as would the case be when the generic product has been shown to be bioequivalent to a nondomestic reference product, a clinical decision is required to declare a "second-tier" generic product "prescribable" for a patient who is naïve to that particular medicine, but certainly interchangeability of such a product is highly questionable.

REFERENCES

1. Pharmaceutical and Analytical: Medicines Control Council, Department of Health, 2.02 P&A, Jun07, v2, 2007. http://www.mccza.com/showdocument .asp?Cat=17&Desc=Guidelines%20-%20Human%20Medicines. Accessed April 29, 2009.
2. Biostudies: Medicines Control Council, Department of Health, 2.06 Biostudies, Jun 07, v2, 2007. http://www.mccza.com/showdocument.asp?Cat=17&Desc=Guidelines% 20-%20Human%20Medicines. Accessed April 29, 2009.
3. Dissolution: Medicines Control Council, Department of Health, 2.07 Dissolution, Jun07, v2, 2007. http://www.mccza.com/showdocument.asp?Cat=17&Desc= Guidelines%20-%20Human%20Medicines. Accessed April 29, 2009.
4. Generic Substitution: Medicines Control Council, Department of Health, December 2003. http://www.mccza.com/showdocument.asp?Cat=17&Desc=Guidelines%20-%20Human%20Medicines. Accessed April 29, 2009.
5. Medicines and Related Substances Control Act, 1965 (Act No. 101 of 1965) as amended by Act No. 90 of 1997 and Act No. 59 of 2002. http://www.mccza.com. Accessed April 29, 2009.
6. Circular 14/95, Data Required as Evidence of Efficacy, Annexure 13, Medicines Control Council, Department of Health, October 2, 1995.
7. Medicines and Related Substances Control Act, 1965 (Act No. 101 of 1965) as amended by Act No. 59 of 2002. http://www.mccza.com. Accessed April 29, 2009.
8. Midha KK, Rawson MJ, Hubbard JW. Commentary: The role of metabolites in bioequivalence. Pharm Res 2004; 21:1331–1344.
9. General Information: Medicines Control Council, Department of Health, 2.01 General Information, February 08, v4, April 2008. http://www.mccza.com/showdocument. asp?Cat=17&Desc=Guidelines%20-%20Human%20Medicines. Accessed April 30, 2009.
10. Gibaldi M, Perrier D. Pharmacokinetics. 2nd ed. New York: Marcel Dekker, 1982.
11. Amidon GL, Lennernas H, Shah V,et al. A theoretical basis for a biopharmaceutics classification: The correlation of in-vitro drug product dissolution and in-vivo bioavailability. Pharm Res 1995; 12:413–420.
12. Guidance form Industry, Waiver of the in vivo bioavailability and bioequivalence studies for immediate-release solid oral dosage forms based on a Biopharmaceutics Classification System, Food and Drug Administration, Rockville, MD, USA, 2000. http://www.fda.gov/cder/guidance/index.htm. Accessed April 30, 2009.

APPENDIX I

The following were the requirements for proof of efficacy of generic medicines as stated in

Circular 14/95 issued on 2nd October, 1995

Section 2. Proof of Efficacy May be Submitted by Using One of the Following Methods, Depending on the Relevancy:

2.1 *Bioavailability*
2.2 *Dissolution*
2.3 *Disintegration*
2.4 *Acid neutralizing capacity*
2.5 *Microbial growth inhibition zones*
2.6 *Proof of release by membrane diffusion*
2.7 *Particle size distribution*
2.8 *Blanching test*
2.9 *Any other method an applicant wishes to submit, provided the rationale for submitting the particular method is included.*

2.1 *Bioavailability*
    2.1.1 *Bioavailability as proof of efficacy may be used in any instance as proof of efficacy, but <u>must</u> be used in the following cases:*
        a) *for a modified release tablet or capsule dosage form e.g.; slow release or sustained release.*
        b) *When a monograph in the USP for an active does not include a method for dissolution.*
        c) *When the active is mentioned in the attached list A, irrespective of a monograph being available in the USP.*
        d) *Suspensions, except when a monograph for dissolution is available in the USP for the active in suspension.*
        e) *Exceptions as indicated under alternative methods of proof of efficacy.*
    2.1.2 *Experimental subjects. Generally speaking 12 subjects are required in a cross-over study for immediate release tablets or capsules, and 20 subjects for modified release tablets or capsules.*
    2.1.3 *When bioavailability studies presented for registration purposes were derived from pilot batches, acceptable data derived from production batches must be submitted before distribution of the production batches can take place.*

Part 2.2.1 of Circular14/95, viz.;
    Dissolution as proof of efficacy may be used in the following instances:

    a) *When a monograph for the active in the USP includes a dissolution requirement and the active is not on the attached list A.*
    b) *When a monograph for a combination of actives in the USP includes dissolution requirement and the actives are not on the attached list A.*
    c) *When a monograph for a combination of actives is not included in the USP, but monographs for the individual actives with dissolution requirements are included in the USP, these individual monographs may be used, provided that the actives are not on the attached list A.*

The "attached list A" contained a number of drug substances listed in alphabetic order and the list was headed as follows:

    "MEDICINES CONTAINING THE FOLLOWING CHEMICAL ENTI-TIES, CERTAIN COMBINATIONS, SPECIFIC GROUPS OF ENTITIES OR MEDICINES FALLING INTO A SPECIFIC PHARMACOLOGICAL GROUP, WHICH WILL USUALLY REQUIRE DATA OTHER THAN COM-PARATIVE DISSOLUTION DATA AS EVIDENCE OF EFFICACY."

The following *proviso* was also included:
    "*Additions and deletions to the following list may be made by Council from time to time.*"
The list contained 96 drugs and drug combinations listed in alphabetic order but no reasons provided why those specific names were on the list.
    Section 2.2.2 stated the following:

a) *Assay results of the reference- and test product must be submitted.*
b) *Content uniformity test results, if it is a requirement for lot release, alternatively mass uniformity of the reference and test product must be submitted.*
c) *Dissolution must be done in 3 media, whereof:*

i) the first medium must be that specified in then USP monograph
ii) the other 2 media shall generally be from the following, spanning a wide pH range, and not the same as the medium specified in the USP:
  ii.i) an acidic medium e.g. Gastric fluid USP
  ii.ii) water
  ii.iii) an alkaline medium e.g. intestinal fluid USP

d) The USP requirements for at least 6 units of the product, the apparatus, medium volume, and rotation speed specified in the monograph, for the media required by the USP, must be followed.
e) For the other 2 media 6 units each must be used for the reference and test product. The USP requirements for the apparatus, medium volume, and rotation speed specified in the monograph, for the media required by the USP, may be followed.
f) The method used must be described in full, including media, media volumes, apparatus, rotation speed, filter size (which must not be larger than 1 micrometer if not specified), assay method and calculations to compensate for dilution of the ingredients due to withdrawal for dissolution media.
g) The study shall be a multipoint study.
h) The results shall be presented in a tabulated and graphical form. The tabulated form shall include the individual results of the dosage forms, the mean, and the standard deviation of the dosage forms. The graphical form will be the mean of the studies of the reference and test products set out on a graph for each medium.
i) If the active is insoluble in the other 2 media a motivation may be submitted to Council for the omission of further testing in the other 2 media."

In addition to the above, the following were also stated as being acceptable data required as evidence of efficacy.

2.3 Disintegration
  2.3.1 Disintegration as proof of efficacy may be used in the following instances:
    a) Vitamins or vitamins and mineral combinations when a claim is made as a supplement.
    b) Phenolphthalein
    c) Sucralfate
  2.3.2 The following are guidelines for the submission of disintegration data:
    a) The disintegration test included for Nutritional Supplements in the USP must be used for the vitamins.
    b) The general disintegration test included in the USP may be used for the other substances.
2.4 Acid neutralising capacity
  2.4.1 Acid neutralizing capacity may be as proof of efficacy may be used in the following instances:
    For all products with an antacid or acid neutralizing claim.
  2.4.2 The following are guidelines for the submission of acid neutralizing data:
    The acid neutralizing capacity test included in the USP must be used.
2.5 Microbial growth inhibition zones
  2.5.1 Microbial growth inhibition zones as proof of efficacy may be used in the following instances:
    For all products with a bacteriostatic/bactericidal/antiseptic claim.

2.6   *Proof of release by membrane diffusion*
    2.6.1   *Release by membrane diffusion as proof of efficacy may be used in the follow-ing instances:*
    For all creams, ointments and gels.
2.7   *Particle size distribution*
    2.7.1   *Particle size distribution as proof of efficacy may be used in the following instances:*
    For inhalations
    2.7.2   *The following are guidelines for the submission of particle size distributions: The Anderson sampler or equivalent apparatus may be used.*
2.8   *Blanching test*
    2.8.1   *The blanching test as proof of efficacy may be used in the following instances: For cortisone containing creams and ointments.*
2.9   *Any other method an applicant wishes to submit, provided the rationale for the particular method is included.*
    Please note that it is the prerogative of the applicant to submit any other data as proof of efficacy, motivating the reason to do so.
    Signed: **REGISTRAR OF MEDICINES** *1995/09/27*

9 South America and Pan American Health Organization

Silvia Susana Giarcovich

Department of Pharmacology, School of Pharmacy and Biochemistry, University of Buenos Aires; and DIFFUCAP-EURAND SACIFI, Buenos Aires, Argentina

Ricardo Bolaños

Department of Pharmacology, School of Medicine, University of Buenos Aires; Department of Projects and Plans, Direction of Planification, National Administration of Drugs, Food and Medical Technology (ANMAT), Ministry of Health; and Working Group of Bioequivalence, Pan American Network of Drug Regulatory Harmonization, PAHO, Buenos Aires, Argentina

INTRODUCTION

Current medicine policies, both in developed and underdeveloped countries, seek the same objectives of efficacy, safety, and quality of accessible products. However, different historical, social, and economic circumstances in such countries have resulted in different scenarios associated with their own internal issues which have been and are being currently resolved according to each particular prevailing situation.

The map of registration requirements for pharmaceutical products in the Americas is heterogeneous. Neither registration of innovator products nor that of noninnovator ones are identical and the documentation requirements for the registration of either are different among those countries.

There is currently a competitive market between innovator products and noninnovator products. The noninnovator products involve both generic products and so-called "similar products." The term "generic product" was initially used in those countries that recognized the intellectual property of pharmaceutical products to refer to products registered by competitors once the patent protection period had expired. Usually generic products are commercialized without a brand name although sometimes a brand name is used (the brand generics). The term "similar product" is used in countries where, historically, no patent protection rights on pharmaceutical products existed. In those cases, different commercial brands of the same active drug substances are simultaneously commercialized by different pharmaceutical laboratories. Definitions for both generic and similar products are included in the multisource product definition (1).

Harmonization efforts seek to ensure medicines availability at both national and international level by agreement of common requirements for registration, inspection, control, and vigilance. In this context different harmonization initiatives has taken place in Europe,[a] Southeast Asia,[b] America as well as ICH.[c] Particularly, the World Health Organization (WHO)[d] has provided a number of guidelines and technical documents related to good manufacturing practices (GMPs), good clinical practice (GCP), comparator products, essential medicines, multisource products registration, and so on. WHO has also promoted the International Conference of Drug Regulatory Authorities (ICDRA). In this context, regional harmonization efforts in the Americas have been carried out by different economic integration groups[e] and by PAHO (Pan American Health Organization) through PANDRH (Pan American Drug Regulatory Harmonization). This network grouped Pan-American regulatory authorities since 1997. Its work has been organized through continuous meetings of different working groups around the main issues such as GMP, bioavailability/bioequivalence (BA/BE), GCP, drug registration, phamacopoeia, combat to drug counterfeiting, GLP, drug classification, botanical products, pharmacovigilance, and vaccines. Besides the official regulatory authorities, pharmaceutical industry representatives are invited to participate with one observer in each working group. The Latin-American pharmaceutical companies are thus represented by FIFARMA (*Federación Latinoamericana de la Industria Farmacéutica*) and ALIFAR (*Asociación Latinoamericana de Industrias Farmacéuticas*). The documents, surveys, and recommendations produced by PANDRH are in the public domain at www.paho.org.

Several papers on drug regulations in the Americas have been recently published. Homedes and Ugalde (2), reported the results of a preliminary survey conducted in 10 Latin-American countries. The study aimed to document the experiences of different countries in defining and implementing generic drug policies. It focused on determining the cost of registering different types of pharmaceutical products, the time needed to register them, and uncover incentives that governments have developed to promote the use of multisource drugs products. The survey instrument was administered in person in Chile, Ecuador, and Peru and by e-mail in Argentina, Brazil, Bolivia, Colombia, Costa Rica,

---

[a] EMEA: European Agency for the Evaluation of Medicinal Products; CADREAC: Collaboration Agreement of Drug Regulatory European Authorities.

[b] ASEAN: the Association of Southeast Asian Nations.

[c] ICH is the International Conference of Harmonization of Technical Requirements for Registration of Pharmaceuticals for Human Use. It was established in 1990 and is participated by both regulatory authorities and pharmaceutical industry from the European Union, Japan, and United States. Today, ICH has a Global Cooperation Group, where different harmonization initiatives are represented.

[d] WHO has an international legal mandate from 191 members to establish global standards for the promotion and protection of public health.

[e] The different economic integration groups in the Americas are:

  a) TLCN (Tratado de Libre Comercio de Norteamérica): Canada, Estados Unidos, and Mexico.

  b) MERCOSUR: Argentina, Brazil, Paraguay and Uruguay (Chile and Bolivia participate without being members).

  c) SICA (Sistema de la Integración Centroamericana): Guatemala, El Salvador, Honduras, Nicaragua, and Costa Rica.

  d) CAN (Comunidad Andina de Naciones): Bolivia, Colombia, Ecuador, Peru, and Venezuela.

  e) CARICOM (Caribbean community): Caribbean Islands.

Nicaragua, and Uruguay. There were a total of 22 respondents. Survey responses indicated that countries use the terms generic and BE differently and suggested the need to harmonize definitions and technical concepts. Vacca González et al. (3) characterized the situation and regulatory tendencies relating competitor products in 14 countries of Latin America and the Caribe and concluded that harmonization efforts should consider both definitions adopted by different countries, and local policies promoting generics acquisitions.

In this chapter, an overview is provided of the current status of drug registration requirements in South America and the consensus reached under the sponsorship of PAHO. Associated documents, surveys, proposals, and recommendations from PANDRH working groups (WG) (drug registration and BA/BE) are also discussed. The terminology used in this text is in agreement with WHO definitions unless otherwise stated.

The following documents are referred to in this chapter:

- **Survey conducted by PANDRH Drug Registration WG:** "Diagnostic Study on Common Requirements for Pharmaceutical Products Registration in the Americas."[f]
- **Survey conducted by PANDRH BE WG:** "Diagnostic Study on Implementation of Bioequivalence Studies in the Americas."[g]
- **Document prepared by PANDRH Drug Registration WG:** "Draft Proposal of Harmonized Requirements for Pharmaceutical Products Registration in the Americas,"[h] which is still under discussion.
- **Document prepared by PANDRH BE WG:** "Framework for Implementation of Equivalence Requirements for Pharmaceutical Products,"[i] which has been recently approved at the PANDRH V Conference (17–19 November 2008, Buenos Aires, Argentina).

The discussion is focused specifically on the differences and issues related to the equivalence requirements for registration of noninnovator products.

REGISTRATION

Background

Since neither physicians nor patients are able to asses the efficacy, safety, and quality of the medicines that they prescribe and consume, national health authorities must establish mechanisms to ensure the various foregoing considerations. Most countries in the Americas have laws and regulations regarding the registration, inspection, control, and postmarketing surveillance of medicines. However, differences can be found, for example in the following:

- Classification.
- Documentation to be submitted for registration.
- Meaning of the terms "new drug substance" and "new drug product."
- Assessment criteria.

---

[f] http://www.paho.org/Spanish/AD/THS/EV/rm-hp.htm.

[g] http://www.paho.org/Spanish/ad/ths/ev/BE_ImpletEstudio04-esp.pdf.

[h] http://new.paho.org/hq/index.php?option = com_content&task = view&id = 1060&Itemid = 513.

[i] http://new.paho.org/hq/index.php?option=com_content&task=view&id=1052&Itemid=513&limit=1&limitstart=1.

– Labelling and package insert requirements.
– Time elapsed from registration application until authorization.

The PANDRH WG on Drug Registration has produced two documents, which can be found on the web (www.paho.org) and are further discussed:

– A survey to assess the technical and legal drug registration requirements.[j]
– A proposal of harmonized requirements for drug registration.[k]

Before presenting these two documents, some critical terms are discussed.

Some Critical Terms

A critical issue that each country must decide upon before applying for any policy of registration relates to the definitions of terms and explanation of some particular concepts.

New Drug Substance

Some countries consider a "new drug substance"[l] as a drug substance (API) that has never been commercialized in that particular country. However, others have adopted the concept of "country of reference" and consequently, consider that "new" is a drug substance (API) that has never been commercialized either in the country or in the country of reference. Figure 1 helps to explain these differences.

New Drug Product

Similar to the concept of "new drug substance" some countries consider a "new drug product" as a drug product that has never been commercialized in the country, and others adopt the concept of "country of reference" and consequently consider that "new" is a drug product that has never been commercialized either in the country or in the country of reference. Figure 2 illustrates these differences.

Multisource Pharmaceutical Products

Harmonization efforts have tended to a dual registration concept, that is to say, the product is either "new" or "generic." This has been historically so in the United States and Canada (within PAHO) with the NDA and ANDA criteria for registration. However, the situation in Latin-American countries shows more variety of options. Brazil and Mexico on the one hand allow the registration of either "new," "generic" or "similar" products, where the generic product must demonstrate BE to the selected reference product to be declared interchangeable. However, regulatory authorities have often included similar products in a program to require bioequivalence demonstration for registration renewals.

On the other hand, most of the other Latin-American countries do not use the term "generic" in their registration regulations; usually, they refer to "similar drug products." Also, most of these countries require the demonstration of BE of

---

[j] http://www.paho.org/Spanish/AD/THS/EV/rm-hp.htm.
[k] http://new.paho.org/hq/index.php?option = com_content&task = view&id = 1060&Itemid = 513.
[l] *Author's note*: Drug substance is taken in the context of this article as API (active pharmaceutical ingredient).

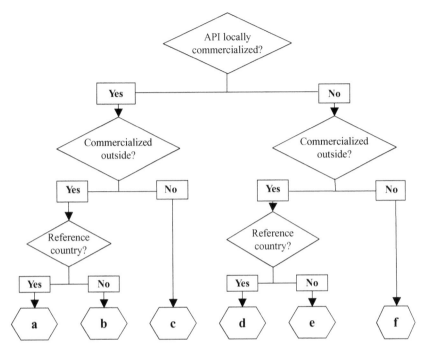

FIGURE 1 **a, b, c:** As the drug substance is already being locally commercialized, it will never be considered a new one. The differences between a, b, and c take into account whether the drug substance is being commercialized outside the country and the category of those countries where reference country categories have been locally established.

d, e, f: As the drug substance is not being locally commercialized, some countries will consider that the drug substance is a new one. However, countries which accept the concept of country of reference will consider that the drug substance is a new one only in cases e and f.

similar/multisource drug product to the reference product when the product is considered to be of high risk.

Lastly, to understand the Latin-American scenario, it is necessary to refer to the WHO definition of a multisource drug product, that is,"*Pharmaceutically equivalent or pharmaceutically alternative products that may or may not be therapeutically equivalent. Multisource pharmaceutical products that are therapeutically equivalent are interchangeable.*" (1)

In practice, both "generic/multisource drug products" and "similar/multisource drug products" that demonstrate bioequivalence to the selected reference product, may be declared interchangeable.

PANDRH Survey: Diagnostic Study of Common Requirements for the Registration of Pharmaceutical Products in the Americas

The survey[m] was carried out by the PANDRH WG on Drug Registration and the full report can be obtained from the PAHO website. The questions were based

---

[m] Some parts of the original document are reproduced (italics) in the present chapter whereas others are summarized.

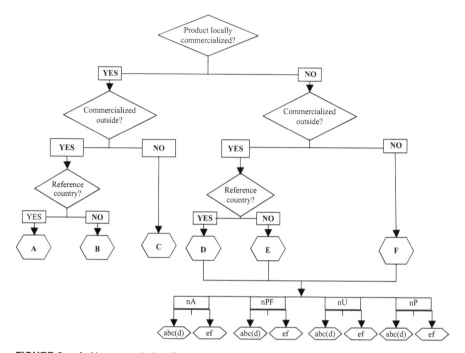

FIGURE 2 nA, New association (fixed-dose combination); nPF, New pharmaceutical formulation; nU, New use; nP, New potency.

A, B, C: As the drug product is already being locally commercialized, it will never be considered a new one. The differences between A, B, and C take into account whether the drug product is being commercialized outside of the country and the category of those countries where reference countries category has been locally established.

D, E, F: As the drug product is not being locally commercialized, some countries will consider that the drug product is a new one. However, countries which accept the concept of country of reference will consider that the drug product is a new one only in cases E and F. Besides, in cases D, E, and F, if the API is an already known one (cases a, b, c, and eventually d of chart shown in "New Drug Substance" section), documentation requirements for registration of the new drug product might be different than when the API is unknown.

on a yes or no response and were sent by the PANDRH Secretariat to invited countries. The percentage of positive responses to the total number of participating countries was informative. The objective was to obtain information from local health authorities of each participating country to know the requirements for medicine registration in the following cases:

- *New molecules*
- *New formulations*
- *New uses*
- *New dosage form*
- *New presentation*
- *New route of administration*
- *New association*
- *Generic equivalent*

- *Similar equivalent*
- *Biologics*
- *Renewal*

The survey implied to answer if the following requirements applied for registration of the different cases mentioned above:

- *General information about the product*
- *Information about the manufacturing facilities*
- *Legal documentation supporting the application*
- *Active Pharmaceutical Ingredient*
- *Final product*
- *Biopharmaceutic technical documentation*
- *Preclinical data*
- *Clinical data*
- *Labelling*
- *Secondary packaging components*
- *Package Inserts*

Observations are summarized as follows mainly taking into account the information relating to requirements for new molecules (with the knowledge that new molecules have the strictest requirements of all studied groups):

General Information About the Product
General information required include commercial name, generic name, presentation, potency, dosage form, route of administration, description of the product, and also cost and time to complete the evaluation of the application. Less than 50% of the countries have already established the required costs and times for evaluation. The percentage is higher for new molecules than for the rest of the groups.

Information About the Manufacturing Facilities
The information required includes name of the license holder, API manufacturer (mostly for new molecules), pharmaceutical manufacturer (100% for new molecules), alternative manufacturing and packing sites (almost all countries for all cases), and responsible person for the release of the batch (15.4–30.8% of the countries).

Legal Documentation Supporting the Application
This item includes manufacturer authorization (>66.7% for all cases), GMP certificate for API manufacturer (<30% of countries), GMP certificate for finished product manufacturer (88.9% is the highest for new molecules), GMP certificate for out-of-the-country manufacturer (<30% for all cases), outsourcing manufacturing contract (maximum of 94.4% for new molecules), patent certificate (maximum of 27.8%), reference standards (<70%), first batch authorization (maximum of 38.9% for new molecules).

Secondary Packaging Components
In a high percentage of the participating countries, the information that has to be present in the secondary packaging material are commercial name, generic

name (INN or DCI), potency, dosage form, expiration date, storage conditions, and prescription conditions. Some countries require registration number, instructions for use, warnings, and contraindications. A fairly low percentage of countries require manufacturing date, legal representative, name of the responsible professional, indications, posology, and interactions. Finally, a very small percentage of countries require the inclusion of restrictions for use and overdose data.

Packaging Insert
A package insert for new APIs is required by 69.2% of the countries. A high percentage of countries require commercial name, generic name, dosage form, instructions of use, indications, dosage, warnings, contraindications, adverse reactions, interactions, and overdose in the insert whereas a smaller number of countries require manufacturer information, expiration date, name of the responsible professional, registration number.

Active Pharmaceutical Ingredient
A high percentage of countries use the INN (94.4% for new molecule), in comparison to those that use IUPAC (44.4% for new molecule) or ATC classification (33.3% for new molecule). Countries require chemical name, molecular and structural formula, molecular weight, organoleptic properties, and physicochemical characteristics (>75% for new molecules). The route of synthesis has to be included for most of the countries (61.1% for new molecules) whereas most of them also require information on impurities and degradation products as well as specifications for batch approval (77.8% for new molecules). A high percentage of countries require stability data and analytical methodology (94.4% for new molecules); however, a smaller percentage require documented validation for the analytical procedures (38.9% for new molecules).

Final Product
A document about the formulation development (44.4% for new molecules), manufacturing method (61.1% for new molecules), and physicochemical properties of the excipients (50.0% for new molecules) is required. Formula and storage conditions are required in 100% of registrations of new molecules. A high percentage of countries require stability data and analytical methodology (94.4% for new molecules) whereas a lower percentage of countries require local analysis for imported products (50.0% for new molecules).

Preclinical Data
Data relating general and special toxicology, pharmacodynamic, and pharmacokinetic studies are required in most countries mainly for new APIs (72.2–94.4%).

Clinical Data
The percentage of countries requiring clinical data are the following:

○ *New chemical entities: 72.2%*
○ *New formulation: 50.0%*
○ *New indication: 66.7%*
○ *New dosage form: 61.1%*

- *New strength: 66.7%*
- *New route of administration: 61.1%*
- *New association: 66.7%*
- *Generics: 38.9%*
- *Similars: 57.1%*
- *Biologics (vaccines, recombinant products, blood products): 50.0%*
- *Renewals: 5.6%*

Due to the observed differences in registration requirements of the different participant countries, it was concluded that a harmonization proposal was needed.

PANDRH Proposal: Proposal of Harmonized Requirements for the Registration of Pharmaceutical Products in the Americas

Since some of the differences in the registration requirements were considered to be critical to ensure public health, a harmonization proposal[n] was elaborated by the Drug Registration WG and discussed along with the PANDRH V Conference although it still remains as a draft document which can be found on the PAHO website (www.paho.org).

The recommended minimum requirements for drug registration information are the following:

1. *Commercial name*
2. *Generic name*
3. *Strength/potency*
4. *Pharmaceutical dosage form*
5. *Technical director*
6. *Legal representative*
7. *Product owner*
8. *API manufacturer*
9. *Finished product manufacturer*
10. *Other laboratories participating in the manufacturing process*
11. *Drug product presentation*
12. *Route of administration*
13. *Storage conditions*
14. *Dispensing conditions*
15. *Qualitative–quantitative formula*
16. *Product legal documents*
17. *Summary of product characteristics*
18. *Labelling and insert*
19. *Labelling of primary packaging*
20. *Labelling of secondary packaging*
21. *Insert*
22. *Health professional information*
23. *Final marketing packaging*
24. *Samples of the finished product*
25. *Certificate of analysis*
26. *Active pharmaceutical ingredient(s)*

---

[n] Parts of the original document are reproduced (italics) in the present chapter.

BIOEQUIVALENCE

Introduction

Bioequivalence requirements for registration of noninnovator products in the Americas are basically divided into three different approaches as follows, United States/Canada, Brazil/Mexico, and the rest of the Spanish-speaking countries.

On the one hand, United States and Canada will always require demonstration of therapeutic equivalence-employing either in vivo or in vitro methodology- to allow health authorities to declare interchangeability between the noninnovator product (the generic product) and the reference product (usually the innovator product). When a noninnovator product fails to demonstrate therapeutic equivalence, it will not be declared interchangeable and consequently it will not be registered as a generic product in either the United States or in Canada.

A different approach is taken in Mexico and Brazil. Both countries have regulations for the registration of generic products from 1999. Mexico was the first Latin-American country with the *Norma Oficial Mexicana* NOM-177-SSA1– 1998 and Brazil was the second with the *Ley* No 9787 from 10.2.99 and the *Res.* No 391 from 9.8.99. However, both of them allow sponsors to choose between two types of registration of noninnovator products, interchangeable generic products and similar products.

Finally, the rest of the Spanish-speaking countries represent the third approach. In general, they do not have regulations for the registration of generic products as such.

However, in some of the countries, a bioequivalence study and/or declaration of interchangeability (through either in vitro or in vivo methodology) is also required as a condition either for registration or commercialization for some of the noninnovator products, mainly chosen according to high health risk criteria. For example, the following countries use this type of registration and/or commercialization requirements[o]:

- **Argentina:** *Disp.* No *3185/1999* from ANMAT.
- **Chile:** *Res. No.* 726 and *727/2005* from Health Ministry.
- **Colombia:** *Res. No.* 1400/2001 from Health Ministry.
- **Costa Rica:** *Pres. Dec. No.32470-S/2005.*
- **Cuba:** Reg. No. 18-07 from Health Ministry.
- **Panama:** *Res. No. 081/2005* from Health Ministry.
- **Venezuela:** *Normas de la Junta Revisora de Productos Farmacéuticos,* from January 1999, Chapter XIV and more recently in *Res. No. 38.499/2006* from Health Ministry.
- **Uruguay:** *Pres. Dec.* of Interchangeability, 2008.

In some countries, such as Peru and Ecuador, expert groups are discussing the way to include therapeutic equivalence studies (either in vitro or in vivo) requirements into their own regulations.

---

[o] The following summary states mainly the first or initial regulations where these requirements can be found. However, in most cases they have been complemented with additional local regulations.

Also, it is important to mention that amongst this group of countries there is a difference which results in two new approaches. Some of these countries include the concept of "country of reference"[p] whereas others do not. When the regulations for registration include the concept of "country of reference," then the introduction of a new product into the local market can sometimes proceed in an abbreviated way. If it can be demonstrated that the new product has already been registered and commercialized in one of the countries of reference, then it can be registered as a similar/multisource product. However, when the concept of "country of reference" is not included in the regulations for registration, then the first pharmaceutical company (usually the innovator), that intends to introduce a new product into the local market, will have to submit complete and comprehensive preclinical and clinical data.

The PANDRH WG on BE (WG/BE) has produced two documents that are discussed later:

- A survey to assess the implementation of BE studies in the Americas[q]
- A proposal for implementation of equivalence criteria requirements[r]

PANDRH Survey: Diagnostic Study on Implementation of Bioequivalence Studies in the Americas

The present survey was carried out by the PANDRH WG/BE and the full report (2005), written in Spanish,[s] can be found on the PAHO website (www.paho.org). Since the WG/BE of PANDRH makes recommendations about the type of equivalence studies (in vitro/in vivo), prioritization of drugs for BE (in vivo) studies and criteria to choose a reference product, it was necessary to assess the real feasibility of carrying out BE studies in Latin America from both regulatory and operational aspects. Hence, detailed information was obtained from health agencies of each participant country based on a yes-or-no questionnaire regarding the following:

- Current GCP and BE regulations
- Registration requirements for BE studies
- Prioritization criteria[t]
- Availability of centers to carry out the studies
- Expertise of the regulatory agency to assess, control, and audit the studies
- Quantity of BE studies already carried out
- Training

The survey was carried out in 2003/2004 and was answered by 20 countries, 15 Spanish-speaking countries, 4 English-speaking countries, and 1

---

[p] Usually the term "country of reference" refers to countries with high surveillance health agencies.

[q] http://www.paho.org/Spanish/ad/ths/ev/BE_ImpletEstudio04-esp.pdf.

[r] http://new.paho.org/hq/index.php?option=com_content&task=view&id=1052&Itemid=513&limit=1&limitstart=1.

[s] Some parts of the original document are reproduced (italics) in the present chapter whereas others are summarized.

[t] Prioritization criteria are intended to be useful mainly in countries where BE studies are not included into registration requirements and refer to criteria to choose which APIs have priority mainly considering sanitary risk.

Portuguese-speaking country (Argentina, Brazil, Bahamas, Barbados, Bolivia, Canada, Colombia, Costa Rica, Dominican Republic, Ecuador, El Salvador, Guatemala, Honduras, Mexico, Nicaragua, Panama, Peru, Surinam, United States, and Venezuela). Since the number of inhabitants of the various countries largely differ, answers were interpreted considering both the number of countries that answered positively (% of the total number of participant countries) and the impact it had on the regional Latin-American/Caribbean number of inhabitants[u] (% of the total population affected belonging to either the Americas or Latin America). Full results of this survey are published on the web and are summarized as follows:

GCP and BE Current Regulations
Fifty percent of participating countries (~22% of Pan-American countries) have some type of GCP regulations applicable to BE studies, and this impacts on ~76% of Latin-American/Caribbean population.

Registration Requirements for BE Studies
Forty-five percent of participating countries (~20% of Pan American countries) require BE studies for registration and this impacts on ~75% of Latin-American/Caribbean population.

Prioritization Criteria
Thirty-five percent of participating countries (~16% of Pan American countries) have already established prioritization criteria with variable number of drugs in positive lists (countries usually list locally prioritized drugs) and this impacts on 66% of Latin-American/Caribbean population.

Availability of Centers to Carry Out the Studies
Forty percent to fifty-five percent of participating countries (~18–25% of Pan-American countries) recognize the participation of local experts, the availability of adequate hospitals/clinics and the requirement of protocol approval by independent ethical committees, and this impacts on ~70% of Latin-American/Caribbean population.

Expertise of Regulatory Agency to Assess, Control, and Audit the Studies
Forty percent to forty-five percent of participating countries (~18% of Pan American countries) declare having adequate resources to assess and authorize studies and this impacts on 72% of Latin-American/Caribbean population whereas only 20% have adequate resources to inspect the different steps of BE studies. In several countries study evaluators have to fulfill other assignments as well. Furthermore, in some cases the health authorities hire external organizations/personnel in order to complete the evaluation.

---

[u] There were 20 participant countries of the survey of 46 countries within all the Americas. Those 20 countries correspond to 93.6% of the total population of the Americas and 18 of them correspond to 89.8% of the Latin-American and Caribbean population.

Number of BE Studies Concluded

Only 25% of participating countries (~11% of Pan American countries) declare the larger number of BE studies already carried out, and this impacts on 60% of Latin-American/Caribbean population.

Training

Forty percent of participating countries (~18% of Pan-American countries) report having carried out BE training courses and this impacts on 70% of Latin-American/Caribbean population.

From the date of the survey until the time of writing this chapter, additional countries have developed new BE regulations, for example, Chile, Costa Rica, Cuba, Panama, and Uruguay. Hence, since the start of the present survey, an increasing number of countries have included BE studies as a registration requirement affecting about 80% of Latin-American population and about 50% of the population of the Americas.

PANDRH-Approved Document: Framework for Implementation of Equivalence Requirements for Pharmaceutical Products

This document was prepared by the WG/BE[v] and was approved by the PANDRH V Conference[w] in November 2008 (full document can be found on PAHO website). The objective of the document is to recommend harmonized criteria for the equivalence of drugs and consists of two parts.

The first part refers to the **scientific criteria for implementing therapeutic equivalence**. The WG/BE decided unanimously to endorse the WHO document "Multisource (generic) pharmaceutical products: Guidelines on registration requirements to establish interchangeability" (1) and to promote its implementation in the Americas.

The second part of the document refers to the **strategic framework for the implementation of studies for drug equivalence.** This part describes the current situation of the Region of the Americas, with particular attention to the special features of Latin America considering that most of the multisource products marketed in the region have been approved in accordance with drug registration requirements of each country at the time of its registration. The gradual implementation of BE demonstration requirements based on health risk of a particular product is recommended. Furthermore, it presents a flow chart integrating both the requirements of fulfillment of GMP, the validity and reliability of the reference product, and the concept of gradualism in prioritization according to health risk and biowaivers.

Equivalence Demonstration and GCP

The WHO document (1) states that multisource pharmaceutical products must be shown, either directly or indirectly, to be therapeutically equivalent to the

---

[v] Some parts of the original document are reproduced in the present chapter (italics) whereas others are summarized.
[w] First version of the document was considered during the IV PANDRH Conference, March 2005, Dominican Republic.

comparator product in order to be considered interchangeable. Suitable test methods to assess equivalence are

(a) *comparative pharmacokinetic studies in humans, in which the active pharmaceutical ingredient and/or its metabolite(s) are measured as a function of time in an accessible biological fluid such as blood, plasma, serum or urine to obtain pharmacokinetic measures, such as AUC and C_{max} that are reflective of the systemic exposure;*
(b) *comparative pharmacodynamic studies in humans;*
(c) *comparative clinical trials; and*
(d) *comparative in vitro tests.*

The document prepared by the WG/BE recognizes that implementation has to be decided by the local health regulatory authority when establishing equivalence requirements for drug registration and commercialization and that *a strategy based on the health risk criteria of each product will facilitate harmonization of implementing equivalence requirements in the region.*

Prioritization According to Sanitary Risk

As part of the proposal, an example of classification in accordance with health risk has been designed with the aim of helping health authorities only when needed. Health risk categories have been established and a score from 1 to 3 has been assigned according to sanitary risk. The health risk concept has been defined in relation to health impact if drug plasma concentrations fall under or above the therapeutic window (the margin whose limits are the nontoxic maximum and effective minimum drug concentrations) hence, three risk levels have been characterized as follows:

HIGH HEALTH RISK: This is the probability of the appearance of threatening complications of the disease for the life or the psychophysical integrity of the person and/or serious adverse reactions (death, patient hospitalization, extension of the hospitalization, significant or persistent disability, disability or threat of death), when the blood concentration of the active ingredient is not within the therapeutic window. For purposes of the selection, this risk level has been assigned a score of 3 (three).

INTERMEDIATE HEALTH RISK: This is the probability of the appearance of non-threatening complications of the disease for the life or the psychophysical integrity of the person and/or adverse reactions, not necessarily serious, when the blood concentration of the active ingredient is not found within the therapeutic window. For purposes of the selection, this risk level has been assigned a score of 2 (two).

LOW HEALTH RISK: This is the probability of the appearance of a minor complication of the disease and/or mild adverse reactions, when the blood concentration of the active ingredient is not within the therapeutic window. For purposes of the selection, this risk level has been assigned a score of 1 (one).

The table below lists the active ingredients[x] and their classification according to their health risk, and the table should be continuously updated:

---

[x] Taken as an example from API′s listed in: "Multisource (generic) pharmaceutical products: guidelines on registration requirements to establish interchangeability" WHO Technical Report Series (1996), No. 863, 114–155.

Active Ingredient	Health Risk
Carbamazepine	3
Cyclosporine	3
Digoxin	3
Ethambutol	3
Ethosuximide	3
Griseofulvin	3
Lithium Carbonate	3
Oxcarbazepine	3
Phenytoin	3
Procainamide	3
Quinidine	3
Theophylline	3
Tolbutamide	3
Valproic Acid	3
Verapamil	3
Warfarine	3
6-mercaptopurine	2
Amiloride	2
Amitriptyline	2
Amoxicillin	2
Atenolol	2
Azathioprine	2
Biperiden	2
Chloramphenicol	2
Cimetidine	2
Ciprofloxacin	2
Clofazimine	2
Clomipramine	2
Clorpromazine	2
Co-Trimoxazole	2
Cyclophosphamide	2
Dapsone	2
Diethylcarbamazine	2
Doxycycline	2
Erythromycin	2
Ethinylestradiol	2
Etoposide	2
Flucytosine	2
Fludrocortisone	2
Furosemide	2
Haloperidol	2
Hydrochlorothiazide	2
Indometacin	2
Isoniazid	2
Ketoconazole	2
Levodopa + Inhib. DDC	2
Levonorgestrel	2
Levotiroxine	2
Methotrexate	2
Methyldopa	2
Metoclopramide	2
Metronidazole	2
Nitrofurantoin	2

(Continued)

Active Ingredient	Health Risk
Noretisterone	2
Oxamniquine	2
Paracetamol	2
Penicillamine	2
Piperazine	2
Piridostigmine	2
Procarbazine	2
Promethazine	2
Propranolol	2
Propylthiouracil	2
Pyrimethamine	2
Quinine	2
Rifampicin	2
Salbutamol, sulphate	2
Spironolactone	2
Tamoxifen	2
Tetracycline	2
Acetazolamide	1
Allopurinol	1
Calcium Folinate	1
Captopril	1
Clomifene	1
Cloxacillin	1
Dexamethasone	1
Diazepam	1
Folic Acid + Ferrous Sulfate	1
Ibuprofen	1
Isosorbide Dinitrate	1
Levamisole	1
Mebendazole	1
Mefloquine	1
Nalidixic Acid	1
Niclosamide	1
Nifedipine	1
Nystatin	1
Phenoxymethylpenicillin	1
Phytomenadione	1
Pirantelo	1
Praziquantel	1
Pyrazinamide	1
Sulfasalazine	1
Aminophylline (see Theophylline)	
Sulfadoxine (see Pirimetam)	

The same document also shows a list of APIs subject to BE in vivo study requirements in different countries in the American region.

Decision Tree for Implementing Equivalence Studies in the Region
The following flow chart integrates GMP criteria, prioritization according to health risk, biowaivers according to the BCS (4) and a reliability requirement for reference products was designed:

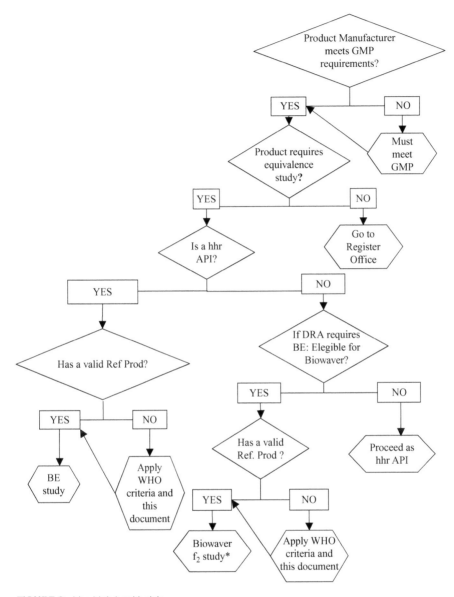

FIGURE 3 hhr, high health risk.
*If a product does not meet f_2 specifications, DRA will decide case by case.

How to Select the Comparator Product in the Region

Generally, local health authorities are responsible for the identification of the most appropriate reference product[y] for each drug. Usually most health

---

[y] *Note of Authors*: "Reference product" corresponds to the term "comparator product" as defined by WHO.

authorities will accept that the innovator product is the most logical comparator product for a multisource pharmaceutical product because its quality, safety,
and efficacy would have been comprehensively assessed and documented. In
Latin America, however, despite the availability of innovator products in the
local market, national requirements of reliability for a reference product still has
to be assessed. Accordingly, in the light of this complex scenario, the PANDRAH
WG/BE has included a recommended mechanism to help the national health
authority to assess the local particular situation and to establish a reliable reference product.

Summary of Regulatory Requirements in Argentina

The main characteristics of the Argentinean regulations (5) regarding multi-
source/similar products registration are summarized below:

a) The decision to categorize a new application for registration as an abbreviated submission or not is primarily related to the commercialization
antecedents of the pharmaceutical product.

b) If the product to be registered is already being commercialized either locally
or in at least one of the countries of reference,[z] the product will be categorized as "similar."

c) There is a list of countries of reference in order to allow an abbreviated procedure for an imported product registration.[aa]

d) Noninnovator products are registered as "similar" products. A registration
procedure for generic products as such, does not exist.

e) The requirements for the registration of noninnovator products mainly
involve GMP and equivalence studies (in vitro and/or in vivo).

f) The requirements of either in vitro or in vivo equivalence studies are
not related to different nomenclature (generics, similars, multisource, etc.)
but to the type of API present in the formulation which is submitted for
registration.

g) The corresponding equivalence studies (in vitro and/or in vivo) have been
mainly established on sanitary risk criteria.

h) The criteria used to decide between in vitro and in vivo equivalence studies
are clearly established in the local regulations.

i) The in vivo equivalence studies are referred to as BA and BE studies.

j) Every registered product has to have demonstrated equivalence (in vitro
and/or in vivo) with respect to a reference product.

k) When a noninnovator product has demonstrated Bioequivalence to a reference product, it can be declared bioequivalent. However, an interchangeability declaration does not exist.

l) Injectables and oral aqueous solutions, gases and powders for reconstitution
into solutions are exempted from the need to demonstrate equivalence.

m) Regulations for prescriptions require the use of the generic name of drugs
instead of the brand name when physicians prescribe medicines.

---

[z] United States, Japan, Sweden, Swizerland, Israel, Canada, Austria, Germany, France, United
Kingdom, The Netherlands, Belgium, Denmark, Spain, and Italy.

[aa] Australia, Mexico, Brazil, Cuba, Chile, Finland, Hungary, Irland, China, Luxenburg, Norway,
and New Zealand.

The authors declare that the contents of this chapter represent their personal point of view without involving any organizations where they are associated.

ABBREVIATIONS

ALIFAR	Asociación Latinoamericana de Industrias Farmacéuticas
ANDA	Abbreviated New Drug Application
ANMAT	Administración Nacional de Medicamentos, Alimentos y Tecnología Médica (Argentinian health authority)
API	Active Pharmaceutical Ingredient
ASEAN	Association of Southeast Asian Nations
BA/BE	Bioavailability/Bioequivalence
BCS	Biopharmaceutic Classification System
CADREAC	Collaboration Agreement of Drug Regulatory European Authorities
CAN	Comunidad Andina de Naciones
CARICOM	Caribbean community
DRA	Drug Regulatory Authority
EMEA	European Agency for the Evaluation of Medicinal Products
FDA	Food and Drug Administration
FIFARMA	Federación Latinoamericana de la Industria Farmacéutica
GCP	Good Clinical Practice
GMP	Good Manufacturing Practice
GLP	Good Laboratory Pactice
ICDRA	International Conference of Drug Regulatory Authorities
ICH	International Conference of Harmonization of Technical Requirements for Registration of Pharmaceuticals for Human Use
INN	International Nonproprietary Name
IUPAC	International Union of Pure and Applied Chemistry
MERCOSUR	Mercado Común del Sur
NDA	New Drug Applications
NAFTA	North American Free Trade Agreement
PAHO	Pan American Health Organization
PANDRH	Pan American Network for Drug Regulatory Harmonization
SICA	Sistema de la Integración Centroamericana
USA	United States of America
WG	Working Group/s
WHO	World Health Organization

ACKNOWLEDGMENTS
We acknowledge the dedication and professional work of members of PANDRH Working Groups, Drug Registration and Bioequivalence, as well as their Coordinators: Esperanza Briceño (Venezuela), María Teresa Ibarz (Venezuela), and Justina Molzon (United States). In addition, with special deep gratitute and respect, we recognize the inestimable help and support of Rosario D'Alessio, (PANDRH/PAHO secretariat). We specially acknowledge the continuous support and expertise from Nelly Marín (PANDRH/PAHO secretariat) and the careful revision of this chapter and precise advice from Horacio Pappa.

REFERENCES

1. Annex 7 "Multisource (generic) pharmaceutical products: guidelines on registration requirements to establish interchangeability. WHO Technical Report Series, No. 937, 2006.
2. Homedes N, Ugalde A. Multisource drug policies in Latin America: Survey of 10 countries. Bull World Health Organ 2005; 83(1):64–70.
3. Vacca González CP, Fitzgerald JF, Bermúdez JAZ. Definición de medicamento genérico ¿un fin o un medio? Análisis de la regulación en 14 países de la Región de las Américas. Rev Panam Salud Pública/Pan Am J Public Health 2006; 20(5):314–323.
4. Annex 8. Proposal to waive in vivo bioequivalence requirements for WHO Model List of Essential Medicines immediate-release, solid oral dosage forms. WHO Technical Report Series, No. 937, 2006.
5. Argentinean regulations relating medicines registration:

   - Dec. ANMAT 150/92

   Argentinean regulations relating BE and clinical studies:

   - Disp. ANMAT No 5330/97
   - Disp. ANMAT No 3436/98
   - Disp. ANMAT No 3185/99
   - Disp. ANMAT No 3112/00
   - Res. SPRS No 229/00
   - Res. SPRS No 40/01
   - Disp. ANMAT No 3311/01
   - Disp. MIN. SALUD-SPRRS-ANMAT No 1383/02
   - Disp. ANMAT No 1277/02
   - Disp. ANMAT No 2807/02
   - Disp. ANMAT No 2814/02
   - Disp. ANMAT No 3598/02
   - Disp. ANMAT No 4290/02
   - Disp. ANMAT No 5318/02
   - Disp. ANMAT No 7062/02
   - Disp. ANMAT No 7307/02
   - Res. Min. Salud No 60/03
   - Res. SPRS No 25 y 19/03
   - Res. SPRRS No 46/03
   - Disp. ANMAT No 3712/04
   - Disp. ANMAT No 3757/04
   - Disp. ANMAT No 4218/04
   - Disp. ANMAT No 690/05
   - Disp. ANMAT No 2124/05
   - Disp. ANMAT No 2749/05
   - Disp. ANMAT No 4844/05
   - Disp. ANMAT No 4457/05
   - Disp. ANMAT No 5040/06
   - Disp. ANMAT No 1746/07
   - Res. MS No 35/07
   - Res. MS No 1490/07
   - Res. MS No 1673/07
   - Disp. ANMAT No2446/07
   - Disp. ANMAT No 1067/08
   - Disp. ANMAT No 6550/08
   - Disp. ANMAT No 1861/08
   - Disp. ANMAT No 1862/08
   - Disp. SP No 6/08
   - Disp. MS No 4691/08

- Res. MS No1062/08
- Disp. ANMAT No 4541/08
- Dec. PEN No 253/08
- Res. MS No 102/09
- Disp. ANMAT No 1310/09

Argentinean regulations relating the use of generic names in prescriptions:

- Res. SP No 326/02
- Ley Esp. Med. No 25.649/02
- Dec. No 987/03

Argentinean regulations relating GMP and current further regulations:

- Disp. ANMAT No 2819/04
- Disp. ANMAT No 3477/05
- Disp. ANMAT No 2372/08

10 | Taiwan

Li-Heng Pao
School of Pharmacy, National Defense Medical Center, Taipei, Taiwan, Republic of China

Jo-Feng Chi
Bureau of Pharmaceutical Affairs, Department of Health, The Executive Yuan, Taipei, Taiwan, Republic of China

Oliver Yoa-Pu Hu
National Defense Medical Center, Taipei, Taiwan, Republic of China

INTRODUCTION

The Bureau of Pharmaceutical Affairs (BPA) within the Department of Health (DOH) is responsible for the regulation of medicinal products in Taiwan. The mission of the BPA is to ensure that medicinal products that are available for the people in Taiwan are of the highest quality, safety, and efficacy. On the basis of BPA statistics, there are over 27,000 active licensed drugs currently available in the Taiwan market (1). Approximately 20% of Taiwan's drug products are imported from abroad, the remainder being manufactured by domestic pharmaceutical companies that operate in compliance with current good manufacturing practice (cGMP) standards. Therefore, the need to ensure bioequivalence of generic drug products with corresponding innovator products is critical in Taiwan. The goal of this chapter was to provide an overview of the generic drug review process and bioequivalence requirements in Taiwan.

Currently the BPA is organized into five sections. Section I is responsible for the law and regulation of pharmaceutical affairs, pharmacy-related compliance, and advertising. Section II is involved with the registration of medical devices and cosmetics. New drug applications (NDAs), clinical trials, and postmarketing surveillance are reviewed and approved according to Section III. Section IV covers the review of abbreviated new drug applications (ANDAs) and their approval for marketing whereas Section V deals with registration of biological and plasma products, radiopharmaceuticals, as well as in vitro diagnostic kits. The Taiwan Food and Drug administration (TFDA) will be formally inaugurated Jan. 1, 2010. Additional information on the organization of the BPA can be found at www.doh.gov.tw.

LEGISLATIVE AND REGULATORY ISSUES

Drug regulation has become more challenging since the advancement of new technologies, cost pressures, as well as trends of globalization. The drug regulatory process and development of the pharmaceutical industry in Taiwan has changed substantially over the past two decades (2–4). The fundamental law regulating medicinal products is based on the *Law for Control of Medicaments and Pharmaceutical Firms* promulgated in 1970. The Law was extensively revised and renamed the *Law of Pharmaceutical Affairs* in 1993. The current *Law of Pharmaceutical Affairs*, with its subsequent amendments in 2004, is the basic law for the regulation of medicinal products in Taiwan. Implementation of good manufacturing practice (GMP) and cGMP had a dramatic impact on Taiwan's pharmaceutical industry. Before advent and implementation of GMP requirements in 1982, there were over 400 domestic pharmaceutical manufactures. Six years later in 1988, the number of manufactures was reduced to 230. To further improve and meet the international standards of domestic production of medicinal products, cGMP regulations were promulgated in 1999. There are currently approximately 160 cGMP pharmaceutical manufactures in Taiwan. Therefore, when applicants seek approval of generic products in Taiwan, the drug products must comply with cGMP standards and requirements.

Following successful implementation of the GMP program in Taiwan, academia, government, and the pharmaceutical industry recognized that data on bioavailability and bioequivalence (BA/BE) were critical to further improve the quality of medicinal products in Taiwan. Consequently, the BPA contracted university professors in 1984 to draft BA/BE guidelines. Through lengthy discussions and amendments via discourse with academics, scientists from industry, government officers, as well as overseas Chinese scientists, the first Taiwan guidelines on BA/BE for generic medicines were issued in 1987. A system of post-marketing safety (PMS) surveillance for new medicinal products was implemented in 1983. All generic products containing new chemical entities (NCEs) approved under the PMS surveillance were required to provide BA/BE data when submitting marketing authorization applications since the implementation of BA/BE guideline in 1987. Since that time, several revisions of the guidelines have been made on the basis of current BA/BE guidelines used in various countries and also on relevant practical situations and issues prevailing in the pharmaceutical industry in Taiwan.

Currently, "generic substitution" is not compulsory in Taiwan.

DRUG QUALITY SURVEILLANCE PROGRAM

Local health authorities conduct regular and unscheduled inspections of manufacturing facilities and sample testing of medicinal products that are manufactured, imported and sold in their respective area. The quality of medicinal products and samples are analyzed by the Bureau of Food and Drug Analysis (BFDA) and action is taken against manufacturers and agents if their products do not meet the required standards.

REGULATION AND INSPECTION OF MANUFACTURERS

After implementation of cGMPs in 1999, 160 pharmaceutical plants were evaluated and recognized by the DOH as manufacturers that were in compliance with cGMP. As mentioned earlier, the DOH operates a system of regular and unscheduled follow-up inspections to ensure that these plants continue to manufacture and operate in compliance with cGMP standards. A pharmaceutical plant is inspected at least once every two years. An unscheduled inspection will be undertaken if the following issues arise:

Concerns about documents relating to manufacturing control for registration.
A severe defect is found in the marketed products.
A plant has previously violated cGMP regulations.

The on-site overseas inspection program was established in 2002. Over 100 foreign on-site cGMP inspections in 28 countries have been completed since the program was implemented in 2003. For imported products, the DOH reviews the plant master files (PMFs) submitted by the pharmaceutical company as a substitute for on-site inspection. The requirements for a PMF to be submitted with marketing authorization applications apply only for those plants without a PMF approved in Taiwan. The DOH is trying to establish a mutual recognition policy (PIC/S) amongst countries to ensure the quality and efficacy of foreign products and then the submission of the PMF and on-site inspection can be waived. PIC/GMP standards have been implemented since 2008. All pharmaceutical manufactures in Taiwan will need to comply with PIC/GMP requirements after year 2012. However, translation of the PMF into English requires a huge effort and is very time consuming. Consideration of the confidentiality of the PMF submitted to another country is also considered to be an issue of concern for some pharmaceutical companies in Taiwan.

The PMF for any imported products generally involves the site of manufacture and the product dosage form. Marketing authorization applications may be waived in Taiwan if the imported product consists of the same dosage form and has been manufactured at the same site with a previously approved PMF.

APPLICATION FOR REGISTRATION OF GENERIC PRODUCTS

In Taiwan, generic medicinal products are reviewed and approved under Section IV within the BPA at the DOH. Figure 1 illustrates the generic drug products review process in the BPA. A generic drug product is defined as one that contains the same active ingredient(s) and is comparable to an approved innovator's medicinal product in dosage form, strength, route of administration, quality characteristics, and intended use. The staff in Section IV serve as the application project manager responsible for the filing review. The applicant needs to certify to the DOH that the patent in question is not infringed by the generic product. The current requirements for the authorization of generic products in Taiwan are listed in Table 1.

The submission of a generic drug product must be filed by the pharmaceutical manufacturer or its representative (agent). The generic sponsor must provide pertinent patent information and a statement to the BPA. The drug product manufacturer must be in compliance with cGMP. This is verified by on-site inspections for domestic products and a review of PMFs for imported products without preinspection. Furthermore, to import a medicinal product into Taiwan, a certificate of origin and a certificate of free sale (CFS) are indispensable documents for acceptance of the submission.

Once the BPA staff have completed the filing review of the submission and have verified that the application has met all the necessary regulatory requirements, the application is then assigned to technical review that focuses on BE data and chemistry and manufacture quality.

Labeling Review Process

After the generic drug application has been accepted for filing by the staff in Section IV, the labeling information is reviewed by the BPA in-house reviewer. Generic products must be properly labeled to provide comprehensive information regarding the pharmaceutical product. The product information should include the name of the product, components and quantity, indications, usage

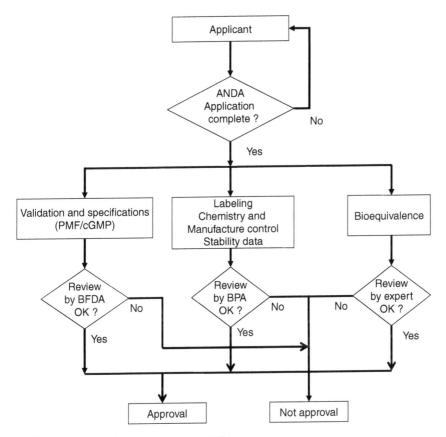

FIGURE 1 Generic drug review process in Taiwan.

TABLE 1 Checklist of Registration Requirements for Generic Medicinal Products in Taiwan

A signed application form with a description of the components and composition of the product
Patent certificate and statement
Comparison of generic and reference listed brand name products
Labeling
BE data
Finished product specifications
Laboratory test methods and results
Specification of raw material and finished products
Description of manufacturing facility and processing instruction
Details of in-process control
Batch records
Packaging and labeling procedures
Characteristics of containers, and
Stability data of finished dosage form.

and dosage, legal status, permit license number, expiry date, batch number, name/address of manufacturer or agent, adverse drug reactions, warning, contraindications, and storage conditions. The labeling review is based on the reference drug product labeling to ensure that the essential information already described is properly labeled.

Chemistry and Manufacture Quality Review Process

Following acceptance for filing by the staff (Section IV), the chemistry and manufacture control and stability of the finished dosage form are reviewed by a BPA reviewer and the relevant specifications are forwarded to the BFDA, an independent institute of the DOH.

Data such as manufacturing process validation, analytical method validation, and stability of the finished dosage form and PMFs are reviewed by the BFDA to ensure that the generic product will be manufactured in compliance with the standards of cGMP and relevant pharmacopoeial specifications when applicable. In addition, a number of samples of the finished product must also be supplied for analysis according to its own product specifications by BFDA to further ensure the quality of the drug.

Bioequivalence Review Process

Subsequently, the BE document is randomly assigned to one external independent expert on BA/BE to review. If additional information is required or problems are raised during the BE review, the reviewer will inform the staff in the BPA resulting in a document that is issued by the DOH to the applicant who is required to resolve the deficiency or specified problems and the response time limit is 70 days. Once the BE review has been completed and all BE requirements are fulfilled based on the current BE guidelines, an official document, accepting the BE results, is issued to the applicant by the DOH.

To ensure the quality and credibility of the BE study, the BA/BE inspection program was established in 2002. During the BE review process, the clinical and analytical sites are randomly inspected on a case- or problem-oriented basis. BE studies of imported drug products performed outside Taiwan are currently acceptable (2008). The certification of foreign BE laboratories and facilities has not yet been possible. When the review of an ANDA is completed, an official document is sent to the applicant as notification of marketing approval of the generic drug product in Taiwan.

BA/BE GUIDELINES IN TAIWAN

Establishing bioequivalence is an important part of the generic review process and is closely linked to the concepts of essential similarity and product interchangeability. Bioequivalence is defined as no significant difference with respect to the rate and extent of active ingredients or active moiety in pharmaceutical equivalents or pharmaceutical alternatives that is available at the site of drug action when given in the same molar dose under similar study conditions. Principally, in a BE study, pharmacokinetic end points are used to establish equivalence between the test and reference drug product. If this is not feasible, pharmacodynamic as well as clinical endpoints may be used only if they can be scientifically justified.

Additional information can be found at www.doh.gov.tw.

In vivo bioequivalence may be exempted if the drug product meets any of the following criteria:

The drug product that
  1. is administered intravenously,
  2. is an oral solution and if its excipients will not affect the absorption of the active ingredient,
  3. is an intramuscular, subcutaneous or intradermal injection solution where the pH value and all the ingredients are essentially the same as the innovator product, except for the preservative and the buffer,
  4. is a preparation intended for application to the skin (topical dermatological product) and which is not intended to be absorbed into the systemic circulation,
  5. is an ophthalmic or otic preparation (usually only "true" solutions are exempt from BE studies), and the drug is not intended to be absorbed into the systemic circulation (i.e., the interior of the eye) to produce the response, such as miotics and mydriatics.

Design and Evaluation of BE Study

Reference and Test Products
The reference product to be used in a BE study is a medicinal product that has been placed under PMS surveillance, generally the innovator/brand drug product marketed in Taiwan or the first approval drug product in Taiwan should be used as reference products.

If medicinal products are not under PMS surveillance (i.e., those approved in Taiwan before 1983), the following is the preference order for the choice of reference products for use in a BE study:

1. Innovator/brand drug product on Taiwan market.
2. The first approved drug product whose BA and quality, safety, and efficacy have been established in Taiwan. If the innovator product is no longer marketed (withdrawn) after approval in Taiwan, an approved domestic generic product can then be used as a reference product in BE study.
3. A drug product whose BE with an innovator product has been confirmed or whose clinical safety and efficacy has been documented and can be justified.

The three points mentioned above are only for medicinal products approved before 1983 (old products and **not placed under PMS surveillance)** in Taiwan. This means that the innovator product is sometimes hard to identify and in some cases no innovator product is available. Therefore, the innovator product either from Taiwan or from another country is the first choice. If not, then the first approved product whose efficacy and safety have been established in Taiwan should be considered as reference product. When no innovator product is available neither in Taiwan nor in other countries, then last option (point 3), can be chosen.

Hence, either an innovator product which is manufactured domestically (e.g., in Taiwan) or if not and it is imported and has marketing approval and is sold in Taiwan, it can be used as the reference product. If an innovator product is not available on the Taiwan market, then the product purchased from the country where the innovator product has been granted a national marketing authorization can be considered with approval by the DOH in Taiwan.

The test product used in the BE study should be from a batch of at least 1/10 of the production batch size intended for market release, or 10,000 units, whichever is higher unless otherwise justified. Quality control based on the product specification and test method (i.e., potency and content uniformity) and dissolution data of the finished products for both reference and test products should be submitted as well. A side-by-side comparison of the compositions of test and reference products is recommended. Dissolution testing should be conducted in accordance with pharmacopoeial requirements. Alternative methods can be considered with scientific justified.

Study Design

The study protocol should be approved by an ethics review board and written informed consent should be obtained from all subjects prior to participation in the study. In general, the typical randomized cross-over design with appropriate washout periods is recommended for the BE study. Washout periods between each treatment should be more than five elimination half-lives of the parent drug or active metabolite. A parallel study design can be used for oral drug products with a long half-life, which should be justified by the applicant. In general, subjects should fast for at least 10 hours prior to administration of the drug except for a food-effect BE study. The drug products should be administered with at least 200 mL of water and water is not allowed at least one hour before and after drug administration. A standard meal should be provided no less than four hours after drug administration. The subjects should abstain from other medicines or alcohol for at least two weeks before and during the study.

Single- Versus Multiple-Dose Studies

A single-dose fasting pharmacokinetic study is generally recommended for conventional immediate-release drug products to demonstrate BE. Multiple-dose studies may be appropriate for the BE of immediate-release dosage form as described.

Study subjects are patients.
The assay sensitivity precludes the possibility to adequately characterize the pharmacokinetic of drug after single-dose administration.
Drugs that exhibit nonlinear kinetics following multiple administrations

For modified-release drug products, either multiple-dose or single-dose studies under both fasting and fed conditions are required. Omission of either fasting or fed study should be justified by the applicant. If a multiple-dose study design is employed, attainment of steady state is essential. At least three consecutive trough drug concentrations on three consecutive days should be obtained to demonstrate that steady-state has been reached.

Number of Subjects

A sufficient number of subjects, which is based on sample-size estimation with appropriate power calculation should be included in the BE study. Generally, the minimum number of subjects should be not less than 12 for the assessment of bioequivalence. An add-on study is not encouraged unless this is anticipated and should be stated clearly in the protocol as well as details of the proposed statistical treatment.

Study Population

In general, healthy adult (males, females, or both) volunteers should be employed and the inclusion of subjects from the target population that the drug product is intended to be used is encouraged. The inclusion and exclusion criteria should be clearly stated in the protocol. Generally, subjects should be between 20 and 55 years and whose weights should be within the normal range (i.e., ±20% of ideal body weight). Subjects should be screened for their eligibility through medical history, physical examination (electrocardiogram and chest X ray), blood biochemistry test (including AST, ALT, γ-GT, alkaline phosphate, total bilirubin, glucose, uric acid, BUN, creatinine, albumin, total cholesterol, and triglyceride), complete blood count test (including hemoglobin, hematocrit, red blood cells, white blood cells, plates, and differential white blood cells), as well as a routine urine test (including pH, glucose, blood, and protein). Patients should be considered for enrollment in BE studies when safety considerations make it unacceptable to use healthy volunteers. Phenotyping and/or genotyping of subjects is permissible for BE studies on certain drugs.

Bioanalytical Methodology

An accurate, precise, and specific analytical method should be applied for assaying the drug and/or its metabolites in plasma or other suitable matrix. The US FDA guidance, Bioanalytical Methods Validation for Human Studies 2001 is recommended for the establishment and validation of bioanalytical methods for BE studies (5).

Sampling Times and Measurement of Analyte(s)

Generally, measurement of parent drug concentrations in plasma or serum, rather than in urine, is recommended. If urine samples are used, the reason should be justified. Measurement of a metabolite may be considered if the concentration of the parent drug is too low to be accurately measured in the biological matrix. When measurement of a metabolite is used for BE the reason should be justified. For example, the measurement of the active moiety instead of the parent may be appropriate for a prodrug. In the case of chiral drugs, the measurement of a specific enantiomer is acceptable when the enantiomers have different pharmacodynamic or pharmacokinetic characteristics. Use of a stereoselective assay may also be acceptable when systemic availability of different enantiomers is nonlinear.

For fixed-combination products, BE is assessed against either an existing approved combination product as the comparator or if the latter is not available, then comparisons should be made by dosing with the individual components.

Blood samples should be taken at appropriate times and sufficient frequency to adequately characterize the maximum concentration of the drug in the blood (C_{max}), the area under the drug concentration versus time curve (AUC) as well as the terminal elimination rate constant (k_{el}). The sampling of biological fluid should continue for longer than three terminal half-lives ($t_{1/2}$) of the drug. For long half-life drugs with low intrasubject variability (CV% < 30%) in distribution and clearance, an AUC truncated at 72 hours may be used for BE evaluation. In such cases, the applicants should consult the DOH prior to undertaking studies.

Assessment of BE and Data Submission

Parameters To Be Assessed

For single-dose studies, including a fasting or a food-effect study, C_{max}, T_{max}, the area under the curve to the last quantifiable concentration (AUC_{0-t}) and to infinity ($AUC_{0-\infty}$) are used for the evaluation of BE. For multiple-dose studies, area under the curve of a dosing interval (AUC_τ), C_{max}, C_{min}, average concentration (C_{ave}), and fluctuation at steady state should be assessed for BE evaluation. Parameters such as mean residence time (MRT), k_{el}, and $t_{1/2}$ are recommended for submission as reference data. These latter secondary parameters help justify the consistency and quality of the BE data.

When urine samples are used, the urinary excretion rate and cumulative amount of parent drug can be used to evaluate BE. All results from the BE study should be presented clearly and all individual data and results should be reported including dropouts, subject withdrawals, and exclusion of data. Any exclusion of data including outliers should be scientifically justified.

Chromatograms of at least one-third of the subjects and chromatograms of the full validation of the analytical method should be submitted as well.

Statistical Analysis

A 90% confidence interval (CI) or two one-sided test with a significance level of 5% should be used for assessing BE. Analysis of variance (ANOVA) should be applied to the study data to determine the 90% CI. In general, logarithmic transformation should be provided for measures used for BE demonstration. The pharmacokinetic parameters, except for T_{max}, obtained from the study should be analyzed using ANOVA after logarithmic transformation. The variables such as subject, period, sequence, and treatment should be included in the model of ANOVA and the results of the ANOVA analysis should also be reported. In addition to the 90% CI for BE assessment, a summary of the statistics of all the relevant pharmacokinetic parameters should be given. If a BE dossier is submitted where the biostudy was performed some years previously, the review is considered on a case-by-case basis depending on the nature of the drug, drug product, and scientific evidence.

BE Acceptance Criteria

The acceptable range for the declaration of BE is that the 90% CI of the ratios of $AUC_{0-\infty}$ and C_{max} of the test to reference product should be within 0.8 to 1.25 limits. Under the current BE guidelines, the BE limit remains unchanged for narrow therapeutic range drugs. For drugs with no safety or efficacy concerns (i.e., for a drug with a wide therapeutic window or flat dose–toxicity relationship and with clinical justification), a wider range, for example, 0.75 to 1.34, for C_{max} may be acceptable whereas the AUC criterion remains at 0.8 to 1.25. In such cases, scientific justification should be provided by the applicant and consultation with the DOH is required. Currently, there is no special provision for the BE acceptance criteria for highly variable drugs.

Special Pharmaceutical Dosage Forms

Currently, there are no special recommendation for drug products not intended to be absorbed into the systemic circulation such as nasal sprays, nasal aerosols and inhalers, creams, ointments, gels, lotions, etc. The assessment of these products is considered on a case-to-case basis. The human skin blanching assay, also known as the vasoconstrictor assay, for BE of topical corticosteroid products as provided by FDA's guidance is acceptable.

WAIVER OF IN VIVO BE STUDIES

For drug products in the same dosage form and which are proportionally similar in active and inactive ingredients but in a different strength, in vivo BE can be waived under the following circumstances:

If the BE of the higher strengths have been approved, the lower strengths can be waived based on dissolution profile comparisons.

If the BE of lower strengths have been approved and linear pharmacokinetics over the clinical dose range can be demonstrated, the higher strengths can be waived by acceptable dissolution comparisons.

The similarity of dissolution profiles between higher and lower strengths is based on the f_2 test in at least three dissolution media (e.g., pH 1.2, 4.5, and 6.8) or in media that can mimic gastrointestinal physiological pH changes for both immediate-release and modified-release products. An f_2 value of ≥ 50 suggests a similar dissolution profile.

A waiver of in vivo BE studies on the basis of the biopharmaceutical classification system (BCS) has not been adopted in the guidelines at present.

VARIATIONS AND LICENCE RENEWALS

Marketing authorizations are valid for five years and the marketing authorization holder must request extension/renewal of the license before the expiry date. No modifications or alterations of the original registration are allowed without prior approval of the BPA. Regulations regarding scale-up and postapproval changes (SUPAC) of medicinal products have been revised according to the international SUPAC guidance (6).

The BPA in Taiwan strives toward deregulation, transparency of regulations, and the streamlining of the review process. The current BE approval process and requirements are aimed at ensuring the quality of generic products. The BPA works with scientists and pharmaceutical companies to provide safe and effective generic drugs to the people of Taiwan.

REFERENCES

1. Bureau of Pharmaceutical Affairs, Department of Health, Executive Yuan, Taiwan, Republic of China, 2009.
2. Hsieh YY, Hunag WF. The drug regulatory process of the Republic of China. J Clin Pharmacol 1998; 28:200–203.
3. Hu O YP, Hsiao ML, Liu LL. Drug regulatory process of the Republic of China-Taiwan's experiences in bioavailability and bioequivalence. Drug Inf J 1995; 29:1049–1054.
4. Tzou MC, Chi JF, Hu O YP. Generic drug in Taiwan. Regul Aff J 1999; 11:554–560.
5. Guidance for Industry: Bioanalytical method validation, 2001. http://www.fda.gov/cder/guidance/4252fnl.htm. Accessed May 9, 2009.
6. SUPAC-IR: Immediate-Release Solid Oral Dosage Forms: Scale-Up and Post-Approval Changes: Chemistry, Manufacturing and Controls, In Vitro Dissolution Testing, and In Vivo Bioequivalence Documentation, 1995. http://www.fda.gov/cder/guidance/cmc5.pdf. Accessed May 9, 2009.

11 Turkey

Ilker Kanzik

IDE Pharmaceutical Consultancy Ltd. Co., Istanbul, Turkey

A. Atilla Hincal

IDE Pharmaceutical Consultancy Ltd. Co., Ankara, Turkey

INTRODUCTION

Registration

Marketing authorization (registration decisions) for medicinal products for human use is issued by the Ministry of Health (MoH), but the entire procedure is managed by the General Directorate of Drug and Pharmacy (GDDP). On the other hand, for veterinary medicinal products and dietary supplements such as vitamins and minerals, herbal extracts, and other substances that are generally found in foods, the Ministry of Agriculture is the relevant authority.

Turkish licensing regulations (regulation) for all (i.e., innovator/brand and generic drug products) pharmaceutical products published on January 19, 2005 (1), came into force on June 30, 2005. This new legislation brought Turkish law in line with that of the European Union (EU) and covered all aspects of the registration procedure. According to this legislation the MoH will take all appropriate measures to ensure that the procedure for granting an authorization to place a medicinal product on the market is completed within 210 days of the submission of a valid application.

Applicants intending to register any drug product must submit an application for authorization to the GDDP. Applications for marketing authorization of an innovator/brand drug product must be accompanied by the appropriate fees as well as various documents and particulars, including the manufacturer's product development methodology and processes, control methods (sterility tests, stability tests, etc.), and the results of physicochemical, biological or microbiological, pharmaco-toxicological tests, and clinical trials (1).

Prescription (generic or innovator/brand) and nonprescription medicines (over the counter, OTC) follow the same marketing authorization procedure. It is likely that after initial consideration, additional questions will be put to the applicant and the answers would then be further considered by the Registration Approval Committee leading to a final opinion on the application. If an application is refused, the reasons for the decision will be set out in full. The applicant has the right to object in written or orally. A marketing authorization will be refused if it appears that the medicinal product is harmful under normal conditions of use, that it has no or very little therapeutic effect or that it does not have the stated composition in terms of quality and quantity. Authorization will also be refused if the information requested is incomplete or not given. Marketing authorization may be suspended or withdrawn on the same grounds as those invoked for refusal to grant marketing authorization. Marketing approvals are valid for a period of five years, and can be renewed for further five-yearly periods. To do this, the licensee must submit an application to the MoH at least six months before the license expires.

European Union Relations and Obligations

The rules governing the registration of medicinal products in Turkey have been considerably modified to approach the EU requirements. In this context, approval procedures in Turkey are designed to be similar to those existing in the EU, for example, administrative requirements, structure of the dossier including use of the common technical document (CTD) (2). All new Turkish regulations will be, as far as possible, compatible with those of the EU.

It should always be remembered that, Turkey started the Customs Union with the EU on January 1, 1996, in accordance with the Association Council Decisions No: 1/95 and 2/97 (3,4).

In essence the Customs Union aimed to provide free circulation of industrial products, including pharmaceuticals, between Turkey and the EU. Under that agreement, Turkey abolished all tariffs on imports by the end of 1996, introduced patent protection by 1999 and was obliged to respect and adopt all EU international property legislation listed in Annex 8 of the Association Council Decision 1/95 (4). This included an undertaking to implement the GATT-TRIPS (5) provisions no later than three years after the agreement was enforced, and to adopt domestic legislation in line with the EU and its member states.

In summary, Turkish relations with the EU by means of the Customs Union provided the elimination of physical and technical trade barriers in addition to harmonizing the policies and standards of the framework governing the pharmaceutical industry.

Generic Drug Products

A generic medicinal product is defined in the regulations as

> Medicinal product for human use which has the same qualitative and quantitative composition in active substances and the same pharmaceutical form as the original product, and whose bioequivalence with the reference medicinal product has been demonstrated by appropriate bioavailability studies.

After the publication of a regulation on May 27, 1994 (6) bioequivalence (BE) became compulsory in order for a generic drug product to receive a marketing license. It is obvious, therefore, that BE is the key requirement for an abridged (generic) application of a drug product in Turkey.

Abridged Application

For abridged applications, without prejudice to the provisions of the Decree on the Protection of the Patent Rights dated 24.06.1995 no: 551 (7), applicants shall not be required to provide the results of toxicological and pharmacological tests or the results of clinical trials if they can demonstrate: either that the medicinal product is similar to a medicinal product authorized in Turkey and that the holder of the marketing authorization for the original medicinal product has consented to the toxicological, pharmacological, and/or clinical references contained in the file on the original medicinal product being used for the purpose of examining the application in question; or that the constituent or constituents of the medicinal product have a well-established medicinal use, with recognized efficacy and an acceptable level of safety, supported by means of a detailed scientific bibliography.

The generic or similar biological medicinal product, once authorized, can however only be placed on the market six years after the authorization of the reference medicinal product, depending on the exclusivity period applicable for the reference medicinal product.

However, in the event of a different therapeutic indication, route of administration, dosage being envisaged compared to the innovator/brand products which have been introduced into the market, it shall be necessary to provide the results of appropriate clinical trials and where necessary the results of toxicological, pharmacological studies.

With regard to new medicinal products containing known constituents, but which have not yet been used in combination for therapeutic purposes, it is necessary to present the results of the relevant toxicological and pharmacological tests and clinical trials relating to that combination. However, it shall not be obligatory to provide references relating to each constituent.

Application and Approval of the Trial

Applications for a BE study can be made either by the sponsor or a legal representative (Contract Research Organization, CRO) of the sponsor according to the *Regulations Regarding Clinical Trials* published on December 23, 2008 (8). The necessary documents including the "Study Protocol" prepared according to the GCP and GLP Guidelines are required to be submitted to both ethics committee and the Pharmaceutical General Directorate of the MoH. Any amendments made to the protocol should be reported to the general directorate and the relevant ethics committee by the sponsor or the investigator. Rules encompassing topics such as the design, conduct, monitoring, budgeting, assessment and reporting of trials, protection of all the rights and bodily integrity of volunteers, preservation of the safety of trial data must be complied with by all parties participating in the trial so as to ensure that the trials are conducted at international scientific and ethical standards are evaluated by the ethics committees. Time frames for the evaluation by the ethics committee and the MoH are 45 days and 60 days (total), respectively.

All studies should be monitored in accordance with the regulations. The general directorate and the relevant ethics committee should be informed by the sponsor of the trial about any serious, unexpected adverse effects during the clinical trial, which may have resulted in death or are life threatening within a maximum of seven days the receipt of the referred information. Monitoring reports including additional information regarding these cases should be within eight days as of the receipt of the information.

Centers where bioavailability (BA) and BE studies can be conducted are evaluated by the MoH and approved if the sites meet the standards. The Ministry has the right to inspect with or without prior notice, trials conducted in Turkey and abroad as well as the trial sites, the sponsor and the CRO, the sites where the investigational products are manufactured, the laboratories where the analyses relating to the trial are conducted, for their compliance with the provisions of this regulation and other relevant legislations. Depending on the result of the inspection, the trial may be stopped if necessary by the Ministry.

DESIGN AND CONDUCT OF BIOEQUIVALENCE STUDIES FOR ORALLY ADMINISTERED DRUG PRODUCTS

The design and conduct of the study should follow Turkish and/or EU regulations on good clinical practice (GCP), including the requirements of an ethics committee, OECD (Organization for Economic Co-Operation and Development), good laboratory practices (GLP), and ISO 17025 (communication of the MoH 52072/December 03, 2001).

A BE study is basically a comparative BA study designed to establish equivalence between test and reference products. Several in vivo and in vitro methods may be appropriate to document BA and BE. In descending order of preference, the Turkish regulations require pharmacokinetic, pharmacodynamic, clinical, and in vitro studies.

Definitions

The following definitions are specifically described in the Turkish BE guidelines/regulations/laws. However, those BE regulations are relatively old (1994), hence the relevant definitions in EU Guideline (2001) are also valid for Turkey.

Bioavailability

BA means the rate and extent to which the active substance or its active moiety is absorbed from a pharmaceutical dosage form and becomes available at the site of action or biological fluids (usually plasma or plasma) representing the site.

Bioequivalence

Two medicinal products are bioequivalent if they are pharmaceutically equivalent and if their BAs after administration in the same molar dose are similar to such a degree that their effects, with respect to both efficacy and safety, will be identical.

Pharmaceutical Equivalence

Medicinal products are pharmaceutically equivalent if they contain the same amount of the same active substance(s) in the same dosage forms and administered by the same route and that meet the same or comparable standards.

Therapeutic Equivalence

A medicinal product is therapeutically equivalent with another product if it contains the same active substance or therapeutic moiety and, clinically, shows the same efficacy and safety as that product, whose efficacy and safety have been established.

Study Design

The study should be designed in such a way that any formulation effect can be distinguished from other effects. For oral suspensions and immediate-release tablets and capsules, a single-dose in vivo fasting study is usually sufficient. To reduce the variability, a cross-over design is generally the first choice. When a test formulation (generic) is to be compared with a reference formulation (innovator/brand), a two-period, two-sequence cross-over design is often considered to be the design of choice. Other designs and methods may be considered in special cases, but it should be well justified in the study protocol. Volunteers (healthy human subjects) should be randomly included in study periods. Although, single-dose studies for immediate-release products are generally deemed to be sufficient, in certain cases and for modified-release products, steady-state studies may be required under the following conditions:

a) If problems of sensitivity preclude sufficiently precise plasma concentration measurements after single-dose administration.
b) If intraindividual variability in the plasma concentrations and elimination rate is intrinsically high. A drug product is called highly variable if its intraindividual (i.e., within-subject) variability is greater than 30%. A high CV as estimated from the ANOVA model is thus an indicator for high within-subject variability.
c) Dose- or time-dependent pharmacokinetics is prevalent.
d) For prolonged-release products and enteric-coated tablets (in addition to single-dose investigations).

In these types of steady-state investigations, the drug administration scheme should comply with the common dosage recommendations.

The number of subjects required to be included in BE studies is determined by the following:

a) The error variance associated with the primary characteristic to be studied as estimated from a pilot experiment, or from previous studies or from published data.
b) The significance level desired. For the determination of the minimum number of subjects (n) it is recommended that n is generally based on the criterion to achieve an 80% probability of concluding BE if the difference between test and reference treatment is $\leq 5\%$. The use of a larger α level with the fewer variables is also recommended.
c) The expected deviation from the reference product compatible with BE (delta, i.e., percentage difference from 100%).
d) The required statistical power of the study.

The clinical and analytical standards imposed may also influence the statistically determined number of subjects. However, generally the minimum number of subjects should not be less than 12.

Subsequent treatments should be separated by adequate washout periods (at least three times the terminal half-life). In steady-state studies, washout of the previous treatment last dose can overlap with the build-up of the second treatment, provided the build-up period is sufficiently long (at least three times the terminal half-life). The sampling schedule should be planned to provide an adequate estimation of C_{max} and to cover the plasma concentration–time curve long enough to provide a reliable estimate of the extent of absorption. This is generally achieved if the area under the curve (AUC) derived from measurements is at least 80% of the AUC extrapolated to infinity. To get a reliable estimate of the terminal half-life, it should be obtained by collecting at least three to four samples during the terminal log linear phase. For drugs with a long half-life (12–24 hours or longer) relative BA can be adequately estimated using truncated AUC as long as the total collection period is appropriately justified. In this case the sample collection time should be adequate to ensure comparison of the absorption process. C_{max} and a suitably truncated AUC can be used to characterize peak and total drug exposure, respectively. For drugs that demonstrate low intrasubject variability in distribution and clearance, an AUC truncated at 72 hours (AUC_{0-72} hour) can be used in place of AUC_{0-t} or $AUC_{0-\infty}$. For drugs demonstrating high intrasubject variability in distribution and clearance, AUC truncation warrants caution.

Subjects

Selection of Subjects

The studies should normally be performed using healthy volunteers. The inclusion/exclusion criteria should be clearly stated in the protocol. In general, volunteers suitable for inclusion should be within the age limits (between 18 and 55 years) and of weight within the normal range according to accepted normal values for the body mass index (BMI), where BMI = mass (kg)/height (m)^2. Subjects should be asked for informed consent and thereafter screened for suitability by means of clinical laboratory tests, an extensive review of medical history, and a comprehensive medical examination with respect to inclusion and exclusion criteria. Subjects should preferably be nonsmokers and without a history of alcohol or drug abuse. If moderate smokers are included (less than 10 cigarettes per day) they should be identified as such and the consequences for the study results should be discussed.

Standardization of the Study

The test conditions should be standardized to minimize the variability of all factors involved except any formulation effects. Therefore, standardization of subject diet, fluid intake and exercise is recommended. Subjects should preferably be fasting at least during the night prior to administration of the products. If the summary of product characteristics (SmPC) of the reference product contains specific recommendations in relation to food intake and implications for interaction with food, the study should be designed accordingly. In other words, if it is stated that the presence of food will affect the absorption, then the study should be performed also under fed conditions.

The time of day for ingestion should be specified and as fluid intake may profoundly influence gastric passage for oral administration dosage forms, the volume of fluid (at least 240 mL) should be standardized. All meals and fluids taken after the treatment should also be standardized in regard to composition and time of administration during the sampling period. The subjects should not take other medicines during a suitable period before and during the study and should abstain from food and drinks, which may interact with circulatory, gastrointestinal, liver or renal function (e.g., alcoholic or xanthine-containing beverages or certain fruit juices). As the BA of an active moiety from a dosage form could be dependent upon gastrointestinal transit times and regional blood flows, posture and physical activity should be standardized.

1

248

Kanzik and Hincal

Genetic Phenotyping

Phenotyping and/or genotyping of subjects should be considered for exploratory BA studies and all studies by using parallel group design. It may also be considered in cross-over studies (e.g., BE, dose proportionality, food interaction studies, etc.) for safety or pharmacokinetic reasons. If a drug is known to be subject to major genetic polymorphism, studies could be performed in panels of subjects of known phenotype or genotype for the polymorphism in question.

Characteristics To Be Investigated

Moieties to be measured in BE studies are active drug substances or the active moiety in the administered dosage form (parent drug). In some situations, however, measurements of an active or inactive metabolite may be necessary instead of the parent compound. Such situations include cases where the use of a metabolite may be advantageous to determine the extent of drug input, for example, if the concentration of the active substance is too low to be accurately measured in the biological matrix (e.g., major difficulty in analytical method, product unstable in the biological matrix or half-life of the parent compound too short) thus giving rise to significant variability.

In BA studies, the shape of, and the area under the plasma concentration versus time curves are mostly used to assess extent and rate of absorption. The use of urine excretion data may be advantageous in determining the extent of drug input in case of products predominately excreted renally, but has to be justified when used to estimate the rate of absorption. Sampling points or periods should be chosen such that the time–drug concentration profile is adequately defined so as to allow the estimation of relevant parameters.

From the primary results, the desirable BA characteristics are estimated, namely AUC_t, AUC_∞, C_{max}, t_{max}, A_{et}, Ae_∞ as appropriate, or any other justifiable characteristics. The method of estimating AUC values should be specified. For additional information $t_{1/2}$ and MRT can be estimated. For studies at steady state, AUC_τ, C_{max}, C_{min}, and fluctuation [$(C_{max} - C_{min})/C_{av}$] should be provided.

Pharmacodynamic studies are not recommended for orally administered drugs when the drug is absorbed into the systemic circulation and a pharmacokinetic approach can be used to assess systemic exposure and establish BE. However, in some cases the quantitative measurement of a drug and/or metabolite in plasma or urine cannot be made with sufficient accuracy and/or reproducibility. In those cases, studies in healthy volunteers or patients using pharmacodynamic parameters may be used for establishing equivalence between the test and reference products. Demonstration of a dose–response relationship may become necessary. Measurements should be made with sufficient frequency to permit a reasonable estimate of the total area under the effect–time curve. The baseline values in each period should be comparable and the complete effect curve should remain below the maximum physiological response. The methodology used for carrying out the study should be validated for precision, accuracy, specificity and reproducibility. Nonresponders should be excluded from the study by prior screening. A correction for potential nonlinearity of the relationship between the dose and area under the effect–time curve should be made and also baseline corrections should be considered during data analysis.

Sample Analysis (Bioanalysis)

The bioanalytical part of BE trials should be conducted in accordance with applicable principles of OECD—GLP, since Turkey is one of the OECD countries. Bioanalytical methods used to determine the active moiety and/or its biotransformation product(s) in blood, plasma, serum, or urine or any other suitable matrix should comply with the requirements to confirm and validate specificity, accuracy, sensitivity, and precision A full validation report together with the relevant data should be reported.

It is a primary requirement to demonstrate the stability of the active ingredient and/or its biotransformation product(s) (in biological fluids to obtain reliable data).

Reference and Test Products

Test products (generic) in an application for approval of a generic product are normally compared with the corresponding dosage form of an innovator medicinal product (reference product). The reference product can be chosen from any drug product of the innovator registered in any EU member state and/or the United States. The choice of reference product from other regions should be justified by the applicant. A pharmaceutical product can be used as a reference product if it is an imported product of the innovator company from EU or United States. Products of innovator companies which are manufactured in Turkey are not permitted to be used as a reference product. In addition, in the process of choosing a reference product, most of the generic companies run their own dissolution studies with the innovator company's products that were manufactured and/or distributed in different EU countries, to see if there is any difference between the dissolution profiles among these products. The best product that gives similar dissolution profiles with their own products is chosen as the reference product to be used in the BE study.

Test products used in biostudies must be prepared in accordance with GMP regulations. Since many parameters, such as hardness of the tablet, friability, disintegration, dissolution, and deviation of tablet weight in relation to the declared tablet average weight, related substances and microbiological limits, etc., may have minor or major effects on dissolution results, batch control results of the test product should be reported within the analysis certificate in detail.

Batch Size

In the case of oral solid forms for systemic action, the test product should usually originate from a batch of at least 1/10 of production scale or 100,000 units, whichever is greater, unless otherwise justified. The production of batches used should provide a high level of assurance that the product and process will be feasible on an industrial scale; in case of a production batch smaller than 100,000 units, a full production batch will be required. If the product is subjected to further scale-up this should be properly validated.

Samples of the product from full production batches should be compared with those of the test batch, and should show similar in vitro dissolution profiles when employing suitable dissolution test conditions.[see Appendix II of Note for Guidance on the Investigation of Bioavailability and Bioequivalence, 2001 (9)].

The study sponsor must retain a sufficient number of all investigational product samples used in the study for one year after the product's shelf life or two years after completion of the trial or until approval whichever is longer to allow re-testing, if it is requested by the authorities.

Data Analysis
The main goals and objectives of a BE study is to demonstrate equivalency within a clinically significant acceptance range (80–125%) and to limit the risk of false acceptance decisions of BE. In certain cases, however, a wider interval may be acceptable. The interval must be prospectively defined, for example, 0.75 to 1.33, and justified addressing, in particular any safety or efficacy concerns for patients switched between formulations. The possibility offered here by the guideline to widen the acceptance range of 0.80 to 1.25 for the ratio of C_{max} (not for AUC) should be considered exceptional and limited to a small widening (0.75–1.33). Furthermore, this possibility is restricted to those products for which at least one of the following criteria applies:

1. Data regarding pharmacokinetic/pharmacodynamic (PK/PD) relationships for safety and efficacy are adequate to demonstrate that the proposed wider acceptance range for C_{max} does not affect pharmacodynamics in a clinically significant way.
2. If pharmacokinetic/pharmacodynamic data are either inconclusive or not available, clinical safety and efficacy data may still be used for the same purpose, but these data should be specific for the compound to be studied and persuasive.
3. The reference product has a highly variable within-subject BA.

A post hoc justification of an acceptance range wider than defined in the protocol is not acceptable. Information that would be required to justify results lying outside the conventional acceptance range at the post hoc stage should be utilized at the planning stage, either for a scientific justification of a wider acceptance range for C_{max}, or for selecting an experimental approach that allows the assessment of different sources of variability.

When a parametric approach is used, the statistical method for testing relative BA (e.g., BE) should be based upon the 90% confidence interval for the ratio of the population means (test/reference), for the parameters under consideration. This method is equivalent to the corresponding two one-sided test procedures with the null hypothesis of bioinequivalence at the 5% significance level. In the case of concentration or concentration-related characteristics (e.g., AUC) the data should be transformed prior to analysis using a logarithmic transformation. In the parametric approach, if it is doubtful to assume log normal (AUC, C_{max}) or normal distribution (t_{max}), a nonparametric approach is recommended. This approach may also be selected as a general statistical approach to evaluate all BA characteristics during a specific investigation.

Currently, Turkey follows this approach, even though it is not mentioned in its regulation.

In Vitro Dissolution
In vitro dissolution test data obtained with the batches of test and reference preparations that were used in the BA and BE studies should always be reported. The similarity of dissolution profiles between the test product and reference product should be demonstrated in each of the three buffers within the range

of pH 1 to 8 (preferably at or about pH 1.2, 4.5, 6.8). The similarity/difference of the in vitro dissolution profiles may be compared by the similarity/difference factors (e.g., f_2 and f_1). A f_2 value between 50 and 100 suggests that two dissolution profiles are similar and a f_1 value between 1 and 15 suggests that two dissolution profiles are within the acceptable differences. In cases where more than 85% of the drug is dissolved within 15 minutes, dissolution profiles may be accepted as similar without further evaluation (see Appendix II of Note for Guidance on the Investigation of Bioavailability and Bioequivalence, 2001) (9).

Reporting of Results

The report of a BA or a BE study should contain complete documentation including the study protocol, conduct and evaluation and confirmation (QA) that the study was in compliance with GCP rules. The authenticity of the whole of the report is attested by the signature of the principal investigator. Furthermore, responsible person(s)/investigator involved with any particular section of the study should sign their respective sections of the report. Names and affiliations of the responsible person(s)/investigator, site of the study and duration should be stated. The names and batch numbers of the products used in the study as well as the composition(s), finished product specifications and comparative dissolution profiles should be provided. In addition, the applicant should submit a signed statement confirming that the test product is the same as the one that is submitted for registration.

All results should be clearly presented and the method used to derive the pharmacokinetic parameters (e.g., AUC) from the raw data should be specified. If pharmacokinetic models are used to evaluate the parameters, the model and computing procedure used should be justified. The reason for not submitting any data should be explained. All individual subject data should be included and individual plasma concentration/time curves presented on a linear/linear and log/linear scale. All the data from subjects who dropped-out should be included. Dropouts and withdrawal of subjects should be fully documented and accounted for. A representative number of chromatograms or other raw data should be included covering the whole concentration range for all, standard and quality control samples as well as the specimens analyzed together with the analytical validation report. Currently, the MoH requests 100% of the chromatograms.

Applications for Generic Products Containing Approved Active Substances

If it is intended to register a generic product which contains an active substance of a currently authorized product (original), its BE with this product should be demonstrated.

If bioinequivalence may pose a risk of therapeutic failure, then BA studies must be carried out. In other words, if the active substance of a product has, for example, a narrow therapeutic range and bioinequivalence may pose serious adverse effects, BE studies will be mandatory.

Steady-State Studies

There are no specific recommendations for this type of study. However, the EU guidance on Modified Release Oral and Transdermal Dosage Forms: Section II:

(Pharmacokinetic and Clinical Evaluation), 1999 (10), is the guideline that is followed by the Turkish MoH.

Outliers
Only data for clinically proven outliers can be excluded from the statistical evaluations (11,12).

Add-On Studies
Add-on studies are not permitted.

EXEMPTION FROM BIOEQUIVALENCE STUDIES (WAIVERS)
In general, it is not necessary to submit BA studies if

a) products differ only in the strength of active substance provided that the conditions below are fulfilled:

  linear pharmacokinetics
  the same qualitative composition
  same ratio between active ingredient and excipients or (in the case of low strength) same ratio of excipients
  the two drug products of two different generic companies are manufactured by the same manufacturer and at the same manufacturing site
  - BA and BE studies have been performed with the highest strength of the product against the original product
  similar in vitro dissolution rates under the same testing conditions

b) the product is reformulated with minor changes or the manufacturing process has been changed by the same manufacturer and justification is provided that these changes will not affect BA to any significant degree. In case of exemption from BE studies, in vitro data should demonstrate the similarity of dissolution profile between the test product and the reference product in each of three buffers within the range of pH 1 to 8 at 37°C (preferably at or about pH 1.0, 4.6, 6.8).

c) the product is to be administered as an aqueous intravenous solution containing the same active substance in the same concentration as the currently authorized original product.

d) the product contains the same active substance at the same concentration as a currently authorized product in the same pharmaceutical dosage form and does not contain any excipients that could affect the gastric passage of the active substance such as a solution dosage form for oral administration (elixir, syrup, or others).

e) an acceptable correlation can be shown between in vivo absorption and in vitro dissolution and the in vitro dissolution rate of the new product that is equivalent to the currently authorized generic product under the same conditions that are used to determine the correlation.

f) dermal products (including corticosteroids) for local use, that is, intended to act without systemic absorption. However, if there exists a certain degree of undesired partial absorption, safety evaluations may be requested. For other products (oral, nasal, inhalation, ocular, dermal, rectal, vaginal, etc., administration) intended to act without systemic absorption, BE and/or clinical studies are required.

g) a gas for inhalation.

Suprabioavailability

If *Suprabioavailability* is found, that is, if the new product displays an extent of absorption appreciably larger than the approved product, reformulation to lower dosage strength should be considered. In this case, the biopharmaceutical development should be reported and a final comparative BA study of the reformulated new product with the old approved product should be submitted.

In case reformulation is not carried out the dosage recommendations for the suprabioavailable product will have to be supported by clinical studies. Such a pharmaceutical product should not be accepted as therapeutically equivalent to the existing reference product. If marketing authorization is obtained, the new product may be considered as a new medicinal product.

Finally, it is emphasized that where no specific recommendations are provided for in the Turkish guidelines, the Turkish MoH requests sponsors to follow current EU guidelines and recommendations (see Chapter 5 of this book), viz.:

> *Note for Guidance on the Investigation of Bioavailability and Bioequivalence, 2001; "Questions & Answers on the Bioavailability and Bioequivalence Guideline, 2006* and *Note for Guidance on Modified Release and Transdermal Dosage forms: Section II (Pharmacokinetic and Clinical Evaluation), 2000.*

REFERENCES

1. Regulation on Licensing for Medicinal Products for Human Use, Official Gazetta No: 25725/19 January 2005.
2. The European Parliament and the Council of the European Union. Directive 2001/83/EC of the European Parliament and of the Council of November 6, 2001 on the Community code relating to medicinal products for human use, CPMP/EWP/QWP/1401/98. Accessed July 26, 2001.
3. Decision No 1/95 of the EC-Turkey Association Council of 22 December 1995 on implementing the final phase of the Customs Union (96/142/EC). http://www.mfa.gov.tr/data/AB/EUAssociationCouncilDecision195CustomsUnionDecision.pdf. Accessed December 22, 1995.
4. Decision No 2/97 of the EC-Turkey Customs Cooperation Committee. http://www.avrupa.info.tr/Files/File/RECOURCE_CENTRE/key_links/DECISION_No19.doc. Accessed May 30, 1997.
5. GATT-TRIPS Implementation. http://www.wto.org. Accessed April 1994.
6. Regulation on the Evaluation of Bioavailability and Bioequivalence of Medicinal Products Official Gazetta of Turkey No: 21942. Accessed May 27, 1994.
7. Decree on the Protection of the Patent Rights dated 24.06.1995 no: 551; http://www.tpe.gov.tr/portal/default.jsp. Accessed December 7, 1995.
8. Regulations Regarding Clinical Trials. Official Gaetta of Turkey No: 27089, December 23, 2008
9. EU Note for Guidance on the Investigation of Bioavailability and Bioequivalence. CPMP/EWP/1401/98, 2001.
10. EU Note for Guidance on Modified Release Oral and Transdermal Dosage Forms: Section II: (Pharmacokinetic and Clinical Evaluation). CPMP/EWP/280/96, 1999.
11. Turkish MoH Communication no. 058123 dated 16 November 2006.
12. Turkish MoH Communication no. 038031, June 10, 2008.

Barbara M. Davit and Dale P. Conner

Office of Generic Drugs, Center for Drug Evaluation and Research, United States Food and Drug Administration, Rockville, Maryland, U.S.A.

INTRODUCTION

All prescription and over-the-counter generic drugs marketed in the United States must meet standards established by the U.S. Food and Drug Administration (FDA). In approving a new generic drug for marketing, the FDA concludes that the generic product is therapeutically equivalent to its corresponding reference product (usually the innovator product, but sometimes another generic product if the innovator product was withdrawn). The FDA believes that therapeutically equivalent drug products can be substituted with the full expectation that both products will produce the same clinical response (1). A generic drug is approved by the FDA if it is (*i*) pharmaceutically equivalent to an approved safe and effective reference product in that it (a) contains identical amounts of the same active drug ingredient in the same dosage form and route of administration, and (b) meets compendial or other applicable standards of strength, quality, purity, and identify; (*ii*) bioequivalent to the reference product in that it (a) does not present a known or potential bioequivalence problem, and it meets an acceptable in vitro standard (usually dissolution testing), or (b) if it does present such a known or potential problem, it is shown to meet an appropriate bioequivalence standard; (*iii*) adequately labeled; and (*iv*) manufactured in compliance with current good manufacturing practice regulations (1). It is important to note that the regulatory oversight of generic drug chemistry, manufacturing, and controls is identical to that imposed upon innovator drug products (2).

BIOEQUIVALENCE

No topic seems so simple but stimulates such intense controversy and misunderstanding as the topic of bioequivalence. The apparent simplicity of comparing in vivo performance of two drug products is an illusion that is quickly dispelled when one considers the difficulties and general public misunderstanding of the accepted regulatory methodology. In the United States, one often hears members of the public and medical experts alike stating various opinions on the unacceptability of approved generic drug products based on misconceptions regarding the determination of therapeutic equivalence of these products to the approved reference. These misconceptions include the belief that the U.S. FDA approves generic products that have mean differences from the reference product of 20% to 25% and that generic products can differ from each other by as much as 45%. In addition, some incorrectly assume that, since most bioequivalence testing is carried out in normal volunteers, it does not adequately reflect bioequivalence and therefore therapeutic equivalence in patients. When the current bioequivalence methods and statistical criteria are clearly understood it becomes apparent that these methods constitute a strict and robust system that provides assurance of therapeutic equivalence. In this chapter we discuss the history, rationale, and methods used for the demonstration of bioequivalence for regulatory purposes in the United States. In addition, we will touch on some of the controversial issues and difficulties in demonstrating bioequivalence for locally acting drug products.

OBJECTIVES OF BIOEQUIVALENCE STUDIES

The most important concept in the understanding of bioequivalence is that the sole objective is to measure and compare formulation performance between two or more pharmaceutically equivalent drug products. Formulation performance is defined as the release of the drug substance from the drug product leading to bioavailability of the drug substance and eventually leading to one or more pharmacologic effects, both desirable and undesirable. If equivalent formulation performance from two products can be established then the clinical effects, within the range of normal clinical variability, should also be equivalent. This is the same principle that leads to an equivalent response from different lots of the brand-name product.

When generic drugs are submitted for approval through the abbreviated new drug application (ANDA) process in the United States, they must be both pharmaceutically equivalent and bioequivalent to be considered therapeutically equivalent and therefore approvable. To be considered pharmaceutically equivalent, two products must contain the same amount of the same drug substance and be of the same dosage form with the same indications and uses. Thus, an immediate-release tablet would not be considered pharmaceutically equivalent to an oral liquid suspension, capsule or modified-release tablet. Bioequivalence means the absence of a significant difference in the rate and extent to which the active ingredient becomes available at the site of drug action when administered at the same molar dose under similar conditions in an appropriately designed study. Two drug products are considered therapeutically equivalent if they are pharmaceutical equivalents and if they can be expected to have the same clinical effect and safety profile when administered to patients under the conditions specified in the labeling. The FDA believes that products classified as therapeutically equivalent can be substituted for each other with the full expectation that the substituted product will produce equivalent clinical effects and safety profile as the original product.

HISTORY OF BIOEQUIVALENCE EVALUATION IN THE UNITED STATES

Enactment of the Food, Drug, and Cosmetic Act

In 1938, the U.S. Congress enacted the Federal Food Drug and Cosmetic Act. The new law required, among other things, that a "new drug" product would need to provide proof of safety before it could be marketed. The new drug application (NDA) was established to provide a mechanism for proof of safety of drugs to be submitted to the FDA. Regulations were promulgated as to the form and content of the data to be submitted for an NDA. Originally, only toxicity studies were required along with informative labeling and adequate manufacturing data. These early requirements have since evolved into the comprehensive regulations found in Title 21 of the Code of Federal Regulations Part 300, Sub-chapter D: Drugs for Human Use (21 CFR Part 300).

The Drug Efficacy Study Implementation

In 1962, The Kefauver–Harris Amendment to the Food Drug and Cosmetic Act mandated that all new drug products subsequently approved for marketing must have adequate evidence of effectiveness, as well as safety (3). The FDA was assigned the responsibility for receiving, reviewing, and evaluating required data submissions, and enforcing compliance with the law. An applicant submit-ting an NDA was now required to submit "substantial evidence" in the form of "adequate and well-controlled studies" to demonstrate the effectiveness of the drug product under the conditions of use described in its labeling. The new drug effectiveness provision of the law also applied retrospectively to all drugs approved between 1938 and 1962 on the basis of safety only. The FDA contracted with the National Academy of Sciences/National Research Council (NAS/NRC) to review this group of drugs for effectiveness. The NAS/NRC appointed 30 panels of experts and initiated the Drug Efficacy Study. The panels reviewed approximately 3400 drug formulations and classified them either effective or less than effective (4). The FDA reviewed the reports and any supporting data, and published its conclusions in the Federal Register as Drug Efficacy Study Imple-mentation (DESI) notices. The DESI notices contained the acceptable marketing conditions for the class of drug products covered by this notice.

Many drug products had active ingredients and indications that were iden-tical or very similar to those of drug products found to be effective in the DESI review but lacked NDAs themselves. Initially, in implementing the DESI pro-gram, the FDA required that each of these duplicate drug products should have its own approved NDA before it could be legally marketed. Later, the FDA con-cluded that a simpler and shorter drug application was adequate for approving duplicate DESI drugs for marketing, and, in 1970, created the ANDA procedure for the approval of duplicate DESI drug products (5‾-7). The FDA believed that it was not necessary for firms seeking approval of duplicate DESI drug prod-ucts to establish the safety and efficacy of each new product identical in active ingredient and dosage form with a drug product previously approved as safe and effective. However, many of the DESI notices included, as a requirement for approval of the duplicate drug application, presentation of evidence that the "biologic availability" of the test product was similar to that of the innovator's product.

Development of FDA's Bioavailability/Bioequivalence Regulations

Identification of the Need for Regulatory Bioequivalence Studies
Introduction in the late 1960s and early 1970s of the sophisticated bioanalytical techniques made possible measurements of drugs and metabolites in biologi-cal fluids at concentrations as low as a few nanograms per milliliter. By using these methods, the disposition of drugs in the human body could be charac-terized by determining pharmacokinetic profiles. The rate processes of drug absorption, distribution, metabolism, and excretion could now be quantified and related to formulation factors and pharmacodynamic effects. As these techniques were applied to investigate the relative bioavailability of various marketed drug

products, it became apparent that many generic formulations were more bioavailable than the innovator products, while others were less bioavailable.

In the late 1960s and early 1970s, many published studies documented differences in the bioavailability of chemically equivalent drug products, notably chloramphenicol (8), tetracycline (9), phenylbutazone (10), and oxytetracycline (11). In addition, a number of cases of therapeutic failure occurred in patients taking digoxin. These patients required unusually high-maintenance doses and were subsequently found to have low plasma digoxin concentrations (12). A cross-over study conducted on four digoxin formulations available in the same hospital at the same time revealed striking differences in bioavailability. The peak plasma concentrations, following a single dose, varied by as much as sevenfold among the four formulations. These findings caused considerable concern because the margin of safety for digoxin is sufficiently narrow that serious toxicity or even lethality can result if the systemically available dose is as little as twice that needed to achieve the therapeutic effect.

Creation of an Office of Technology Assessment

To address this problem of bioinequivalence among duplicate drug products, the U.S. Congress in 1974 created a special Office of Technology Assessment (OTA) to provide advice on scientific issues, among which was the bioequivalence of drug products. The OTA formed the drug bioequivalence study panel. The basic charge to the panel was to examine the relationships between chemical and therapeutic equivalence of drug products, and to assess whether existing technological capability could assure that drug products with the same physical and chemical composition would produce comparable therapeutic effects. Following an extensive investigation of the issues, the panel published its findings to the U.S. Congress in a report, dated July 15, 1974, entitled Drug Bioequivalence (13,14). Notably, the panel concluded that variations in drug bioavailability were responsible for some instances of therapeutic failures and that analytical methodology was available for conducting bioavailability studies in man. Several recommendations pertained to in vivo bioequivalence evaluation. The panel recommended that efforts should be made to identify classes of drugs for which evidence of bioequivalence is critical, that current law requiring manufacturers to make bioavailability information available to the FDA should be strengthened, and that additional research aimed at improving the assessment and prediction of bioequivalence was needed.

Publication of the 1977 Bioavailability and Bioequivalence Regulations

In 1977, the FDA published its *Bioavailability and Bioequivalence* regulations under 21 CFR. The regulations were divided into subpart A—*General Provisions*, subpart B—*Procedures for Determining the Bioavailability of Drug Products*, and subpart C—*Bioequivalence Requirements* (15). The regulations greatly aided the rational development of dosage forms of generic drugs, as well as the subsequent evaluation of their performance. With the publication of these regulations, a generic firm could file an ANDA that provided demonstration of bioequivalence to an approved drug product in lieu of clinical trials. Subpart B defined bioavailability in terms of rate and extent of drug absorption, described procedures for determining bioavailability of drug products, set forth requirements for submission of in vivo bioavailability data, and provided general guidelines for the conduct of in vivo bioavailability studies. Subpart C set forth requirements for marketing a drug product subject to a bioequivalence requirement. ANDAs were generally still restricted to duplicates of drug products approved prior to October 10, 1962 and determined to be effective for at least one indication in a DESI notice. A duplicate drug product had to meet bioequivalence requirements if well-controlled trials showed that it was either not therapeutically equivalent or bioequivalent to other pharmaceutically equivalent products. Narrow therapeutic index (NTI) drugs also had to meet bioequivalence requirements, as did drugs with low aqueous solubility, poorly absorbed drugs, drugs with nonlinear pharmacokinetics, drugs that underwent extensive first-pass metabolism, drugs which were unstable in the gastrointestinal (GI) tract, and drugs for which absorption was limited to a specific portion of the GI tract. Finally, a duplicate drug product had to meet bioequivalence requirements if competent medical determination concluded that a lack of bioequivalence would have a serious adverse effect in the treatment or prevention of a serious disease or condition.

An important feature of the 1977 regulations was the provision for waiver of in vivo bioequivalence study requirements (biowaivers) under certain circumstances. Applicants could file waiver requests for parenteral solutions, topically applied preparations, oral dosage forms not intended to be absorbed, gases or vapors administered by the inhalation route, and oral-solubilized dosage forms. Waivers could be granted for duplicate DESI-effective parenteral drug products (suspensions excluded) and duplicate DESI-effective immediate-release oral drug products, which were not on the list of FDA pharmacological classes and drugs for which in vivo bioequivalence testing was required. Biowaivers could also be granted for drug products in the same dosage form, but a different strength, and proportionally similar in active and inactive ingredients to a drug product from the same manufacturer for which in vivo bioavailability had been demonstrated. Both drug products were required to meet an appropriate in vitro test (generally dissolution) approved by the FDA.

Availability of the Paper NDA Route for Duplicate Drug Products
The FDA did allow some duplicate drug versions of post-1962 drug products to be marketed under a "paper NDA" policy (16). Under this policy, in lieu of conducting their own tests, manufacturers of such duplicate drug products could submit safety and effectiveness information derived primarily from published reports of well-controlled studies. However, such reports of adequate and well-controlled studies in the literature were limited, and the FDA staff effort involved in reviewing paper NDAs became a substantial and often inefficient use of resources.

Present-Day Bioequivalence Requirements

The 1984 Hatch–Waxman Amendments
In 1984, the Drug Price Competition and Patent Term Restoration Act (the Hatch–Waxman Amendments) amended the Federal Food, Drug, and Cosmetic Act by creating Section 505(j) of the Act [21 USC 355 (j)], which established the present ANDA approval process (17). Section 505(j) extended the ANDA process to duplicate versions of post-1962 drugs, but also required that an ANDA for any new generic drug product shall contain information to show that the

generic product is bioequivalent to the reference listed drug product. Evidence of bioequivalence was now required for all dosage forms: tablets, capsules, suspensions, solutions, topical ointments and creams, transdermal patches, ophthalmics, injectables, and so on. The new law stated that a drug shall be considered to be bioequivalent to a listed drug if "the rate and extent of absorption of the drug do not show a significant difference from the rate and extent of absorption of the listed drug when administered at the same molar dose and of the therapeutic ingredient under similar experimental conditions in either a single dose or multiple doses. . ."

The 1992 Revisions to FDA's Bioavailability/Bioequivalence Regulations
In 1992, the FDA revised the *Bioavailability and Bioequivalence Requirements* of 21 CFR Part 320 to implement the Hatch–Waxman Amendments (18). In its present form, 21 CFR Part 320 consists of subpart A, *General Provisions*, and subpart B, *Procedures for Determining the Bioavailability and Bioequivalence of Drug Products*. Subpart A describes general provisions including definitions of bioavailability and bioequivalence. Subpart B states the basis for demonstrating in vivo bioavailability or bioequivalence and lists types of evidence to establish bioavailability or bioequivalence, in descending order of accuracy, sensitivity, and reproducibility. Subpart B also provides guidelines for the conduct and design of an in vivo bioavailability study and lists criteria for waiving evidence of in vivo bioequivalence. The present biowaiver regulations now apply to solutions of all parental solutions, including intraocular, intravenous, subcutaneous, intramuscular, intraarterial, intrathecal, intrasternal, and interperitoneal, but no longer permit automatic biowaivers for all topical and nonsystemically absorbed oral dosage products (18). Waivers of in vivo testing can now be granted for ophthalmic, otic, and topical solutions. A DESI-effective immediate-release oral drug product can be granted a waiver of in vivo testing, provided it is not listed in the FDA's *Approved Drug Products with Therapeutic and Equivalence Evaluations* as having a known or potential bioequivalence problem (1). Other aspects of the present regulations governing biowaivers are similar to the 1977 regulations.

STATISTICAL EVALUATION OF BIOEQUIVALENCE DATA

Introduction
Statistical evaluation of bioequivalence studies of systemically active drugs is based on analysis of drug blood or plasma/serum concentration data. The area under the plasma concentration versus time curve (AUC) is used as an index of the extent of drug absorption. Generally, both AUC determined until the last measurable blood sampling time (AUC_{0-t}) and AUC extrapolated to infinity (AUC_{∞}) are evaluated. Drug peak plasma concentration (C_{max}) is used as an index of the rate of drug absorption.

Early Days of FDA's Bioequivalence Review Process
Criteria for approval of generic drugs have evolved since the 1970s (19). In the early 1970s, approval was based on mean data. Mean AUC and C_{max} values for the generic product had to be within ±20% of those of the brand-name product. In addition, plasma concentration–time profiles for immediate-release products had to be reasonably superimposable. Beginning in the late 1970s, the 75/75

(or 75/75–125) rule was added to the criteria. According to the 75/75 rule, the test/reference ratios of AUC and C_{max} had to be within 0.75 to 1.25 for at least 75% of the subjects. This was an attempt to consider individual variability in rate and extent of absorption. In the early 1980s, the power approach was applied to AUC and C_{max} parameters in conjunction with the 75/75 rule. The power approach consisted of two statistical tests: (*i*) a test of the null hypothesis of no difference between formulations using the *F* test; and (*ii*) the evaluation of the power of a test to detect a 20% mean difference in treatments.

Statistically, the power approach and the 75/75 rule have poor performance, and the FDA discontinued the use of these methods in 1986. The problems with both the 75/75 rule and power approach methods arose from the fact that they were based on the conventional null hypothesis test of no difference. Conventional hypothesis testing does not assess the evidence in favor of the conclusion that the test and reference means are equivalent, but rather assesses the evidence in favor of a conclusion that the test and reference means are different, which is not the question of interest in bioequivalence analysis (20⁻-22). That is, the objective of bioequivalence analysis is to establish whether the test and reference means are equivalent—in other words, is the difference between the two means an acceptable difference?

THE TWO ONE-SIDED TESTS PROCEDURE

The two one-sided tests procedure, used by the FDA since 1986 for bioequivalence analysis, resolved the problems of hypothesis testing (20). The two one-sided tests procedure tests two conditions. Stated simply, the first condition tests if the test product is significantly less bioavailable than the reference product. The second condition tests if the reference product is significantly less bioavailable than the test product. A significant difference is defined as 20% at the alpha equals 0.05 level. As per these statistical criteria, the mean test/reference ratio of the data is usually close to one. The criteria above may be re-stated to illustrate the rationale for the 0.80 to 1.25 (or 80–125%) confidence interval criteria. In the first case illustrated above, the bioequivalence limit for the test/reference ratio = 0.80. In the second case, the bioequivalence limit for the reference/test ratio = 0.80. Since by convention, bioequivalence ratios are expressed as test/reference, the second bioequivalence limit = 1.25, that is, the reciprocal of 0.80. This may be stated in clinical terms as follows. If a patient is currently receiving a brand-name reference product and is switched to a generic product, the generic product should not deliver significantly less drug to the patient than the brand-name product. Conversely, if a patient is currently receiving the generic product and is switched to the brand-name reference product the brand-name product should not deliver significantly less drug to the patient than the generic product.

Logarithmic Transformation of Bioequivalence Data

Until 1992, for bioequivalence statistical analysis, the FDA generally recommended that applicants perform analysis of variance (ANOVA) on untransformed AUC and C_{max} data to determine the 90% confidence limits of the differences. Following a 1991 meeting of the Generic Drugs Advisory Committee, which focused on statistical analysis of bioequivalence data, the FDA began to recommend that applicants perform ANOVA on log-transformed data.

The Generic Drug Advisory Committee recommended log transformation for bioequivalence analysis for two reasons. First, the ANOVA used to conduct the bioequivalence statistics is based on a linear statistical model (23,24). However, the form of expression for AUC suggests a multiplicative model, since $AUC = (F \times D)/(V \times K_e)$, where F is the fraction of drug absorbed, D the dose, V the volume of distribution, and K_e the elimination rate constant. For this reason, FDA statisticians concluded that effects on AUC are not additive if the data are analyzed on the original scale of measurement. Thus, since $\ln(AUC)$ is equal to $\ln(F) + \ln(D) - \ln(V) - \ln(K_e)$, logarithmic transformation of AUC allows it to be analyzed using the ANOVA, which assumes a linear statistical model. A similar argument can be made for C_{max}.

The second reason for log transformation is that C_{max} and AUC, like much biological data, correspond more closely to a log-normal distribution than to a normal distribution (23). Plasma concentration data and derived pharmacokinetic parameters tend to be skewed, and their variances tend to increase with the means. Log transformation generally remedies this situation, and makes the variances independent of the means. In addition, skewed frequency distributions are often made more symmetrical by log transformation.

To summarize, since 1992, the FDA has formally recommended that applicants perform ANOVA on the geometric mean test/reference AUC and C_{max} ratios and determine the 90% confidence interval for the ratios in performing bioequivalence analysis (25). To obtain geometric means, the data are log-transformed prior to conducting an ANOVA, then back-transformed before calculating the test/reference ratios. The 90% confidence interval encompasses the two one-sided tests, each carried out at the $\alpha = 0.05$ (5%) level. The FDA requires that bioavailability of the generic formulation relative to the brand name should be within 0.80 to 1.25 and must be known with a 90% confidence. The determination of bioequivalence using this approach is termed "average bioequivalence" (20).

The Role of T_{max} in Bioequivalence Analysis

The FDA does not ask ANDA applicants to use statistical procedures to compare the time to drug peak plasma concentrations (T_{max}) for the test and reference products. Although theoretically a relatively sensitive measure of absorption rate, T_{max} is thought to have shortcomings as an indirect measure of the rate of drug absorption (26,27). For example, ANOVA analysis cannot be applied to T_{max}. Unlike C_{max} and AUC, which are continuous variables, T_{max} is a discrete measure (28). In addition, most pharmacokinetic studies typically employ irregular sampling schemes to collect T_{max} data, and as a result these data are not routinely amenable to proper statistical evaluation (29). For these reasons, the FDA has decided not to impose bioequivalence acceptance criteria on the parameter T_{max} (30). Nonetheless, the FDA believes that T_{max} should be considered in bioequivalence decision making, and routinely examines T_{max} data in bioequivalence studies as supportive data to verify that the test and reference products have the same rate of absorption (31).

CURRENT METHODS AND CRITERIA FOR
DOCUMENTING BIOEQUIVALENCE

FDA's General Recommendations for In Vivo
Bioequivalence Study Design

Introduction
Part 320 of Title 21 of the Code of Federal Regulations (21 CFR) contains FDA's regulations on procedures for determining bioavailability or bioequivalence of drug products (32). CDER's Guidance for Industry: *Bioavailability and Bioequivalence Studies for Orally Administered Drug Products—General Considerations* ("BA/BE Guidance") contains advice on how to meet the bioavailability and bioequivalence requirements of 21 CFR Part 320 (30). The BA/BE Guidance also applies to nonorally administered drug products where reliance on systemic exposure measures is suitable for documenting bioavailability and bioequivalence (e.g., transdermal systems, certain rectal and nasal drug products).

General Bioequivalence Study Design Recommendations
There are several types of designs suitable for in vivo bioequivalence studies. The preferred design for most orally administered dosage forms is a two-way cross-over, two-period, two-sequence single-dose study, in healthy subjects, performed under fasting conditions. In this design, each study subject receives each treatment, test and reference, in random order. Plasma or blood samples are collected for approximately three pharmacokinetic half-lives for determination of the rate and extent of drug release from the dosage form and absorption by each subject. A washout period is scheduled between the two periods to allow the subjects to completely eliminate the drug absorbed from the first dose before administration of the second dose. Although this design is carried out for most orally absorbed drug products, it may become impractical for drugs with long pharmacokinetic half-lives, that is, longer than 30 hours (e.g., nevirapine). In this case a single-dose parallel design may be used instead (33). For drugs with very long half-lives, concentration sampling may be carried out for a period of time corresponding to two times the median T_{max} (time to C_{max}) for the product. For drugs that demonstrate low intrasubject variability in distribution and clearance, an AUC truncated at 72 hours may be used in place of AUC_{0-t} or AUC_{∞} (30).

Number of Subjects: Single-Dose Versus Steady-State Bioequivalence Studies
The FDA recommends that investigators enroll a minimum of 12 subjects (25). Most bioequivalence studies submitted in support of ANDAs enroll from 24 to 36 subjects. The FDA asks investigators to conduct single-dose bioequivalence studies because it has been shown that these are more sensitive to detecting differences in formulation performance than multiple-dose studies (30,34–38).

Appropriate Drug Product Strength for Bioequivalence Studies
Most bioequivalence studies are conducted on the highest strength of a drug product line, unless it is necessary to use a lower strength for safety reasons. Use of the highest strength is particularly critical for drugs that display nonlinear kinetics because of nonlinear (usually capacity-limited) elimination or presystemic metabolism, with the result that plasma concentrations increase more than

proportionally with an increase in dose (39). For such drugs, small differences in the rate or extent of absorption can potentially have substantial effects on the AUC (40). Thus, using the highest strength in bioequivalence studies, or, in some cases, the highest starting dose—so that drug pharmacokinetics are potentially in the "nonlinear range" ensures that a generic formulation will not pass bioequivalence acceptance criteria unless it is formulated to provide nearly the same rate and extent of exposure as the corresponding reference product. For drugs for which rate and/or extent of absorption increases less than proportionally with an increase in dose (41), the bioequivalence study will be most discriminating if conducted at the lowest strength or, if only one strength is marketed, at the lowest recommended dose.

Fed Bioequivalence Studies

Because food can influence the bioavailability of orally administered drugs, the FDA recommends that applicants conduct bioequivalence studies under fed conditions in most cases. FDA's Guidance for Industry, *Food-Effect Studies and Fed Bioequivalence Studies* ("Food Guidance"), contains recommendations about how to design, and how to decide whether it is necessary to conduct fed bioequivalence studies (42). Fed bioequivalence studies are generally conducted using meal conditions expected to provide the greatest effects on formulation performance and GI physiology such that systemic drug bioavailability may be maximally affected. Typically, the drug is administered to subjects within 30 minutes of consuming a high-fat, high-calorie meal. The FDA recommends that these studies use a randomized, balanced, single-dose, two-treatment (fed vs. fasting), two-period, two-sequence cross-over design. The acceptance criteria for fed bioequivalence studies is the same as for fasting bioequivalence studies—the 90% confidence interval of the geometric mean test/reference AUC and C_{max} ratios must fall within the limits of 80% to 125%.

The decision tree for determining when it is necessary to conduct fed bioequivalence studies differs for immediate-release and modified-release products (42). For immediate-release solid oral dosage forms, fed bioequivalence studies are recommended whenever the FDA-approved package insert (or label) for the reference drug contains statements about the effect of food on absorption or administration. If the label states that the product should be taken only on an empty stomach, fed bioequivalence studies are not recommended; in such cases it is necessary to evaluate bioequivalence only under fasting conditions (e.g., risedronate sodium) (43).

For modified-release solid oral dosage forms, fed bioequivalence studies are always conducted in addition to the fasting bioequivalence studies, with few exceptions. This is primarily because many excipients used to get modified-release behavior depend on the environmental conditions of the GI tract to provide either delayed or extended delivery of the drug (44). Thus, by changing conditions such as gastric pH, gastric emptying rate, intestinal motility, and intestinal secretions, food can interact with components of the modified-release formulation, thereby influencing drug bioavailability. Therefore, for modified-release products, the generic applicant must demonstrate that food effects on test product oral bioavailability does not differ from food effects on reference product oral bioavailability; in other words, that the test and reference products are bioequivalent under fed conditions. If the label warns that the modified release product must be given on an empty stomach for reasons of safety (45) or efficacy (46), then the FDA may conclude that bioequivalence need only be determined under fasting conditions.

In very few cases, bioequivalence is evaluated only under fed conditions because there are safety concerns associated with administration of the product on an empty stomach (47).

Study Population in Bioequivalence Studies

The FDA recommends that in vivo bioequivalence studies be conducted in individuals that are representative of the general population, taking into account age, sex, and race factors (30). For example, if a drug product is to be used in both sexes, the sponsor should attempt to include similar proportions of males and females in the study; if the drug product is to be used predominantly in the elderly, the applicant should attempt to include as many subjects of 60 years of age or greater as possible. Restrictions on admission into the study should generally be based solely on safety considerations.

Bioequivalence studies should be conducted in the intended patient population when there are significant safety concerns associated with use in healthy subjects. For example, bioequivalence studies of drugs used for cancer chemotherapy are generally conducted in cancer patients (48,49). These studies should be conducted in patients who are already stabilized on the medication of interest.

TYPES OF EVIDENCE TO ESTABLISH BIOAVAILABILITY AND BIOEQUIVALENCE

What FDA's Regulations Say

Subpart B of the *Bioavailability and Bioequivalence Requirements* in 21 CFR Part 320 lists the following in vivo and in vitro approaches to determining bioequivalence in descending order of accuracy, sensitivity, and reproducibility (32):

- In vivo measurement of active moiety or moieties in biologic fluid;
- In vivo pharmacodynamic comparison;
- In vivo limited clinical comparison;
- In vitro comparison;
- Any other approach deemed appropriate by FDA.

Bioequivalence Studies with Pharmacokinetic Endpoints

Figure 1 illustrates, for a model of oral dosage form performance, why the most sensitive approach is to measure the drug in biological fluids, such as blood, plasma, or serum. The active ingredient leaves the solid dosage form and dissolves in the GI tract, and following absorption through the gut wall, appears in the systemic circulation. The step involving dissolution of the drug substance prior to absorption is the critical step that is determined by the formulation. Other steps illustrated in the diagram are patient- or subject-determined processes not directly related to formulation performance. Variability of the measured endpoint increases with each additional step in the process. Therefore, variability of clinical measures is quite high compared to blood concentration

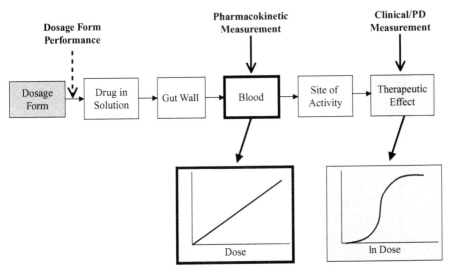

FIGURE 1 The most sensitive approach in evaluating bioequivalence of two formulations is to measure drug concentration in biological fluids, as illustrated in this diagram showing the relationship between dosage form performance and therapeutic response. Following oral dosing, the active ingredient leaves the solid dosage form, dissolves in the gastrointestinal tract, and, following absorption through the gut wall, appears in the systemic circulation. Formulation performance is the major factor determining the critical steps of dosage form disintegration and drug substance dissolution prior to absorption. All other steps following in vivo drug substance dissolution are patient- or subject-determined processes not directly related to formulation performance. The variability of the measured endpoint increases with each additional step in the process, such that variability of clinical measures is quite high compared to that of blood concentration measures. As a result, a pharmacodynamic or clinical approach is not as accurate, sensitive and reproducible as an approach based on plasma concentrations.

measures. Figure 2 shows that the blood concentration of a drug directly reflects the amount of drug delivered from the dosage form.

Most bioequivalence studies submitted to the FDA are based on measuring drug concentrations in plasma. In certain cases, whole blood or serum may be more appropriate for analysis. Measurement of only the parent drug released from the dosage form, rather than a metabolite, is generally recommended because the concentration–time profile of the parent drug is more sensitive to formulation performance than a metabolite, which is more reflective of metabolite formation, distribution, and elimination (30). Measurement of a metabolite may be preferred when parent drug concentrations are too low to permit reliable measurement. In this case, the metabolite data are subjected to a confidence interval approach for bioequivalence demonstration. Both the parent and metabolite are measured in cases where the metabolite is formed by presystemic or first-pass metabolism and contributes meaningfully to safety and efficacy. In this case, only the parent drug data are analyzed using the confidence interval approach. The metabolite data are not subjected to confidence interval analysis but rather used to provide supportive evidence of comparable therapeutic outcome.

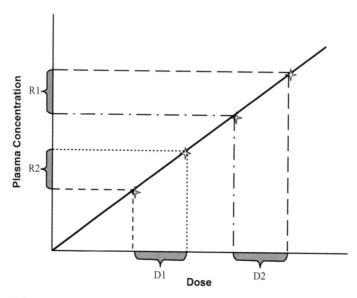

FIGURE 2   The blood concentration of a drug directly reflects the amount of drug delivered from the dosage form. The corresponding responses over a wide range of doses will be of adequate sensitivity to detect differences in bioavailability between two formulations. This is illustrated for two widely different doses, D_1 and D_2. Any differences in dosage form performance are reflected directly by changes in blood concentration (R_1 and R_2).

Urine measurements are not as sensitive as plasma measurements, but are necessary for some drugs such as orally administered potassium chloride (50) and alendronate sodium (51), because serum concentrations are too low to allow for accurate measurement of drug absorbed from the dosage form. Both the cumulative amount of drug excreted (A_e) and maximum rate of urinary excretion (R_{max}) are evaluated statistically in bioequivalence studies that rely on urine concentrations.

Bioequivalence Studies with Pharmacodynamic Endpoints

In situations where a drug cannot be reliably measured in blood, it may be appropriate to base bioequivalence evaluation on an in vivo test in humans in which an acute pharmacologic (pharmacodynamic) effect is measured as a function of time. Generally, the pharmacodynamic response plotted against the logarithm of dose appears as a sigmoidal curve, as shown in Figure 3. It is assumed that, after absorption from the site of delivery, the drug or active metabolite is delivered to the site of activity and, through binding to a receptor or some other mechanism, elicits a quantifiable pharmacodynamic response. Since additional steps contribute to the observed pharmacodynamic response, a pharmacodynamic assay is not as sensitive to drug formulation performance as are blood drug concentrations. In developing a pharmacodynamic assay for bioequivalence evaluation, it is critical to validate the assay by selecting the correct dose. The dose should be in the range that produces a change in response, as shown in the midportion of the curve. In other words, the pharmacodynamic assay should be sensitive to

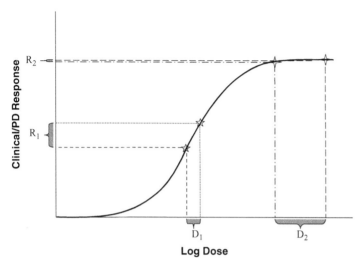

FIGURE 3 In evaluating bioequivalence in a study with pharmacodynamic or clinical endpoints, it is critical to select a dose that falls on the middle ascending portion of the sigmoidal dose–response curve. The most appropriate dose for a study based on pharmacodynamic or clinical endpoints should be in the range that produces a change in response (R_1), as shown in the midportion of the curve (D_1). A dose that is too high will produce a minimal response at the plateau phase of the dose–response curve, such that even large differences in dose (D_2) will show little or no change in pharmacodynamic or clinical effect (R_2). Thus, two formulations that are quite different may appear to be bioequivalent.

small changes in dose. A dose that is too high will produce a minimal response at the plateau phase of the dose–response curve, such that even large differences in dose will show little or no change in pharmacodynamic effect. A pharmacodynamic study can be conducted in healthy subjects. The pharmacodynamic response selected should directly reflect dosage form performance but may not necessarily directly reflect therapeutic efficacy.

The FDA accepts bioequivalence studies with pharmacodynamic endpoints for locally-acting drug products. To be adequately sensitive to distinguish between two products that are not bioequivalent, the dose used in the pivotal bioequivalence study should be on the linear portion of the dose–response curve. In such cases, it is necessary to conduct a pilot pharmacodynamic study by using the reference product to determine the optimal dose for the pivotal bioequivalence study. Topical corticosteroids are examples of a drug product class for which the pharmacodynamic approach is suitable (52). In this case, the pharmacodynamic endpoint is based on the ability of corticosteroids to produce blanching or vasoconstriction in the microvasculature of the skin. Acarbose is another example of a drug product for which bioequivalence can be determined using a pharmacodynamic approach (53). Acarbose lowers blood glucose by inhibiting the activity of α-glucosidase within the GI tract following ingestion of food or other sources of sugar. In this case, the pharmacodynamic endpoint is based upon the ability of acarbose to lower serum glucose after administration of a sucrose load

Bioequivalence Studies with Clinical Endpoints

If it is not possible to develop reliable bioanalytical or pharmacodynamic assays, then it may be necessary to evaluate bioequivalence in a well-controlled trial with clinical endpoints. This type of bioequivalence study is conducted in patients and is based on evaluation of a therapeutic, that is, clinical, response. The clinical response follows a similar dose-response pattern to the pharmacodynamic response, as shown in Figure 3. Thus, in designing bioequivalence studies with clinical endpoints, the same considerations for dose selection apply as for bioequivalence studies with pharmacodynamic endpoints. As with a pharmacodynamic study, the appropriate dose for a bioequivalence study with clinical endpoints should be on the linear rising portion of the dose–response curve, since a response in this range will be the most sensitive to changes in formulation performance. Due to high variability and the subjective nature of clinical evaluations, the clinical response is often not as sensitive to differences in drug formulation performance as a pharmacodynamic response. For these reasons, the clinical approach is the least accurate, sensitive and reproducible of the in vivo approaches to determine bioequivalence.

Bioequivalence studies with clinical endpoints generally employ a randomized, blinded, balanced, parallel design. Studies compare the efficacy of the test product, innovator product, and placebo to determine if the two products containing active ingredient are bioequivalent. The placebo is included to assure that the two active treatments in the clinical trial actually are being studied at a dose that affects the therapeutic response(s). Failure to assure that the treatments are clinically active in the trial would show that the trial has no sensitivity to differences in formulation, i.e. the response is on the flat bottom of the dose–response curve (Fig. 3). A generic equivalent of the innovator product should be able to demonstrate bioequivalence for selected clinical endpoint(s) that adequately reflect drug appearance at the site(s) of activity and therefore formulation performance. Fluticasone propionate nasal spray is an example for which such an approach is suitable. Fluticasone propionate is a corticosteroid, which, when formulated as a nasal spray, is indicated to treat the nasal symptoms of nonallergic rhinitis. For this product, the clinical endpoint is based on total nasal symptom score (TNSS), which is a composite score of patient self-rated symptoms, expressed as a mean change from baseline of the TNSS (54). FDA considers two fluticasone propionate nasal products to have an equivalent clinical response when the 90% confidence interval for the point estimate (mean ratio between test and reference product for the change in TNSS relative to baseline) is within an 80% to 125% acceptance interval.

Bioequivalence Studies with In Vitro Endpoints

With suitable justification, bioavailability and bioequivalence may be established by in vitro studies alone. This approach is suitable for some types of locally acting resins, such as cholestyramine (55) and sevelamer (56), which produce their corresponding therapeutic responses by forming nonabsorbable complexes in the intestine with bile acids and phosphate salts, respectively. For such products, the in vitro measures of bioequivalence are based on binding rate studies. The 90% confidence interval of the test/reference ratios of the equilibrium binding constants should fall within the limits of 0.80 to 1.25.

Two Bioequivalent Products Will Produce the Same Response in Patients
As already described, most studies determining bioequivalence between generic products and the corresponding brand-name products are conducted in healthy subjects. It is true that drug pharmacokinetic profiles may differ between healthy subjects and particular types of patients. This is because some disease states affect different aspects of drug substance absorption, distribution, metabolism, and elimination. However, the effects of disease on relative formulation performance, that is, release of the drug substance from the drug product, are rare. Bioequivalence studies are designed to measure and compare formulation performance between two drug products within the same individuals. It is expected that any difference between in vivo drug release from the two formulations will be the same whether the two formulations are tested in patients or normal subjects. Thus, generic and brand-name products which are bioequivalent can be substituted in patients because they will produce the same effect(s). This is illustrated by findings from a recent observational cohort study comparing effectiveness and safety in patients switched from brand-name warfarin sodium tablets to generic warfarin sodium tablets (57). The generic product was approved on the basis of standard bioequivalence studies in normal volunteers. The observational cohort study showed that the two products had no difference in clinical outcome measures.

WAIVERS OF IN VIVO BIOEQUIVALENCE BASED ON IN VITRO DISSOLUTION TESTING

Introduction
Under certain circumstances, product quality bioavailability and bioequivalence can be documented using in vitro approaches (58). In vitro dissolution testing to document bioequivalence for nonbioproblem DESI drugs remains acceptable. In vitro dissolution characterization is encouraged for all product formulations investigated, including prototype formulations, particularly if in vivo absorption characteristics are well-defined for the different product formulations. Such efforts may enable the establishment of an in vitro–in vivo correlation. When an in vitro–in vivo correlation is available (18), the in vitro test can serve as an indicator of how the product will perform in vivo.

Immediate-Release Drug Products
For immediate-release products, an in vivo bioequivalence demonstration of one or more lower strengths can be waived based on acceptable dissolution testing and an in vivo study on the highest strength (30). All strengths should be proportionally similar in active and inactive ingredients. For reasons of safety of study subjects, it is sometimes appropriate to conduct the in vivo study on a strength that is not the highest. In these cases, the FDA will consider a biowaiver request for a higher strength if elimination kinetics are linear over the dose range, if the strengths are proportionally similar, and if comparative dissolution testing on all strengths is acceptable. Examples of drug products for which an in vivo study is not recommended on the highest strength due to safety include aripiprazole tablets (59) and lamotrigine tablets (60).

Modified-Release Drug Products

For modified-release oral drug products, application of dissolution waivers varies depending on whether the product is formulated as a beaded capsule or tablet. For capsules in which the strength differs only in the number of identical beads containing the active moiety, it is not necessary for the applicant to conduct in vivo testing on lower strengths provided that dissolution testing is acceptable and that bioequivalence is demonstrated in an in vivo study for the highest strength. For tablets, it may not be necessary to conduct in vivo studies on lower strengths provided that (*i*) the lower strengths are proportionally similar in its active and inactive ingredients to the strength that underwent acceptable in vivo bioequivalence testing; and (*ii*) the dissolution profiles of the lower strengths in at least three media (e.g., pH 1.2, 4.5, and 6.8) are similar to the profiles of the strength that underwent acceptable in vivo testing.

The Biopharmaceutics Classification System

Applicants can request biowaivers for immediate-release products based on an approach termed the biopharmaceutics classification system (BCS) (61). The BCS is a framework for classifying drug substances based on solubility and intestinal permeability. With product dissolution, these are the three major factors governing rate and extent of absorption from immediate-release products. The BCS classifies drug substances as

Class 1: high solubility, high permeability
Class 2: low solubility, high permeability
Class 3: high solubility, low permeability
Class 4: low solubility, low permeability

The FDA believes that demonstration of in vivo bioequivalence may not be necessary for immediate-release products containing BCS Class 1 drug substances, as long as the inactive ingredients do not significantly affect absorption of the active ingredient(s). This is because, when a drug dissolves rapidly from the dosage form (in relation to gastric emptying) and has high intestinal permeability, the rate and extent of its absorption is unlikely to depend on dissolution and/or GI transit time.

The CDER Guidance for Industry: Waiver of In Vivo Bioavailability and Bioequivalence Studies for Immediate Release Solid Oral Dosage Forms Based on a Biopharmaceutics Classification System (61), recommends methods for determining drug solubility and permeability for applicants who wish to request biowaivers based on BCS. The drug solubility class boundary is based on the highest dose strength of the product that is the subject of the biowaiver request. The permeability class can be determined in vivo (mass balance, absolute bioavailability, or intestinal perfusion approaches) or in vitro (permeation studies using excised tissues or a monolayer of cultured epithelial cells). Test and reference dissolution profiles should be compared in three media: 0.1 N HCl or simulated gastric fluid without enzymes; pH 4.5 buffer, and pH 6.8 buffer or simulated intestinal fluid without enzymes.

HIGHLY VARIABLE DRUGS

Issues with In Vivo Bioequivalence Studies of Highly Variable Drugs

The width of the 90% confidence interval is proportional to the estimated drug variability (in particular, within-subject variability for a cross-over design) and inversely proportional to the number of subjects participating in the study. Historically, the FDA has applied the bioequivalence limits of 80% to 125% to almost all drug products regardless of the size of within-subject variability. As a result, the number of subjects required for a study of highly variable drugs or drug products can be much greater than normally needed for a typical BE study. Highly variable drugs are defined as drugs for which the within-subject variability in the bioequivalence measures, AUC and $C_{max} \geq 30\%$. At a 2004 meeting of the Pharmaceutical Sciences Advisory Committee, government, academic, and industry scientists expressed concern that applying the conventional bioequivalence criteria to highly variable drugs/products may unnecessarily expose a large number of healthy subjects to a drug when this large number of subjects is not needed for assurance of bioequivalence (62).

It is believed that drugs with high within-subject variability generally have a wide therapeutic window; in other words, despite high variability, these products have been demonstrated to be both safe and effective (63). For these reasons, scientists and statisticians at the FDA investigated various approaches available for determining bioequivalence that would reduce the sample size required for a bioequivalence study, but prevent therapeutically inequivalent products reaching the market.

Reference-Scaled Average Bioequivalence Approach

For drugs with an expected within-subject variability of 30% or greater, the FDA recommends using a reference-scaled average bioequivalence approach (64,65). By using this approach, the bioequivalence study uses a three-period, reference-replicated, cross-over design with sequences of TRR, RTR, and RRT. Specifically, subjects receive a single dose of the test product once and reference product twice on separate occasions with random assignment to the three possible sequences of product administration. This partial replicate design allows for the estimation of within-subject variability of the reference product. The three-period design was selected over a four-period design because of efficiency. The only advantage of the four-period design is that it allows the calculation of the variability of the test product. The test product variability is not used in the proposed statistical method. The minimum number of subjects that would be acceptable is 24.

To analyze the bioequivalence study data by using this approach, measurements of both C_{max} and AUC are first log-transformed and the averages, μ_T and μ_R, of the test and reference products are calculated. The within-subject variability of the reference product, σ^2_{WR}, is also calculated.

Scaled average bioequivalence for both AUC and C_{max} is evaluated by testing the following null hypothesis

$$H_0: \quad \frac{(\mu_T - \mu_R)^2}{\sigma^2_{WR}} > \theta \tag{1}$$

(for given $\theta > 0$) versus the alternative hypothesis

$$H_1 : \quad \frac{(\mu_T - \mu_R)^2}{\sigma_{WR}^2} \leq \theta \tag{2}$$

where μ_T and μ_R are the averages of the log-transformed measures C_{max} and AUC for the test and reference products, respectively; usually testing is done at level $\alpha = 0.05$; and θ is the scaled average bioequivalence limit. Furthermore,

$$\theta = \frac{(\ln \Delta)^2}{\sigma_{W0}^2} \tag{3}$$

where Δ is 1.25, the usual average BE upper limit for the untransformed test/reference ratio of geometric means, and $\sigma_{W0} = 0.25$. Rejection of the null hypothesis H_0 supports the conclusion of equivalence.

A 95% upper confidence bound for $(\overline{Y}_T - \overline{Y}_R)^2 / s_{WR}^2 \leq \theta$ determined in a BE study must be $\leq \theta$, or equivalently, a 95% upper confidence bound for $(\overline{Y}_T - \overline{Y}_R)^2 - \theta s_{WR}^2$ must be ≤ 0. The scaling is mixed. If s_{WR} is less than 0.294, then the two one-sided tests procedure is used to determine bioequivalence. If s_{WR} is greater than or equal to 0.294, then the reference-scaled procedure is used to determine bioequivalence. The value of 0.294 is set by the FDA. Additionally, the point estimate (test/reference geometric mean ratio) must fall within [0.80, 1.25].

Thus there are two parts to the proposed bioequivalence criteria for highly variable drugs, the scaled average bioequivalence evaluation and a point estimate constraint. The test product must pass both conditions before it is judged bioequivalent to the reference product.

The FDA believes that the reference-scaled average bioequivalence approach addresses many of the concerns about the bioequivalence of highly variable drugs that have been raised for the past several years. The approach adjusts the bioequivalence limits of highly variable drugs/products by scaling to the within-subject variability of the reference product in the study. For drugs and products that are highly variable, reference-scaling effectively decreases the sample size needed for demonstrating bioequivalence. The additional requirement of a point-estimate constraint will impose a limit on the difference between the test and reference means, thereby eliminating the potential that a test product would enter the market based on a bioequivalence study with a large mean difference.

NARROW THERAPEUTIC INDEX DRUGS

There are no additional approval requirements for generic versions of NTI drugs versus non-NTI drugs. The FDA does not set specific standards based on therapeutic index (30). The bioequivalence criteria, using the 90% confidence interval approach, are quite strict; there is no need to apply stricter criteria for NTI drugs. The current FDA position is that any generic product may be switched with its corresponding reference listed drug.

FAILED BIOEQUIVALENCE STUDIES

Reasons Why Products Fail to Meet Bioequivalence Limits

There are several reasons why bioequivalence studies fail. Figure 4 shows various scenarios of bioequivalence results for several hypothetical formulations

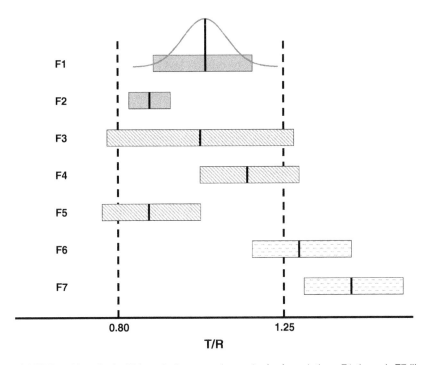

FIGURE 4 Hypothetical bioequivalence study results for formulations F1 through F7 illustrate various scenarios of passing and failing bioequivalence criteria. The width of each 90% confidence interval (CI) is shown as a bar, although in actuality, the log-transformed test/reference (T/R) ratios are distributed as a bell-shaped curve. F1 and F2 represent results of studies in which the 90% CIs of the test/reference ratios (T/R) fall between 0.80 and 1.25 (pass bioequivalence criteria). For F1, the ratio of T/R means (point estimate) is near 1.00. For F2, the point estimate is less than 1.00, but because of low variability, the 90% CI still falls within acceptable limits. F3 through F7 show ways in which studies fail to pass CI criteria. With F3, the point estimate is near 1.00, but because of high variability, the 90% CI is very wide and the drug does not pass bioequivalence criteria. F3 may pass CI criteria if the number of study subjects is increased. By contrast, F4 through F7 have variability comparable to F1. F4 represents a failure on the low side (T is less bioavailable than R), and F5 represents a failure on the high side (R is less bioavailable than T). Since the point estimates for F3 and F4 are still within the 0.8 to 1.25 range, these formulations may also meet CI criteria if a greater number of subjects are dosed. F6 does not meet the upper bound of the 90% CI, and the point estimate exceeds 1.25. For F7, the entire CI is outside the acceptance criteria (bioinequivalence). Formulations F6 and F7 are so different from the reference that both will still fail CI criteria even if the number of subjects is increased.

(labeled F1 through F7). For simplicity, the width of the 90% confidence interval is shown as a bar, although it is important to remember that the results are truly not an even distribution but a normal or log-normal distribution. The log-transformed test/reference ratios from a bioequivalence study are distributed as a bell-shaped curve, with most of the subjects' ratios centered around the center or mean, and fewer subjects' ratios falling at the edges. The top bar in Figure 4 (F1) represents a study with a 90% confidence interval of the test to reference ratio falling between the limits of 0.80 to 1.25 and the test/reference ratios centered around 1.00. This is what most applicants would like to achieve with the to-be-marketed formulation for a given product. The second bar (F2) also represents a 90% confidence interval of test/reference ratios falling within 0.8 to 1.25. Although the mean test/reference ratio is less than 1.00, the variability is very low with the result that this product also meets the 90% confidence interval criteria. The remaining bars in Figure 4 show various scenarios of failure to demonstrate bioequivalence. The third bar (F3) from the top depicts a situation where the test/reference ratios are still centered around 1.00, but because of high variability and probably inadequate sample size, the 90% confidence interval is very wide. This example illustrates how highly variable drugs often need more subjects to attain sufficient statistical power to pass bioequivalence criteria. It is very likely that a new study on the same formulation would pass if more subjects were enrolled, thereby increasing the power of the study and decreasing the resulting width of the 90% confidence interval. The next two bars (F4 and F5) show results of studies which fail to meet bioequivalence criteria, one a failure on the low side (test product has lower bioavailability than reference product), the other a failure on the high side (reference product has lower bioavailability than test product). These two formulations have comparable variability to the formulation that passed (F1), but fail because there is a difference between the test and reference formulations. Because the ratio of means, or point estimate, is still within the 0.80 to 1.25 limits in each of these two cases, it is also possible that these two formulations may pass another study if many more subjects were enrolled. The product represented by the bar that is second from the bottom (F6) does not meet the upper bound of the 90% confidence interval, and also the point estimate exceeds 1.25. It is likely that this product will not pass even if the power of the study is increased by enrolling more subjects. The bottom bar (F7) represents a very extreme case in which the entire confidence interval is outside the acceptance criteria. In this extreme case, the two products are bioinequivalent. The two products are so different that it is highly improbable that repeat studies would ever demonstrate bioequivalence.

In bioequivalence studies of generic products, the most common reason for failure is that the study was underpowered with respect to the number of subjects in the dataset. The width of the confidence interval is controlled by the number of subjects and by the variability of the pharmacokinetic measures. Studies may be underpowered for various reasons. The applicant may have failed to enroll an adequate number of subjects. There may be an excessive number of withdrawals, or there may be missing data because of lost samples. Sometimes a study may fail because of subjects who appear to have an aberrant response on a given dosing day (66). For example, noncompliant subjects may cause the study to fail. The FDA discourages deletion of outlier values, particularly for nonreplicated study designs (25).

Need to Require Submission of Failed Bioequivalence Studies of Generic Drugs

Until recently, applicants did not include failed bioequivalence study data in ANDA submissions. Although applicants submitting New Drug Applications (NDAs) are required to submit data from all clinical studies to the FDA, this was not the case for ANDAs. The Food Drug and Cosmetic Act Section 505(b)(1)(A) states that, for NDAs, all human investigations made to show whether or not a drug is safe for use and whether such drug is effective must be submitted. Similar language was not included in Section 505(j), the section covering the submission of ANDAs. Therefore, generic firms for many years interpreted this language to mean that failed bioequivalence studies did not have to be submitted in their ANDAs. In November of 2000, the Advisory Committee for Pharmaceutical Science recommended that generic applicants submit to the FDA results of all bioequivalence studies on the to-be-marketed ("final") formulations (67). The committee expressed the opinion that applicants should submit the results of failed bioequivalence studies as complete summaries, further suggesting that the FDA should do a brief, but careful examination to identify potential problems worthy of requesting additional information.

The "All Bioequivalence Studies" Rule for Generic Drug Submissions

In 2009, the FDA published a new final rule relating to failed bioequivalence studies, "Requirements for Submission of Bioequivalence Data" (68). The new rule amends the Bioavailability/Bioequivalence Regulations of 21 CFR Part 320 to require an ANDA applicant to submit data from all bioequivalence studies that an applicant conducts on a drug product formulation submitted for approval, including studies that do not meet the specified bioequivalence criteria. The final rule also amends portions of 21 CFR Part 314 (Applications for FDA Approval to Market a New Drug) subpart C, "Abbreviated Applications." All bioequivalence studies submitted on the same drug formulation as that submitted for approval must be submitted to the FDA as either a complete study report or a summary report of the bioequivalence data. The term "same drug product formulation" means the formulation of the drug product submitted for approval and any formulations that have minor differences in composition or method of manufacture from the formulation submitted for approval, but are similar enough to be relevant to the agency's determination of bioequivalence.

FDA's Guidance on What Constitutes the "Same Drug Product"

A draft guidance for industry, *Submission of Summary Bioequivalence Data for ANDAs*, is intended to assist applicants who are submitting ANDAs in complying with the Requirements for Submission of Bioequivalence Data rule (69). The guidance provides information on the types of ANDA submissions covered by the final rule; a recommended format for summary reports of bioequivalence studies; and the types of formulations that FDA considers to be the same drug product formulation for different dosage forms based on differences in composition.

SUMMARY

Current bioequivalence methods in the United States and other countries are designed to provide assurance of therapeutic equivalence of all generic drug products with their innovator counterparts. The sole objective of bioequivalence testing is to measure and compare formulation performance between two or more pharmaceutically equivalent drug products. For generic drugs to be approved in the United States, they must be pharmaceutically equivalent and bioequivalent to be considered therapeutically equivalent and therefore approvable. In the United States, a mechanism for submitting ANDAs for generic products was initiated in 1962 and expanded by the Hatch–Waxman amendment of

1984. The requirement that ANDA submissions contain information showing that a generic drug product is bioequivalent to the innovator product is mandated by law, under Section 505(j) of the U.S. Federal Food, Drug, and Cosmetic Act. Additional Federal laws, published under Title 21 of the Code of Federal Regulations, implement Section 505(j). Part 320 of 21 CFR, the *Bioavailability and Bioequivalence Requirements*, states the basis for demonstrating in vivo bioequivalence, lists the types of evidence to establish bioequivalence (in descending order of accuracy, sensitivity, and reproducibility), and provides guidelines for the conduct and design of an in vivo bioavailability study. The FDA publishes Guidances for Industry to advise the regulated industry on how to meet the *Bioavailability and Bioequivalence Requirements* set forth in 21 CFR Part 320. The FDA makes every attempt to update these guidances as the need arises to ensure that they reflect state-of-the art scientific thinking regarding the most accurate and sensitive methods available to demonstrate bioequivalence between two products. Consulting with panels of experts such as Advisory Committees, participating in meetings and workshops with academia and industry (both in the United States and abroad), and inviting public comment on draft guidances are among the mechanisms that the FDA employs to keep guidance development current.

Current statistical criteria for determining acceptability of bioequivalence studies in the United States and in other countries assure that the test product is not significantly less bioavailable than the reference (usually the innovator) product, and that the reference product is not significantly less bioavailable than the test product. The difference for each of these two tests is 20%, with the result that the test/reference ratios of the bioequivalence measures must fall within the limits of 0.80 to 1.25. A generic product which does not meet these criteria is not approved. The FDA holds that the most accurate, sensitive, and reproducible method for determining bioequivalence is to measure drug concentrations in blood/plasma/serum in a single-dose study using human subjects. If it is not possible to accurately and reproducibly measure drug concentrations in such biological fluids, other approaches may be used, such as measuring active metabolite or measuring drug in urine. For locally active drug products with little systemic availability, bioequivalence may be evaluated by pharmacodynamic, clinical-endpoint, or highly specialized in vitro studies. The FDA now requires ANDA applicants to submit results of all in vivo studies on the to-be-marketed formulation, whether these studies meet or fail bioequivalence acceptance limits, to identify potential problems worthy of seeking additional information.

Biowaivers are granted in some circumstances. Conditions under which waivers may be granted are also stipulated in 21 CFR Part 320. For those drug products which are systemically available and have demonstrated acceptable in vivo bioequivalence, the requirement for an in vivo study may be waived for lower strengths only if the strengths are proportionally similar and show acceptable in vitro dissolution by a method approved by the FDA. Biowaivers may also be granted if an applicant satisfactorily demonstrates that a drug dissolves rapidly from the dosage form and has high intestinal permeability (BCS Class 1).

In approving a generic product, the FDA makes a judgment that it is therapeutically equivalent to the corresponding reference product. It should be clear that regulatory bioequivalence evaluation of generic drug products in the United States is quite rigorous. In fact, surveys of bioequivalence data in ANDAs approved since the enactment of the Hatch–Waxman Amendments in 1984 show that rate and extent of drug exposure from generic drugs differ very little from that of their corresponding innovator counterparts (70–72). The FDA believes that a health care provider can substitute an approved generic product for the brand product with assurance that the two products will produce an equivalent therapeutic effect in each patient.

REFERENCES

1. Approved Products with Therapeutic Equivalence Evaluations, 29th ed. Washington, DC: US Department of Health and Human Services, Public Health Service, Food and Drug Administration, Center for Drug Evaluation and Research, Office of Pharmaceutical Sciences, Office of Generic Drugs, 2009. http://www.fda.gov/cder/orange/obannual.pdf. Accessed July 3, 2009.
2. Current Good Manufacturing Practice for Finished Pharmaceuticals. Title 21 Code of Federal Regulations. Part 211. Washington, DC: U.S. Government Printing Office. April 1, 2008.
3. Kefauver-Harris Amendments, Public Law (PL) 87–781, 87th Congress, October. 10, 1962.
4. Drug Efficacy Study: A Report to the Commissioner of Food and Drugs, National Academy of Sciences. Washington, DC: National Research Council, 1969.
5. 34 Fed Regist 2673, February 27, 1969.
6. 35 Fed Regist 6574, April 24, 1970.
7. 35 Fed Regist 11273, July 14, 1970.
8. Glazko AJ, Kinkel AW, Alegnani WC, et al. An evaluation of the absorption characteristics of different chloramphenicol preparations in normal human subjects. Clin Pharmacol Ther 1968; 9:472–483.
9. Barr WH, Gerbracht LM, Letcher K, et al. Assessment of the biologic availability of tetracycline products in man. Clin Pharmacol Ther 1972; 13:97–108.
10. Chiou WL. Determination of physiological availability of commercial phenylbutazone preparations. Clin Pharmacol Ther 1972; 12:296–299.
11. Blair DC, Barnes RW, Wildner EL, et al. Biologic availability of oxytetracycline HCl capsules: A comparison of all manufacturing sources supplying the United States market. JAMA 1971; 215:251–254.
12. Lindenbaum J, Mellow MH, Blackstone MO, et al. Variations in biological availability of digoxin from four preparations. N Eng J Med 1971; 285:1344–1347.
13. Office of Technology Assessment. Drug Bioequivalence. A Report of the Office of Technology Assessment Drug Bioequivalence Study Panel. Washington, DC: U.S. Government Printing Office, 1974.
14. Office of Technology Assessment. Scientific Commentary: Drug bioequivalence. J Pharmacokinet Biopharm 1974; 2:433–466.
15. 42 Fed Regist 1648, January 7, 1977.
16. 57 Fed Regist 17950, April 28, 1992.
17. Drug Price Competition and Patent Term Restoration Act of 1984, Public Law 98–417, 98 Stat 1585–1605, September 24, 1984.
18. 57 Fed Regist 17998, April 28, 1992.
19. Dighe SV, Adams WP. Bioequivalence: A United States regulatory perspective. In: Welling PG, Tse FLS, Dighe SV, eds. Pharmaceutical Bioequivalence. New York, NY: Marcel Dekker, Inc., 1991:347–380.
20. Schuirmann DJ. A comparison of the two one-sided tests procedure and the power approach for assessing the equivalence of average bioavailability. J Pharmacokinet Biopharm 1987; 15:657–680.
21. Westlake WI. Bioequivalence testing: A need to rethink. Biometrics 1981; 37:589–594.
22. Schuirmann D. Statistical evaluation of bioequivalence studies. In: Sharma KN, Sharma KK, Sen P, eds. Generic Drugs, Bioequivalence and Pharmacokinetics. New Delhi, India: University College of Medical Sciences, 1990: 119–125.

23. Proceedings of the September 26, 1991, meeting of the Generic Drugs Advisory Committee. Available from Division of Dockets Management, Food and Drug Administration, 5630 Fishers Lane, Room 1061 (HFA-305), Rockville, MD, 20852.
24. Bolton S. Statistical considerations for establishing bioequivalence. In: Shargel L, Kanfer I, eds. Generic Drug Product Development – Solid Oral Dosage Forms. New York, NY: Marcel Dekker, Inc., 2005:257–279.
25. US Dept of Health and Human Services, Food and Drug Administration, Center for Drug Evaluation and Research. Guidance for Industry: Statistical Approaches to Establishing Bioequivalence, January 2001. http://www.fda.gov/downloads/Drugs/GuidanceComplianceRegulatoryInformation/Guidances/ucm070244.pdf. Accessed July 3, 2009.
26. Endrenyi L, Fritsch S, Yan W. Cmax/AUC is a clearer measure than Cmax for absorption rates in investigations of bioequivalence. Int J Clin Pharmacol Ther Toxicol 1991; 29:394–399.
27. Duquesnoy C, Lacey LF, Keene ON, et al. Evaluation of partial AUCs as indirect measures of rate of drug absorption in comparative pharmacokinetic studies. Eur J Pharm Sci 1998; 6:259–263.
28. Tozer TN, Bois FY, Hauck WW, et al. Absorption rate vs. exposure: which is more useful for bioequivalence testing? Pharm Res 1996; 13:453–456.
29. Basson RP, Ghosh A, Cerimele BJ, et al. Why rate of absorption inferences in single dose bioequivalence studies are often inappropriate. Pharm Res 1998; 15: 276–279.
30. US Dept of Health and Human Services, Food and Drug Administration, Center for Drug Evaluation and Research. Guidance for Industry: Bioavailability and Bioequivalence Studies for Orally Administered Drug Products – General Considerations, March 2003. http://www.fda.gov/downloads/Drugs/GuidanceComplianceRegulatoryInformation/Guidances/ucm070124.pdf. Accessed July 3, 2009.
31. FDA Response to Docket No. 0lP-0546/PSAl& SUP1, November 21, 2003. http://www.fda.gov/ohrms/dockets/dailys/03/dec03/121003/01p-0546-pdn0001-vol1.pdf. Accessed July 3, 2009.
32. Bioavailability and Bioequivalence. Title 21 Code of Federal Regulations. Part 320. Washington, DC: U.S. Government Printing Office. Revised as of April 1, 2008.
33. US Dept of Health and Human Services, Food and Drug Administration, Center for Drug Evaluation and Research. Guidance for Industry: Nevirapine. May 2008. http://www.fda.gov/downloads/Drugs/GuidanceComplianceRegulatory Information/Guidances/ucm088752.pdf. Accessed June 21, 2009.
34. Chen, ML, Shah V, Patnaik R, et al. Bioavailability and bioequivalence: An FDA regulatory overview. Pharm Res 2001; 18:1645–1650.
35. Blume H, McGilveray IJ, Midha KK. BIO-International '94 Conference on Bioavailability, Bioequivalence and Pharmacokinetic Studies and Pre-Conference Satellite on "In Vivo/In Vitro Correlation." Munich Germany, June 14–17, 1994. Eur J Drug Metab Pharmacokinet 1995; 2:3–13.
36. El-Tahtawy AA, Jackson AJ, Ludden TM. Evaluation of bioequivalence of highly-variable drugs using Monte Carlo simulations. I. Estimation of rate of absorption for single and multiple-dose trials using Cmax. Pharm Res 1995; 12:1634–1641.
37. El-Tahtawy AA, Jackson AJ, Ludden TM. Comparison of single and multiple dose pharmacokinetics using clinical bioequivalence data and Monte Carlo simulations. Pharm Res 1996; 11:1330–1336.
38. El-Tahtawy AA, Tozer TN, Harrison F, et al. Evaluation of bioequivalence of highly variable drugs using clinical trial simulations. II. Comparison of single- and multiple-dose trials using AUC and Cmax. Pharm Res 1998; 15:98–104.
39. Wagner JG. Nonlinear pharmacokinetics: Examples of anticancer agents and implications in bioavailability. In: Sharma KN, Sharma KK, Sen P, eds. Generic Drugs, Bioequivalence and Pharmacokinetics. New Delhi, India: University College of Medical Sciences, 1990:111–118.

40. US Dept of Health and Human Services, Food and Drug Administration, Center for Drug Evaluation and Research. Guidance for Industry: Phenytoin Sodium, May 2008. http://www.fda.gov/cder/guidance/bioequivalence/default.htm. Accessed July 3, 2009.

41. Clinical Pharmacology Biopharmaceutics Review(s), September 28, 2000. Mifepristone. Drugs@FDA. http://www.accessdata.fda.gov/drugsatfda_docs/nda/2000/20687_mifepristone.cfm. Accessed July 3, 2009.

42. U S Dept of Health and Human Services, Food and Drug Administration, Center for Drug Evaluation and Research. Guidance for Industry: Food-Effect Bioavailability and Fed Bioequivalence Studies. December 2002. http://www.fda.gov/downloads/Drugs/GuidanceComplianceRegulatoryInformation/Guidances/ucm070241.pdf. Accessed July 3, 2009.

43. US Dept of Health and Human Services, Food and Drug Administration, Center for Drug Evaluation and Research. Draft Guidance for Industry: Risedronate Sodium. December 2008. http://www.fda.gov/downloads/Drugs/GuidanceComplianceRegulatoryInformation/Guidances/ucm089588.pdf. Accessed July 3, 2009.

44. Davit BM, Conner DP. Food Effects on Drug Bioavailability: Implications for New and Generic Drug Development. In: Krishna R, Yu L, eds. Biopharmaceutics Applications in Drug Development. New York, NY: Springer, 2008:317–335.

45. US Dept of Health and Human Services, Food and Drug Administration, Center for Drug Evaluation and Research. Guidance for Industry: Efavirenz. May 2008. http://www.fda.gov/downloads/Drugs/GuidanceComplianceRegulatory Information/Guidances/ucm086238.pdf. Accessed July 3, 2009.

46. US Dept of Health and Human Services, Food and Drug Administration, Center for Drug Evaluation and Research. Guidance for Industry: Didanosine. May 2008. http://www.fda.gov/downloads/Drugs/GuidanceComplianceRegulatory Information/Guidances/ucm085602.pdf. Accessed July 3, 2009.

47. US Dept of Health and Human Services, Food and Drug Administration, Center for Drug Evaluation and Research. Guidance for Industry: Mefloquine. May 2008. http://www.fda.gov/downloads/Drugs/GuidanceComplianceRegulatory Information/Guidances/ucm088650.pdf. Accessed July 3, 2009.

48. US Dept of Health and Human Services, Food and Drug Administration, Center for Drug Evaluation and Research. Draft Guidance for Industry: Imatinib Mesylate. March 2009. http://www.fda.gov/downloads/Drugs/Guidance ComplianceRegulatoryInformation/Guidances/ucm118261.pdf. Accessed June 22, 2009.

49. US Dept of Health and Human Services, Food and Drug Administration, Center for Drug Evaluation and Research. Draft Guidance for Industry: Methotrexate Sodium. April 2009. http://www.fda.gov/downloads/Drugs/GuidanceComplianceRegulatoryInformation/Guidances/UCM148209.pdf. Accessed June 22, 2009.

50. US Dept of Health and Human Services, Food and Drug Administration, Center for Drug Evaluation and Research. Draft Guidance for Industry: Alendronate Sodium. January 2008. http://www.fda.gov/downloads/Drugs/Guidance ComplianceRegulatoryInformation/Guidances/ucm082421.pdf. Accessed July 3, 2009.

51. US Dept of Health and Human Services, Food and Drug Administration, Center for Drug Evaluation and Research. Guidance for Industry: Potassium Chloride Modified-Release Tablets and Capsules – In Vivo Bioequivalence and In Vitro Dissolution Testing. October, 2005. http://www.fda.gov/downloads/Drugs/GuidanceComplianceRegulatoryInformation/Guidances/ucm072890.pdf. Accessed July 3, 2009.

52. US Dept of Health and Human Services, Food and Drug Administration, Center for Drug Evaluation and Research. Guidance for Industry: Topical Dermatologic Corticosteroids: In Vivo Bioequivalence. June 1995. http://www.fda.gov/

downloads/Drugs/GuidanceComplianceRegulatoryInformation/Guidances/ucm
070234.pdf. Accessed July 3, 2009.

53. FDA Center for Drug Evaluation and Research – Final Response Letter,
Docket No. FDA-2007-P-0418. May 7, 2008. http://www.regulations.gov/
fdmspublic/component/main?main = DocumentDetail&o = 09000064805529eb.
Accessed July 3, 2009.

54. FDA Center for Drug Evaluation and Research – Final Response Letter, Dockets No. 2004P-206/CP1; 2004–239/CP1, SUP 1, SUP 2 & PSA 1;
2004P-0348/CP1 & SUP 1; and 2004P-0523/CP1 & PSA1. February 22, 2006.
http://www.regulations.gov/fdmspublic/component/main?main = DocumentDe-
tail&o = 090000648046ffd5. Accessed July 3, 2009.

55. US Dept of Health and Human Services, Food and Drug Administration,
Center for Drug Evaluation and Research. Interim Guidance for Industry:
Cholestyramine Powder In Vitro Bioequivalence. July 1993. http://www.fda.gov/
downloads/Drugs/GuidanceComplianceRegulatoryInformation/Guidances/
ucm070230.pdf. Accessed June 22, 2009.

56. US Dept of Health and Human Services, Food and Drug Administration, Center for Drug Evaluation and Research. Draft Guidance for Industry: Sevelamer Hydrochloride. July 2008. http://www.fda.gov/downloads/Drugs/
GuidanceComplianceRegulatoryInformation/Guidances/ucm089621.pdf. Accessed
July 3, 2009.

57. Swenson CN, Fundak G. Observational cohort study of switching warfarin sodium
products in a managed care organization. Am J Health Syst Pharm 2000; 57:
452–455.

58. 57 Fed Regist 29354, July 1, 1992.

59. US Dept of Health and Human Services, Food and Drug Administration, Center for Drug Evaluation and Research. Draft Guidance for Industry: Aripiprazole. November 2007. http://www.fda.gov/downloads/Drugs/Guidance
ComplianceRegulatoryInformation/Guidances/ucm082554.pdf. Accessed July 3,
2009.

60. US Dept of Health and Human Services, Food and Drug Administration, Center for Drug Evaluation and Research. Guidance for Industry: Lamotrigine. May
2008. http://www.fda.gov/downloads/Drugs/GuidanceComplianceRegulatory
Information/Guidances/ucm086283.pdf. Accessed June 22, 2009.

61. US Dept of Health and Human Services, Food and Drug Administration, Cen-
ter for Drug Evaluation and Research. Guidance for Industry: Waiver of In Vivo
Bioavailability and Bioequivalence Studies for Immediate-Release Solid Oral Dosage
Forms Based on a Biopharmaceutics Classification System. August 2000. http://
www.fda.gov/downloads/Drugs/GuidanceComplianceRegulatoryInformation/
Guidances/ucm070246.pdf. Accessed July 3, 2009.

62. Committee Discussion. FDA Advisory Committee for Pharmaceutical Science
Meeting Transcript. April 13–14, 2004. http://www.fda.gov/ohrms/dockets/ac/
04/minutes/4034M1.htm. Accessed July 3, 2009.

63. Benet L. Therapeutic Considerations of Highly Variable Drugs. FDA Advisory
Committee for Pharmaceutical Science Meeting Transcript. October 6, 2006.
http://www.fda.gov/ohrms/dockets/ac/06/slides/2006–4241s2-index.htm.
Accessed July 3, 2009.

64. Haidar SH, Davit B, Chen ML, et al. Bioequivalence approaches for highly variable
drugs and drug products. Pharm Res 2008; 25:237–241.

65. Haidar SH, Makhlouf F, Schuirmann DJ, et al. Evaluation of a scaling approach for
the bioequivalence of highly variable drugs. AAPS J 2008; 10:450–454.

66. Davit B. Outliers and Inadequate Profiles in Bioequivalence Studies – US FDA
Perspective. Health Canada Scientific Advisory Committee on Bioavailability and
Bioequivalence Record of Proceedings. June 3–4, 2004. http://www.hc-sc.gc.ca/dhp-
mps/alt_formats/hpfb-dgpsa/pdf/prodpharma/sacbb_rop_ccsbb_crd_2004–06-03-
eng.pdf. Accessed July 3, 2009.

67. Committee Discussion. Advisory Committee for Pharmaceutical Sciences Meeting Transcript. November 16, 2000. http://www.fda.gov/ohrms/dockets/ac/00/transcripts/3657t2.pdf. Accessed July 3, 2009.
68. 74 Fed Regist 2849, January 16, 2009.
69. US Dept of Health and Human Services, Food and Drug Administration, Center for Drug Evaluation and Research. Draft Guidance for Industry: Submission of Summary Bioequivalence Data for ANDAs. April 2009. http://www.fda.gov/downloads/Drugs/GuidanceComplianceRegulatoryInformation/Guidances/UCM134846.pdf. Accessed July 3, 2009.
70. Nightingale SL, Morrison JC. Generic drugs and the prescribing physician. JAMA 1987; 258:1200–1204.
71. Henney JE. Review of generic bioequivalence studies (letter from the Food and Drug Administration). JAMA 1999; 282:1995.
72. Davit BM, Nwakama PE, Buehler GJ, et al. Comparing generic and innovator drugs: A retrospective review of 12 years of bioequivalence data from the United States Food and Drug Administration. Ann Pharmacother. 2009; 43:1583–1597.

The World Health Organization[a,b]

John Gordon

Division of Biopharmaceutics Evaluation 2, Bureau of Pharmaceutical Sciences, Therapeutic Products Directorate, Health Canada, Ottawa, Ontario, Canada

Henrike Potthast

Federal Institute for Drugs and Medical Devices, Bonn, Germany

Matthias Stahl and Lembit Rägo

WHO Medicines Prequalification Programme, Quality Assurance and Safety of Medicines, Essential Medicines and Pharmaceutical Policies, World Health Organization, Geneva, Switzerland

BACKGROUND

Much progress has been achieved over the last 50 years in the field of pharmaceuticals, both in terms of introducing new medicines and improving the regulation of medicines. This progress involves mostly highly industrialized countries where citizens can benefit from new innovative drugs and enjoy access to quality assured multisource (generic) medicines as well. Lack of access to quality essential drugs, the majority of which are multisource (generic) medicines, remains a serious health problem and global disequilibrium of quality continues to threaten patients in many parts of the world (1). The overall tendency is that resource-constrained or resource-poor countries are less likely to control the quality of products on the market, enjoy political support for the regulators, or have properly resourced and functioning regulatory authorities (2). It is no wonder that in many resource-poor settings patients do not trust locally authorized multisource (generic) products.

In terms of what is required for regulatory approval of medicines it is important to distinguish between two major groups, innovative new medicines (new chemical entities or NCEs) and multisource (generic) medicines. To launch an innovator product the manufacturer/applicant has to pass rigorous scientific assessment by the competent regulatory authorities and prove its product's quality, safety, and efficacy. The most difficult aspect of these is to prove the safety and efficacy of the new drug since that has to be based on original preclinical

---

[a] The WHO is the directing and coordinating authority for health within the United Nations system. It is responsible for providing leadership on global health matters, shaping the health research agenda, setting norms and standards, articulating evidence-based policy options, providing technical support to countries and monitoring and assessing health trends.

[b] The views stated in this chapter reflect the views of the authors and not necessarily those of the WHO.

and clinical research. Multisource (generic) medicines are formulated when patent and other exclusivity rights expire. These medicines have an important role to play in public health as they are well known to the medical community and are usually more affordable due to competition. The quality of a multisource (generic) medicine must be the same as that of an originator. In the case of safety and efficacy, no original research is carried out and reference is made to the data for the respective innovator product. Thus, the key for using multisource (generic) medicines is their therapeutic interchangeability with originator products. To ensure therapeutic interchangeability, multisource products must be pharmaceutically equivalent and proven to have the same safety and efficacy pattern as the originator product, that is, be therapeutically interchangeable. It should be noted that some multisource products may be pharmaceutically equivalent but may not necessarily be therapeutically interchangeable. The manufacture of multisource (generic) quality medicines may not be an easy task and certainly requires the appropriate skills. To ensure that multisource (generic) products are interchangeable and of good quality, well-resourced regulatory authorities nowadays require that a multisource (generic) medicine must

- contain the same active ingredients as the innovator drug;
- be identical in strength, dosage form, and route of administration;
- have the same indications of use;
- be bioequivalent (as a marker for therapeutic interchangeability);
- be manufactured under the same strict standards of good manufacturing practices (GMP) as required for innovator products.

In the case of multisource (generic) medicines, pharmacopoeial monographs are important as they discourage manufacturers from elaborating their own specifications but rather encourage them to develop the products to meet the requirements of pharmacopoeial standards for active pharmaceutical ingredients (APIs) and finished dosage forms. Pharmacopoeial standards and the work of the World Health Organization (WHO) in this area have been described in-depth elsewhere (3).

It is important to point out that pharmacopoeial standards should be used in the framework of all regulatory measures such as good manufacturing practice inspection of active pharmaceutical ingredient and finished dosage form manufacturing, scientific assessment of all quality specifications, therapeutic interchangeability data and labeling information provided by the manufacturer. The greatest value of pharmacopoeial standards is revealed during postmarketing surveillance of the quality of multisource (generic) medicines. While the use of pharmacopoeial standards is critical to ensure the quality of pharmaceutical products, their use cannot guarantee or predict the performance of these products in vivo.

In the WHO the whole area of work of regulatory standards for medicines, including bioequivalence, is overseen by the WHO Expert Committee on Specifications for Pharmaceutical Preparations. The WHO Expert Committee on Specifications for Pharmaceutical Preparations is the highest level advisory body to the WHO's Director-General and its Member States in the area of quality assurance. They are intended to help national and regional authorities (in particular drug regulatory authorities), procurement agencies, as well as major international bodies and institutions, such as the global fund, and international organizations such as UNICEF, to combat problems of substandard medicines and underpin important initiatives, including the prequalification of medicines, the Global Malaria Program, and Stop TB Program.

In light of the HIV/AIDS pandemic, the WHO was requested to give advice on how to ensure the quality of antiretrovirals including fixed-dose combination (FDC) drugs. The WHO Prequalification Program started in 2001 to assure that medicinal products supplied for procurement meet WHO norms and standards with respect to quality, safety, and efficacy (http://www.who.int/medicines/ and http://www.who.int/prequal). The scope was later expanded and today product dossiers submitted to the program belong to the therapeutic areas of HIV/AIDS, tuberculosis, malaria, reproductive health, influenza, or zinc for diarrhoea and usually involve multisource (generic) products. Such products need to conform to the same standards of quality, efficacy, and safety as required of the originator's (comparator) product. Specifically, the multisource product should be therapeutically equivalent and interchangeable with the comparator product. Testing the bioequivalence between a product and a suitable comparator in a well standardized pharmacokinetic study with a limited number of usually healthy subjects is the most common way of demonstrating therapeutic equivalence without having to perform clinical trials involving many patients to prove safety and efficacy. The bioequivalence study therefore provides indirect evidence of the efficacy and safety of a multisource drug product. It is assumed that if the concentration patterns of two drugs in the blood of healthy volunteers are essentially the same, then their safety and efficacy must also be essentially the same. As mentioned earlier, for multisource drug products bioequivalence is the only evidence required to show that the product is safe and efficacious. It is therefore crucial that the bioequivalence study is performed in an appropriate manner. The following describes the standards for planning, design and analysis of a bioequivalence study as outlined in the WHO guidance documents and as applied in the WHO Prequalification Program.

DEFINITIONS AND GENERAL CONSIDERATIONS

Bioavailability and bioequivalence contribute to the so-called biopharmaceutical characterization of a particular medicinal product. According to Ritschel et al. (4) the term *biopharmaceutics* "deals with the physical and chemical properties of the drug substance, the dosage form, the body and the biological effectiveness of a drug and/or drug product upon administration, i.e., the drug availability to the human body from a given dosage form, considered as a drug delivery system. The time course of the drug in the body and the quantifying of the drug concentration pattern are explained by pharmacokinetics."

Thus, the term bioavailability may be considered a link between product quality and clinical performance as it describes the availability of the drug substance/API from the dosage form or formulation. Needless to say that should the API not be available to the biological system following administration, the most sophisticated dosage form would be of no therapeutic value. According to the definition, bioavailability is the rate and extent to which an active pharmaceutical ingredient or an active moiety is delivered from a pharmaceutical dosage form and becomes available at the site of action. An active moiety is rarely measurable at its site of action but, since it is generally agreed that the API in the systemic circulation is usually in some kind of equilibrium with its site of action, systemic concentrations may serve to evaluate bioavailability. Accordingly, concentrations derived after intravenous administrations define the 100% standard, that is, "absolute" bioavailability. Therefore, comparing systemic concentrations after the administration of an oral dosage form to those obtained following intravenous administration results in "absolute bioavailability" of the oral product. However, it is not possible to distinguish between formulation properties and its impact on the pharmacokinetics of the API by means of investigating absolute bioavailability. In contrast relative bioavailability is meant to be the comparison of dosage forms other than intravenous and one of the most interesting is comparing an aqueous oral solution with a solid oral dosage form. The result of such a comparison may serve to describe the formulation impact of the solid dosage form, for example, possibly formulation-related diminished availability. Hence, comparing bioavailability becomes basic for the assessment of bioequivalence of systemically acting (particularly orally administered) generics [i.e., multisource pharmaceutical drug products (MPPs)]. According to the latest WHO guidance on the topic, Annex 7 of the Technical Series Report No. 937 (5) the term bioequivalence is defined as similar bioavailabilities of pharmaceutical equivalent or pharmaceutical alternative products in terms of peak (C_{max} and T_{max}) and total exposure (AUC) after administration of the same molar dose under the same conditions to such a degree, that their effects can be expected to be essentially the same. In other words, essentially similar products should be interchangeable.

APPLICABILITY AND LIMITATIONS OF THE BIOEQUIVALENCE CONCEPT

As it can be seen from the literature on this topic and according to international guideline documents (5,6), the bioequivalence concept can be understood as a simplified way to demonstrate the safety and efficacy of a drug product containing the same quantity of a known API as in an innovator product. The basic underlying assumption is that in the same subject an essentially similar plasma concentration–time course will result in essentially similar concentrations at the site(s) of action, that is, the therapeutic outcome is essentially the same. Hence, evaluating pharmacokinetic measures may be justified as a surrogate method instead of generating clinical/therapeutic data. Accordingly, as stated by van Faassen et al., (7) "...if the fraction of the dose absorbed is the same, the human body should always do the same with the absorbed compound—even in a disease state, this argument is still a valid statement." Importantly, however, the rate of absorption must also be included as a consideration in such a statement. Therefore, if bioequivalence has been demonstrated between a generic and an innovator product, the clinical/toxicological properties should be valid for both products, that is, safety and efficacy as originally proven for the innovator product can be demonstrated for the MPP by means of (e.g., pharmacokinetic) bioequivalence data. Hence, proof of bioequivalence allows the possibility of seeking/granting marketing authorization without the need for further clinical studies. The importance of meaningful bioequivalence study results is obvious from these explanations.

It is possible that an MPP could exhibit sub- or suprabioavailability compared to the comparator product. In such cases, the clinical/toxicological documentation of the comparator cannot be considered valid for the generic product, but it is necessary that it be treated as a new product [see Section 6.11.2 in Ref. (5)].

Another difficulty may arise in cases when no innovator/originator product is available for some reason. The appropriate selection of the comparator as recommended by WHO is discussed in "Choice of Comparator Products" section (vide infra).

For completeness it should be noted that bioequivalence testing is required not only for new MPP applications but also in the case of major formulation changes (variations). For modified MPPs, bioequivalence has to be demonstrated by comparison against the designated reference comparator product. A study comparing the modified MPP to the originally approved (unchanged) MPP is not sufficient as this could lead to "creep" away from the connection between the reference product and the MPP, the link upon which the evidence for the safety and efficacy of the MPP is based. The latter comparison is an adequate option only for innovator products (see "Modified-Release MPPs" section [vide infra]).

In Vivo Equivalence Studies

In Sections 3 and 5 of Annex 7 of the WHO Technical Report of 2006 (5), the necessity to demonstrate bioequivalence of MPPs with an appropriate comparator is generally stated. Different options to perform in vivo studies are described in Table 1. However, pharmacokinetic measurements can usually be regarded as the most sensitive approach for the assessment of bioequivalence. Actually, the final decision regarding the appropriate approach to demonstrate bioequivalence has to be taken by considering the characteristics of the particular drug substance and drug product. This is particularly important as a positive risk assessment (Table 3) may provide the option to demonstrate bioequivalence by a certain in vitro approach which is discussed in "In Vitro Approaches/Biowaiver Options" section later in the chapter.

In case in vivo equivalence studies are carried out, compliance with good clinical practice (GCP) guidelines including proper ethical review and ethical clearances should be ensured. Historically bioequivalence studies in healthy volunteers were often carried out by universities and/or research institutions having qualified staff for study design, conduct, analytical measurement, and statistical interpretation of data. Today the majority of bioequivalence studies

TABLE 1 Options to Show Therapeutic Equivalence of MPPs

	Experimental setting
Generally regarded most sensitive	Comparative pharmacokinetic studies in humans – evaluation of systemic exposure by means of pharmacokinetic measures, for example, AUC and C_{max} (and T_{max})
Prerequisites should be noted	Comparative in vitro tests—see BCS-based biowaiver
Sensitivity not optimal	Comparative pharmacodynamic studies in humans— evaluation of relevant pharmacodynamic endpoints, for example, lowered blood pressure in mm Hg
Rarely used for oral MPP formulations with systemic actions	Comparative clinical trials—evaluation of, for example, noninferiority

as well as most clinical research are performed by Contract Research Organizations (CROs). To facilitate proper conduct of bioequivalence studies the WHO has issued a specific guidance "Additional Guidance for Organizations Performing In Vivo Bioequivalence Studies" (8).

Formulation-Related Biowaiver

Section 4 of the WHO Technical Report of 2006, Annex 7 (5) outlines those cases when bioequivalence studies are not necessary due to the particular formulation and/or site of administration and/or intended effect. Accordingly, aqueous solutions intended for oral or intravenous/parenteral administration are exempted from bioequivalence testing since no formulation effect is expected even in the case when slightly different excipients have been used (i.e., buffer, preservative, antioxidant). Other soluble formulations (e.g., syrups, tinctures, powders for reconstitution) except suspensions may be treated likewise.

Exemption of bioequivalence testing is also generally acceptable for pharmaceutically equivalent products such as gases, otic or ophthalmic formulations, and topical products if they are not intended to be systemically effective, aqueous nebulizer inhalation products or aqueous-based nasal sprays. Hence, formulation-related biowaivers may be applicable if pharmaceutical product quality is deemed sufficient to assure therapeutic equivalence when no formulation effect is expected. Notwithstanding, however, the investigative products should contain the same API in the same molar concentration, whereas any possible impact of difference in excipients used will have to be appropriately addressed by the applicant (Table 2).

TABLE 2 Formulation-Related Biowaiver

Biowaiver for	Prerequisites	
	Active moiety	Excipients
Aqueous intravenous and parenteral solutions	Same molar concentration as comparator	Same or similar
Pharmaceutical equivalent solutions for oral use—except suspensions	Same molar concentration as comparator	Essentially the same
Pharmaceutical equivalent powders for reconstitution as a solution	Same molar concentration as comparator	Same or similar
Gases	Same molar concentration as comparator	–
Pharmaceutical equivalent otic or ophthalmic products	Same molar concentration as comparator	Essentially the same; preservatives, buffer, tonicity or thickening agents may differ
Pharmaceutical equivalent topicals as aqueous solutions	Same molar concentration as comparator	Essentially the same in comparable concentrations
Pharmaceutical equivalent aqueous solutions for nebulizer inhalation or nasal sprays	Same molar concentration as comparator	Essentially the same device; essentially the same excipients in comparable concentrations

CHOICE OF COMPARATOR PRODUCTS

Once a decision has been made with regard to the type of comparisons that must be made to establish the safety and efficacy of the proposed MPP, for example, an in vivo bioequivalence study or in vitro comparisons for a biowaiver, it is then necessary to identify the correct comparator product against which the proposed MPP must be compared.

In the case of a national regulator employing WHO guidance as a basis for the investigation and approval of MPPs, it is recommended that the innovator pharmaceutical product be employed as the comparator product since its quality, safety, and efficacy should have been established both during premarket assessment and postmarket monitoring. The consistent use of the "nationally authorized innovator" product as the designated comparator product will provide a consistent anchor for multiple MPPs in a market. A MPP (generic) product should not be employed as the comparator product if the nationally authorized innovator product is available, as this process could lead to "creep" away from the product data on which the original approval was based.

Should a "nationally authorized innovator" not be available, the WHO guidance Annex 7 (5) proposes a multistepped process for a national regulatory authority to use to select an alternative comparator product. It is recommended that the national drug regulatory authority select one of the following as their comparator product in order of preference:

1. The "WHO comparator product" as described in The WHO Comparator Product section; or
2. The innovator product currently available on the market in a well-regulated country.

As a standard for the stringency of regulation of a "well-regulated" national market, Annex 7 recommends the markets of members of the "International Conference on Harmonisation of Technical Requirements for Registration of Pharmaceuticals for Human Use" (ICH), that is, the United States, European Union, or Japan, or countries associated with the ICH, for example, countries such as Switzerland or Canada.

Should it not be possible to identify a suitable innovator product via these avenues, a comparator should be selected that has been documented to be safe and effective based on evidence such as approval in ICH and associated countries, "prequalification" by the WHO, extensive use in clinical trials reported in peer-reviewed scientific journals, and/or a long and unproblematic history of postmarket surveillance. These would be rare occurrences, perhaps limited to older medicines where the innovative product is no longer produced and the demand for the medicine in many countries is very limited.

It is important to be aware that should a **"nationally authorized innovator" product not be available as the Comparator Product**, products approved on the basis of comparison to the chosen comparator product may or may not be interchangeable with other products currently available within that particular market. **In other words, actual "generic substitution" cannot be recommended.**

WHO Comparator Products

In an effort to assist national drug regulatory authorities and pharmaceutical companies in selecting appropriate comparator products to which comparisons can be made as a part of the authorization process for MPPs, WHO has published a guidance that includes a list of comparator products derived from information collected from drug regulatory authorities and the pharmaceutical industry (6). These "International Comparator Products" or WHO comparator products can be selected by national authorities when a "nationally authorized innovator" is not available. This guidance also suggests criteria, in a decision-tree format, that can be used in the selection of a comparator product.

Selection of Comparator Products—The WHO Prequalification Program

As the prequalification program (PQ) serves a truly international purpose, the selection of appropriate comparator products for use in comparative studies for MPPs intended for PQ should be based on the WHO International Comparator Product, as already described and identified by the WHO (7,9). However, to serve its international purpose, not all of the recommendations made to national authorities are applicable to the PQ project. For example, the use of a nationally authorized innovator as comparator product would only be acceptable within the PQ if that product is the identified WHO International Comparator Product or, if that product is not defined explicitly, it is the innovator product available in an ICH—or ICH-associated market.

If such a product cannot be identified, a product that was approved in the PQ based on an evaluation of full quality documentation and full clinical trial documentation establishing its safety and efficacy, may serve as a comparator product (9).

Comparator Products for Fixed-Dose Combination Products

Current FDC products should not be employed as the comparator product for a proposed FDC product unless approval of the currently available FDC product was based on full clinical trials establishing the safety and efficacy of the product. The use of a FDC product that was approved based on bioequivalence data to the individual components as a comparator product for other FDCs can lead to "creep" away from the product performance observed in the original clinical trials. If an appropriate FDC comparator is not available, the individual component products should be used as the comparator products.

IN VIVO APPROACHES

As discussed earlier, in vivo documentation of bioequivalence is especially important for certain medicines and dosage forms, and the pharmacokinetic bioequivalence study is considered to be the most effective and sensitive design for achieving this purpose.

There are many issues to be considered when designing a bioequivalence study to compare the in vivo performance of two products, however, the ultimate goal is to design a study that minimizes the variability that is not attributable to the formulations being compared and to eliminate any bias in the study. In general, studies comparing product performance following a single administration of each product, that is, single-dose bioequivalence studies, are considered to be the most effective study designs for the purpose of comparing a MPP to a comparator product.

Single-Dose Bioequivalence Studies

Products to Be Compared

Multisource Pharmaceutical Product

For a bioequivalence study to be meaningful for the registration of a product within a national authority or within the PQ, the batch of the MPP used in the bioequivalence study (the biobatch) should be identical to the MPP proposed for commercial market in terms of composition, quality characteristics, and methods of manufacture, that is, it is not acceptable to employ a developmental batch of product that differs from the final "for-market" product as the biobatch. Further, the product units being tested should be taken from batches of industrial scale. If it is not yet practical to produce production scale batches, the test units should be taken from pilot scale batches that are not smaller than 10% of the planned full production batches, or 100,000 units, whichever is larger, and were produced using methods and equipment of manufacture that is consistent with that proposed for full-scale production.

Comparator Product

The product to which the MPP will be compared in a bioequivalence study should be determined as discussed in Choice of Comparator Products section.

Study Design

Cross-Over Versus Parallel Designs

Although there are several options to be considered, the study design of choice for comparing a MPP to a comparator product should generally be a randomized, two-period, two-sequence, single-dose, cross-over bioequivalence study conducted in healthy subjects under fasted conditions. Given that greater variability is generally observed in pharmacokinetic comparisons made between subjects than those made repeatedly within a subject, a cross-over study design is recommended to take advantage of the lower variability associated with intra-subject comparisons. That is, using an intra-subject comparison to study product performance will help minimize the variability in the data that is attributable to factors other than the products themselves. It is recommended that the washout period between doses in a cross-over design be at least five times the terminal half-life of the active ingredient.

Long Half-Life Drugs

There are situations where a cross-over study design may not be feasible and a parallel study design may be appropriate. For example, the use of a cross-over study design may not be practical when studying products containing a drug with a long terminal elimination half-life due to the long washout interval that would be required between drug administrations. WHO Annex 7 (5) suggests that the interval between drug administrations should not exceed three to four weeks. Establishing bioequivalence by using inter-subject comparisons in a parallel-design study will typically necessitate a greater number of subjects than that required to accomplish the same goal with a cross-over study design because of the increased variability as already mentioned.

When designing a study, it is necessary to ensure that the duration of blood sampling be sufficient to characterize drug absorption throughout gastrointestinal (GI) transit of the drug product. Annex 7 (5) recommends that blood samples be collected for 72 hours following drug administration, unless a shorter period can be justified based on the pharmacokinetics of the API. It should normally be ensured that the sampling protocol will permit the capture of 80% of the complete area under the concentration–time curve, that is, $AUC_T/AUC_I \geq 0.8$. However, for drugs possessing a long half-life, Annex 7 allows the collection of blood samples for 72 hours as being sufficient.

Food Considerations
As mentioned earlier, one of the primary considerations in the design of a bioequivalence study is minimizing the variability in the data that is attributable to factors other than the products themselves. The presence of food in the GI tract at the time of drug administration can introduce considerable variability into pharmacokinetic data because of the multiple and complicated effects food can have on both the physiology of the GI tract, the disintegration of the drug product, and the dissolution of the API. However, although a study conducted under fasted conditions is preferred, there are circumstances under which a study conducted under fed conditions may be accepted. For example, if the active ingredient is known to cause significant GI disturbance when administered under fasted conditions, or if the product labelling clearly restricts administration to the fed state, then a study conducted under fed conditions should be used to assess bioequivalence. In such fed studies, the test meal selected should be designed to account for local custom and diet, should be consumed by subjects within a 20-minute time frame, and drug administration should follow within 30 minutes.

The bioequivalence comparison of a proposed modified-release (MR) formulation is discussed in Modified-Release MPPs section but, briefly with respect to food, the comparison of a MR formulation to the appropriate comparator product must be investigated under both fasted and fed conditions. In such situations, the test meal employed in the fed study should be designed to challenge the robustness of the proposed MR formulation by promoting a maximal perturbation of the GI conditions relative to the fasted state, for example, a high-fat, high-calorie meal.

Participants
(a) Number of Subjects: The number of subjects required for a successful bioequivalence study should be determined based on the standards that must be met (see Section Handling of Study Data below) and the drug products being compared. The probability that a study of a given size will meet the applicable standards will depend on the expected mean difference between the test and reference formulations and variability associated with the drug involved, that is, the anticipated intrasubject coefficient of variation for a cross-over design.

A justification for the number of subjects enrolled in a study, which includes a sample size calculation, should always be provided in the study protocol. A minimum of 12 subjects is required for all studies.
(b) Subject Selection: Bioequivalence studies should normally employ healthy volunteers. The volunteers should be standardized based on characteristics

such as age, height, and weight. If the product under development is proposed for use in both sexes, it is suggested that both male and female volunteers be recruited for the study.

In situations where administration of the study drug to healthy volunteers is not acceptable due to its potency or toxicity, a study employing a patient population may be necessary. Such studies will usually involve a multiple-dose study design, as discussed below.

Bioanalytical Methods

The validity of the study conclusions depends on the reliability and reproducibility of the data collected. Therefore, all analytical methods used to measure the active ingredient in the chosen biological fluid must be well characterized, fully validated, and documented.

Bioanalytical methods must meet the requirements of specificity, sensitivity, accuracy, precision, and reproducibility. Investigators are encouraged to adhere to the recommendations of the Bioanalytical Method Validation Conference (10) with respect to both prestudy method validation and within-study clinical sample analyses and accompanying quality control.

The analytical method, validation procedures, and acceptance/rejection criteria must be clearly defined in the analytical protocol and associated standard operating procedures (SOPs) prior to the conduct of a study.

A dossier should include a complete Method Validation Report, copies of all applicable SOPs, and a complete analytical report. The analytical report should summarize the results of the analysis of the clinical study samples along with complete details of the calibration and quality control sample analyses, repeat analyses as per SOP, and a representative sample of chromatograms from the study.

Monitoring of Metabolites

The purpose of a bioequivalence study is to compare the performance of two or more products. The most sensitive approach to achieving this goal is to monitor the parent compound's disposition in the systemic circulation, that is, the API being released from the products. Therefore, the API is the analyte of choice for a bioequivalence study. If, employing up-to-date analytical methodologies, the API cannot be measured accurately in the biological matrix over a sufficient period of time to properly characterize a concentration–time profile, measurement of the primary, therapeutically active metabolite may be justified. The choice of analyte must be established a priori and stated in the study protocol. Should it be found appropriate to monitor the metabolite, the study design, for example, washout period, should be adjusted appropriately.

Handling of Study Data—Statistical Analysis

Standard pharmacokinetic analysis of the concentration–time data should be used to generate the following pharmacokinetic parameters for each period of data from each subject:

- area under the concentration–time curve up to the last sampling time (AUC_T)
- area under the concentration–time curve extrapolated to infinity (AUC_I)

- maximal concentration observed (C_{max})
- the time following dosing that C_{max} was observed (T_{max})
- elimination half-life ($t_{1/2}$).

Statistical analysis of the concentration-dependent parameters, that is, AUC and C_{max}, should be conducted by ANOVA on log-transformed data. For a standard cross-over study, the ANOVA model should include formulation, period, sequence, and subject factors. On the basis of the ANOVA results, the ratio of geometric means and 90% confidence intervals can be calculated using generally accepted approaches (11,12).

With regard to the parameters T_{max} and $t_{1/2}$, descriptive statistics should be provided. Any statistical analysis of T_{max} data should be based on nonparametric methods applied to untransformed data.

Acceptance Ranges
(a) Area under the curve (AUC): The 90% confidence interval for the relative mean AUC_T of the test to the reference product should be within 0.80 to 1.25. If the therapeutic range is particularly narrow, the acceptance range may need to be reduced based on a clinical justification. A larger acceptance range may be acceptable in exceptional cases if justified clinically.
(b) Maximal Concentration (C_{max}): The 90% confidence interval for the relative mean C_{max} of the test to the reference product should be within 0.80 to 1.25. As the measurement of C_{max} is inherently more variable than the measurement of AUC, Annex 7 (5) suggests that there are certain cases where a wider acceptance range may be justified, for example, see "Highly Variable Drugs" section later in the chapter. The range used must be defined a priori and should be justified, taking into account safety and efficacy considerations.
(c) Time to Maximal Concentration (T_{max}): The WHO guidances indicate that the statistical evaluation of T_{max} is necessary only if there is a clinically relevant claim with respect to time of onset of action or concerns about adverse events. In such a case, the nonparametric 90% confidence interval for the relative T_{max} measure should lie within a clinically relevant range.

Highly Variable Drugs
Highly variable drugs are defined as those drug substances that possess a high intrasubject variability, that is, an intrasubject variability of 30% or greater in AUC and C_{max} as estimated by the intrasubject coefficient of variation from the ANOVA from a repeated-measures study. Establishing the bioequivalence of two products containing a highly variable drug can be problematic because the substantial variability, which is not related to the formulation per se, can cause significant widening of calculated 90% confidence intervals. The end result is that larger numbers of subjects are required to demonstrate the bioequivalence of products containing a highly variable drug. Should a regulatory authority wish to make special provision for highly variable drugs, the WHO Annex 7 (5) offers three approaches that could be employed. An accepted approach, if any, should be specified/defined by the regulatory authority.

The WHO PQ program will consider the widening of the acceptance range for the 90% confidence interval on C_{max} from 0.75 to 1.33 if it can be clearly established that the drug in question is a highly variable drug and that there are no concentration-related concerns with regard to the safety or efficacy profiles of the drug. A widening of the acceptance range for AUC may be considered in exceptional circumstances if justified clinically.

Group Sequential Designs/Add-On Studies

The WHO guidelines do not provide specific advice on the use of "add-on" studies or group sequential design studies; approaches that are sometimes employed to address situations where a large number of subjects may be required to demonstrate bioequivalence between two products due to variability issues. Should a manufacturer be interested in employing such a design, it is recommended that a detailed protocol including a complete description of the proposed data handling and statistics be submitted to the relevant regulatory authority for comment and discussion prior to undertaking such a study.

Outliers

Any statistical methods to be used for the detection of outlier data must be clearly defined in the study protocol. The statistical recognition of data as an outlier is not considered to be a justification for its exclusion but, an indication that a further examination of these data is appropriate. The criteria that can be employed to justify exclusion should be defined a priori and, if an additional examination is undertaken to investigate outlying data, this examination must be extended to all subjects to ensure the uniform treatment of the data.

Within-clinical setting issues, for example, significant protocol violations, and/or physiological/medical explanations for the aberrant data, should be sought and would be a critical element in a justification for excluding outlier data from the main dataset. Should exclusion of the outlier data be proposed, complete statistical analyses with and without the outlier data should be included as an appendix to the study report.

Multiple-Dose Studies

Should it be found necessary to perform a steady-state, multiple-dose study, the basic study design considerations discussed earlier would be applicable for the dosing period during which blood samples are collected. Annex 7 (5) describes situations where a multiple-dose study may be appropriate, for example, a multiple-dose study in patients may be required when the API is too potent or too toxic to be administered in healthy volunteers. Other situations in which multiple-dose studies might be appropriate include the following:

- Drugs for which assay sensitivity is too low to adequately characterize the concentration–time disposition profile after a single dose
- Products involving drugs that exhibit nonlinear kinetics at steady state
- MR products with a tendency to accumulate. In the latter case, the multiple-dose study would be required in addition to single-dose studies.

Each drug product must be administered for a sufficient number of dosing intervals prior to the "sample collection" interval to ensure that the patients have attained steady state on that product. In a cross-over design, the washout period following the last dose of the first product can overlap with the approach to steady state of the second product provided this period between drug administrations is sufficiently long, that is, at least three times the terminal half-life of the API. Blood samples should be collected to establish that steady state has been achieved.

With regard to metrics for the assessment of bioequivalence, the parameters C_{max}, C_{min} (the minimum concentration observed in the dosing interval), the peak-trough fluctuation (the percentage difference between C_{max} and C_{min}) and the AUC of the dosing interval being sampled (AUC_τ) should be calculated. The calculated C_{max} and AUC parameters should meet the standards discussed earlier, while the relative mean C_{min} at steady state of the test to reference product should not be less than 0.80.

FIXED-DOSE COMBINATION PRODUCTS

The general principles outlined for bioequivalence studies are also applicable with respect to FDC MPPs [see Section 6.11.1 in Annex 7 (5)]. However, specific questions may have to be addressed, the most important of which is to define an appropriate comparator. As discussed in "Comparator Products for Fixed-Dose Combination Products" section, in the event that an innovator fixed-dose combination product is available such a product should be the reference for bioequivalence testing. Reference is also made to another WHO guidance document covering in detail particular aspects relating to the registration of FDC products in general (13). In other cases, only mono-component innovator products might be available but used together in clinical practice. These mono-component innovator products should then serve as reference for FDC MPPs together with sound clinical data that justify the particular combinational use. Respective clinical information should also be available for the summary of product characteristics (SPC) of the fixed-dose combination MPP if the innovator products are available as separate dosage forms only. In some cases it may also be necessary to follow the rules for the choice of an acceptable reference product as outlined earlier.

BIOEQUIVALENCE ASSESSMENT OF MPPs

Immediate-Release MPPs

As discussed above, a single-dose study under fasting conditions is generally regarded as the most conservative and sensitive approach to assess the rate and extent of absorption, that is, comparative bioavailability between a test and reference product. However, when the drug product should not be taken on an empty stomach and the innovator's SPC indicates so, it may be appropriate to do the comparison under standardized fed conditions. Similarly in the case when there are tolerability concerns under fasting conditions, the study may need to be conducted under fed conditions. In general, for immediate-release MPPs steady-state studies are only acceptable in rare cases, that is, they should be performed in addition to single dose studies if necessary and/or possible (see "Multiple-Dose Studies" section).

It may be possible to waive the in vivo comparison if the MPP contains a drug substance that has been classified as being eligible in terms of possible therapeutic risks, solubility and permeability according to the

Biopharmaceutics Classification System (BCS) as outlined in "BCS-Based Biowaiver" section.

Furthermore, comparative pharmacodynamic studies or clinical trials may be appropriate options if the classical pharmacokinetic bioequivalence approach is not applicable. However, pharmacodynamic and clinical approaches are generally considered less sensitive for bioequivalence testing.

Modified-Release MPPs

To evaluate the rate and extent of absorption in a most sensitive experimental setting, single dose studies are also relevant to assess bioequivalence for MR products. Multiple-dose studies are sometimes required in addition, particularly if accumulation after multiple dosing is expected or if fluctuation is considered important. In contrast to the considerations for immediate-release formulations, the investigation of food effects is relevant for MR products as it is critical to test the formulation-related performance of the products under fasting and fed conditions, that is, to test both ends of the spectrum of conditions under which the product may be used once approved. Testing under the range of possible GI conditions is necessary to monitor for significant changes in product performance under the varying conditions, in particular to exclude the risk of dose dumping, that is, the unintended immediate drug release from a MR dosage form, due to the influence of concomitant food intake. Therefore, in addition to fasting studies, pharmacokinetic bioequivalence studies conducted under fed conditions are usually required for MR MPPs.

It should be noted that a BCS-based biowaiver is generally not acceptable for modified-release formulations.

Comparative pharmacodynamic studies or clinical trials are alternatives when pharmacokinetic bioequivalence studies cannot be performed, for example, for safety and/or tolerability reasons.

IN VITRO APPROACHES/BIOWAIVER OPTIONS

BCS-Based Biowaiver

The possibility of "in vitro documentation of bioequivalence" for "certain medicines and dosage forms" is specified in Section 9 of the WHO guidance document (5). If the drug substance in question is highly soluble and highly permeable (BCS class I) and is manufactured as an immediate-release dosage form, exemption from an in vivo pharmacokinetic bioequivalence study may be considered provided that relevant dissolution requirements are fulfilled.

The in vitro approach basically refers to the BCS. Accordingly, active pharmaceutical ingredients are classified into classes based on their aqueous solubility and permeability characteristics (Table 3).

It should be noted that solubility is not meant to be the absolute solubility here. In contrast, high solubility refers to the highest single unit dose to be completely soluble in 250 mL aqueous buffer medium within the pH range of 1.2 to 6.8 without any stability problems.

As another related physicochemical characteristic, high permeability should be demonstrated for the particular API demonstrating that the fraction of the dose absorbed amounts to at least 85%. Accordingly, high permeability would stand for almost complete absorption of the compound in humans.

TABLE 3   Drug Substance Classification According to the BCS

BCS classification	Solubility	Permeability
BCS class I	High	High
BCS class II	Low	High
BCS class III	High	Low
BCS class IV	Low	Low

Physicochemical measures needed for BCS classification purposes may be taken from the peer-reviewed literature. The WHO Model List of Essential Medicines has been reviewed based on the BCS concept and active compounds are classified accordingly in WHO Technical Report No 937 of 2006, Annex 8 "Proposal to waive in vivo bioequivalence requirements for WHO Model List of Essential Medicines immediate-release, sold oral dosage forms," Table 1 of that document (14).

Prior to attempting to file a BCS-based biowaiver, a theoretical risk assessment is mandatory whereby risk to falsely waive a necessary in vivo study should be minimized. As an example, narrow therapeutic index drugs are generally not eligible for the BCS-based biowaiver approach. Other reasons mentioned in the WHO guidance (5) are outlined in Table 4 below, finally leading to perform bioequivalence testing in vivo rather than in vitro.

In practice, some of those criteria listed in Table 4 seem to be difficult to assess, for example, the meaning of "critical use" or "bioavailability problems," and are probably not easily defined. However, the published literature provides valuable examples of how to evaluate the applicability of the BCS-based biowaiver approach (14–18).

The in vitro dissolution investigations including experimental conditions and characteristics are outlined in Section 9 of the WHO guideline (5). It is of utmost importance to note that it is not sufficient to demonstrate the in vitro

TABLE 4   Risk Assessment for a BCS-Based Biowaiver

Situations Where BCS-Based Biowaivers are *Not* Applicable	
Immediate-release dosage forms with intended systemic action	*Formulation-related* considerations
Critical use medicines (e.g., hormones)	Nonoral, Nonparenteral products with systemic action [e.g., transdermal therapeutic systems (TTS), suppositories, etc.]
Narrow therapeutic range (steep dose–response curve) drugs	MR products with systemic action
Where there are documented evidence of bioavailability problems (or bioinequivalence)	Fixed-combination products with at least one API requiring an in vivo study, that is, this API is not eligible for the BCS-based biowaiver approach
Polymorphs, certain excipients, or a manufacturing process, which may have implications for bioavailability	Nonsolution products with non-systemic action (and without systemic absorption[a]), for example, topical, locally acting emulsions

[a]Systemic measurements might be necessary for safety reasons in some cases.

dissolution characteristics for the particular multisource product, but to ensure "the similarity of dissolution profiles between the test and comparator products" (5).

The WHO guidance incorporates basic aspects of the U.S. FDA guidance on the biowaiver approach (August 2000) [http://www.fda.gov/cder/guidance/index.htm]; however, the current scientific discussions in terms of so called "biowaiver extensions" are also considered. Accordingly, BCS-based biowaivers may be acceptable for drugs containing BCS class 2 and 3 drug substances manufactured as immediate-release dosage forms. As an example, a biowaiver may be possible for BCS class 3 drug products that are "very rapidly" (i.e., at least 85% dissolution within 15 minutes in all required media) dissolving. The relevant dissolution criteria are outlined in Section 9.2.1 of the WHO guideline (5).

Proportionality-Based Biowaivers

Another approach to waive in vivo bioequivalence testing may be taken when an MPP is intended for marketing in different strengths manufactured in proportionally formulated dosage forms as opposed to an application for a single strength only. It should be noted that dose proportionality is defined in two ways according to Section 9.3.1 of the WHO guidance (5). Either exact proportional composition of excipients is evident or different strengths are obtained by altering only the amount of the API which is solely adequate for high potency drugs. The latter case applies if the amount of API is relatively low (<10 mg per dosage unit), the total weight of the dosage form remains nearly the same for all strengths of the product series, the same excipients are used for all strengths and the change in strength is obtained by altering the amount of API only (5). Biowaivers based on dose proportionality may be considered if one in vivo bioequivalence study was performed, usually on the highest dose strength of the product series in question. Convincing comparative in vitro dissolution data are required to link the tested biobatch of the MPP, which was proven bioequivalent to the comparator and the additional proportional strengths.

Proportionality-based biowaivers differ from BCS-based biowaivers in that the former approach is eligible (in principle) for immediate- and extended-release formulations whilst the latter can be considered for specific drug substances in immediate-release oral dosage forms only.

Applicability of In Vitro–In Vivo Correlations

An in vivo–in vitro (iviv) correlation provides insight regarding the relation between in vitro dissolution and drug input in vivo. It is only applicable if dissolution is the rate-limiting step for drug absorption, hence establishing such a correlation in an early development phase is always appropriate for MR drug products. This is because the formulation controls the API release. International guidelines recommend level A iviv correlations (19) to be most appropriate to waive an in vivo bioequivalence investigation as it represents a point-to-point relationship between in vitro and in vivo dissolution, that is, the in vitro dissolution profile is predictive for the drug input in vivo. In this case the respective dissolution test may be used as control method with in vivo relevance. Several methods have been published (and frequently used), which describe how to calculate correlation characteristics. While details are not discussed in the framework of

this chapter, it should be mentioned that iviv correlations are always product related, that is, they cannot replace bioequivalence investigations between different products (e.g., MPP vs. comparator) from different applicants. In contrast, an iviv correlation established for a particular product may help to set meaningful specifications and also obviate the need for bioequivalence studies for certain postapproval changes.

CONCLUSIONS
The basic concept of bioequivalence testing has been implemented in international guidelines for decades although requirements have become more detailed and specific over time. In vitro approaches are nowadays considered to further facilitate production of affordable medicines worldwide. However, meaningful bioequivalence testing requires thorough planning and a decision regarding the approach which may be applicable considering the specific characteristics of a particular drug substance and/or drug product. Development of science, regulatory practice, and therapeutic experience will guide the way forward.

REFERENCES
1. Rägo L. Global disequilibrium of quality. In: Prince R, ed. Pharmaceutical Quality. Illinois: Davies Health Care International Publishing, 2004:3–21.
2. Ratanwijitrasin S, Wondemagegnehu E. Effective Drug Regulation. A multicountry Study. Geneva, Switzerland: World Health Organization, 2002:1–142.
3. Kopp S, Rägo L. The International Pharmacopoeia in the changing environment. Pharm Policy and Law 2007; 9:357–368.
4. Ritschel WA, Kearns GL. Handbook of Basic Pharmacokinetics — Including Clinical Applications, 5th ed. Washington, DC: American Pharmaceutical Association, 1999.
5. Multisource (generic) pharmaceutical products: Guidelines on registration requirements to establish interchangeability. In: WHO Expert Committee on Specifications for Pharmaceutical Preparations, Fortieth Report. Geneva, Switherland: World Health Organization, 2006 (WHO Technical Report Series, No. 937, Annex 7):347–390.
6. Guidance on the selection of comparator pharmaceutical products for equivalence assessment of interchangeable multisource (generic) products. In: WHO Expert Committee on Specifications for Pharmaceutical Preparations, Thirty-sixth Report. Geneva, Switzerland: World Health Organization, 2002 (WHO Technical Report Series, No. 902):161–180.
7. van Faassen F, Vromans H, et al. Biowaivers for oral immediate-release products: Implications of linear pharmacokinetics. Clin Pharmacokinet 2004; 43:1117.
8. Additional guidance for organizations performing in vivo bioequivalence studies. In: WHO Expert Committee on Specifications for Pharmaceutical Preparations, Fortieth Report. Geneva, Switzerland: World Health Organization, 2006 (WHO Technical Report Series, No. 937, Annex 9):439–461.
9. Information for Applicants on the Choice of Comparator Products for the Prequalification Project. Located on the World Health Organization (WHO) Prequalification of Medicines Web site, Guidance on selection of comparator product page. http://apps.who.int/prequal/info_applicants/info_for_applicants_BE_comparator.htm
10. Shah VP et al. Bioanalytical method validation—A revisit with a decade of progress. Pharm Res 2000; 17:1551–1557.
11. Schuirmann DJ. A comparison of the two one-sided tests procedure and the power approach for assessing the equivalence of average bioavailability. J Pharmacokinet Biopharm 1987; 15:657–680.
12. Westlake WJ. Bioavailability and bioequivalence of pharmaceutical formulations. In: Peace KE ed. Biopharmaceutical Statistics for Drug Development. New York: Marcel Dekker, Inc., 1988:329–352.

13. Guidelines for registration of fixed-dose combination medicinal products. In: WHO Expert Committee on Specifications for Pharmaceutical Preparations, Thirty-ninth report. Geneva, World Health Organization, 2005 (WHO Technical Report Series, No. 929, Annex 5):94–142.

14. Proposal to waive in vivo bioequivalence requirements for WHO Model List of Essential Medicines immediate-release, sold oral dosage forms. In: WHO Expert Committee on Specifications for Pharmaceutical Preparations, Fortieth Report. Geneva, World Health Organization, 2006 (WHO Technical Report Series, No. 937, Annex 8):391–439.

15. Stosik AG, Junginger HE, Kopp S, et al. Biowaiver monographs for immediate release solid oral dosage forms: Metoclopramide hydrochloride [published online ahead of print February 12, 2008]. J Pharm Sci.

16. Becker C, Dressman JB, Amidon GL, et al. Biowaiver monographs for immediate release solid oral dosage forms: Pyrazinamide [published online ahead of print February 12, 2008]. J Pharm Sci.

17. Becker C, Dressman JB, Amidon GL, et al. Biowaiver monographs for immediate release solid oral dosage forms: Ethambutol dihydrochloride [published online ahead of print August 21, 2007]. J Pharm Sci.

18. Becker C, Dressman JB, Amidon GL, et al. Biowaiver monographs for immediate release solid oral dosage forms: Isoniazid. J Pharm Sci 2007; 96:522–531.

19. Emami J. In vitro-in vivo correlations: From theory to applications. J Pharm Pharmaceut Sci 2006; 9:169.

Index